Identifying Wood

Identifying Wood

Accurate results with simple tools

R. Bruce Hoadley

The Taunton Press

Cover photo and inside flap photo: Deborah Fillion
Back-cover photos: R. Bruce Hoadley
Photomacrographs and photomicrographs: R. Bruce Hoadley
Other photos, except where noted: Brian Gulick

Library of Congress Cataloging-in-Publication Data

Hoadley, R. Bruce, 1933-
Identifying wood: accurate results with simple tools / R. Bruce Hoadley.
    p.   cm.
  "A Fine woodworking book" — T.p. verso.
  Includes bibliographical references (p.) and index.
  ISBN 0-942391-04-7 : $39.95
    1. Woodwork.   2. Wood—Identification.   I. Title
TT180.H58 1990   89-40575
684'.08—dc20   CIP

The Taunton Press
Inspiration for hands-on living™

10 9 8 7
Printed in the United States of America

A FINE WOODWORKING Book

FINE WOODWORKING® is a trademark of
The Taunton Press, Inc., registered in the
U.S. Patent and Trademark Office.

The Taunton Press
63 South Main Street
P. O. Box 5506
Newtown, Connecticut 06470-5506
e-mail: tp@taunton.com

FOR BARBARA

CONTENTS

LIST OF TABLES

ACKNOWLEDGMENTS

As soon as the preparation for this book was underway, I knew it was not a project I could do alone. As the book neared completion, I realized that I had called upon a long list of people for assistance ranging from wood specimens and photographs to detailed information — or simply advice. I hope that they will share some satisfaction for their contribution.

During the preliminary research stages of the book, I traveled to England and obtained generous assistance from Dr. Jeffrey Burley of the Forestry Department at Oxford University and also from John D. Brazier of the Forest Products Research Laboratory at Princes Risborough. My greatest reward was my acquaintance with William H. Brown of Amersham, Buckinghamshire, formerly a wood technologist and expert in wood indentification with TRADA (Timber Research and Development Association). Bill gave me valuable coaching on the more important tropical woods in the international commercial market and pointed out countless anatomical features that serve to separate similar species.

Throughout the project I received information and advice from Dr. Regis B. Miller, Director of the Center for Wood Anatomy Research at the U.S. Forest Products Laboratory at Madison, Wisconsin. Donna J. Christensen, also of the Forest Products Laboratory, shared insight into identification procedures.

Wood specimens for study or for photography came from a number of sources. Many of the figured veneers photographed were supplied by the Fine Hardwoods/American Walnut Association; others were donated by Stanley F. Stevens. Wood specimens were submitted by Robert M. Kellogg, Charles J. Kozlik, Stephen Smulski,

Arthur J. Noskowiak, Judy Scher and Francisco N. Tamolang.

I received valuable assistance in procuring wood and in the preparation of materials from Dennis Morin, Daniel Pepin and Ethan V. Howard. Microtome surfacing of the many specimens for macrophotography was done mainly by Philip Westover, Susan M. Smith and Bruce Griffin.

The International Association of Wood Anatomists, through the courtesy of Dr. Pieter Baas, kindly granted permission to use standard definitions for the glossary. Photographs of the Joshua-tree and related information were supplied by Rick Anderson, Superintendent, Joshua Tree National Monument. Many nagging photography problems were solved through the helpful suggestions of Al Carlson of Micro-Tech Optical, Inc.

The most help came from those close to me — my family. My wife, Barbara, through her association with other museum professionals, was instrumental in making many important contacts involving the identification of wood in historic objects. My daughters, Lindsay and Susan, were involved in the many phases of finding and filing information, sorting wood specimens and the typing, retyping and editing of drafts of the manuscript. I thank them not only for their direct assistance with the book project, but especially for their encouragement and their patience in enduring interrupted schedules, neglected holidays and fragmented vacations. Even Ben, our Labrador, lost out on many a promised day in the marsh. And the house needs paint!

— *R. Bruce Hoadley*
Amherst, Mass.
October, 1990

INTRODUCTION

"What kind of wood is that?"

I wonder how many times I have heard that question or asked it myself. Considering the endless array of woody plants on earth and the fact that wood has been and always will be an indispensable part of our human existence, it is easy to imagine that those involved with wood will continue to be challenged by problems of wood identification.

The purpose of this book is as straightforward as its title suggests: It is a guide to the identification of wood. In short, it is designed to answer the question, "What kind of wood is that?" It is written for the beginner and attempts to provide a self-taught approach. It requires no prior experience either with wood itself or with the procedures and equipment used in its identification.

Wood identification is based primarily on the anatomical — that is, cellular — structure of wood, although certain physical properties such as color, odor and density are also sometimes useful. The book explains all the anatomical features and physical properties involved in the routine identification of the species covered. The identification procedure uses the simplest techniques and the most commonly available equipment consistent with reliable results.

This is intended as a serious book for those interested in a serious pursuit of wood-identification skills. No prediction is made as to whether you will find the subject easy or difficult — what seems easy to one person may seem difficult to the next. Although the identification process has been simplified as much as possible, wood is a complex material, and understandably its identification does have its complications at times.

With the unaided eye, characteristic patterns or markings can be seen on the surfaces of wood; these reflect tissues or cells as seen en masse. While these visual features are sufficient for the identification of some woods, the majority of woods cannot be identified reliably on sight. Most woods do have at least some visual characteristics that enable them to be classified into groups of somewhat similar look-alikes, but the final problem comes down to distinguishing the wood in question from other similar woods within the same group.

In a typical unknown wood, individual cells or groups of cells critical to identification are examined initially with at least a low-power magnifier, such as a 10-power hand lens. To ensure that cell structure can be seen clearly, the wood must be exposed cleanly with a very sharp cutting edge, such as that of a single-edged razor blade. You will learn that with these simple tools — a razor blade and a 10-power hand lens — the majority of the hardwoods covered in this book can be identified.

In some woods, however, identification requires examination of individual cells or even details of cell walls, which requires the much higher magnification provided by a microscope. Chapter 7 explains the various tools and techniques used in identification.

AN EXAMPLE OF THE IDENTIFICATION PROCESS

To get a general notion of how the identification process works, let's try a relatively simple but real example: You encounter a board of an unknown wood — perhaps in a bookshelf or piece of furniture. Its surface looks like that shown in the photo above right.

The wood has a medium-brown color and no apparent odor. Looking at its surface, you probably realize that the

Tangential face of an unidentified board.

dominant markings result from the tree's **growth rings**. The surface also shows many crisp dark lines that are quite thin — about 1/64 in. — and up to nearly an inch long. You may already know — or will learn in Chapter 2 — that these lines are the **rays** of the wood.

In this case, the distinctive combination of growth rings and rays might already suggest what kind of wood it is, but for the majority of woods the surface features apparent at a glance are insufficient to enable reliable identification. Routinely, the next step is to cut the end surface of the board cleanly and examine it with a hand lens. This process is described in detail in

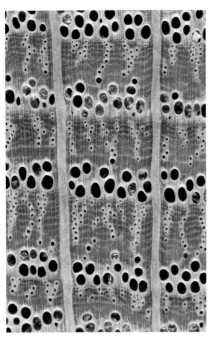

Transverse surface of an unidentified board, as seen with a hand lens. (10x)

Chapter 7. In our example, the wood surface would resemble that shown in the photo above.

This view shows an array of distinctive and meaningful details — cells or cell aggregates that can be used as specific identification features. For example, the relatively straight, light-colored lines that run vertically through the photo are the rays, composed of imperceptibly small ray cells. Most of the rays in this sample are inconspicuous fine lines, but a couple appear by comparison as fat, bold stripes. You probably recognize the repeating bands of cells that run perpendicular to the rays as growth rings.

Within each growth ring some of the cells are large enough to be seen individually as holes, or **pores.** They indicate that this wood is a hardwood, not a softwood, since softwoods do not have pores. Because the largest pores are concentrated in one portion of the growth ring, the wood is called a **ring-porous** hardwood.

Among the pores and rays, other cells are too small to be individually distinct, but their lighter or darker

masses are arranged in characteristic patterns. Chapter 4 systematically describes these and other anatomical features of hardwoods in general. In Chapter 10, these specific features are used as the basis for the identification of each hardwood.

Before reading Chapter 4, which discusses the anatomical structure of hardwoods (and before reading the next paragraph, which gives you the answer!), why not try to identify the sample by making a simple visual match with the woods depicted in the hardwood-identification chapter. Turn to p. 99 and browse through the woods shown in the photographs throughout the chapter.

Did you find a convincing look-alike? It's red oak. You were probably able to identify it correctly based on its obvious visual similarity to the photo at left on p. 103. You will discover that many woods can be recognized reliably with a hand lens. But even with the easy example above you may have noticed that white oak, the photo at right on p. 103, is also ring-porous and also has large rays, leaving some doubt. As with many pairs or groups of similar woods, the uncertain sample can be positively identified with a microscope.

In this procedure, a tiny, ultrathin slice of wood is removed from the edge of the board with a razor blade. It is then placed on a glass microscope slide in a drop of water and topped with a cover glass, as explained in Chapter 7. With the higher magnification of a microscope, you can examine the cells in detail. You would see that in red oak the smaller pores are **solitary,** which means they do not touch other pores; they are also thick-walled. In white oak, however, the smaller pores commonly occur in groups of two or more and are thin-walled. Precisely this comparision is made in the photos of oaks on p. 104. This is about as complicated as it gets.

HAND LENS OR MICROSCOPE?

I prepared this book fully realizing that most readers may not have a microscope and will want to get started in wood identification with only a hand lens. The book is designed to guide you through the separation and identification process as far as possible with just a hand lens. But it does indicate the limits of hand-lens identification, and it explains how microscopic features can be used to make further separations or to confirm hand-lens results.

MICROSCOPIC IDENTIFICATION IS EASIER AND MORE RELIABLE

It is a common but unfortunate misconception that identification by examining thin sections of wood with a microscope is somehow more difficult than hand-lens identification and represents an advanced level of endeavor. I would argue to the contrary, that identification is easier and more reliable with a microscope. I grant that it may take a little more time to prepare sections for microscopic viewing, but the features viewed are far more definitive and more easily judged than the more arbitrary features seen without magnification or with a hand lens. A microscope also gives you much greater flexibility when it comes to the samples you view. You can get accurate results with samples ranging from tiny slivers to painted furniture parts that would defy identification otherwise.

Nor should a microscope be considered an out-of-reach expense. You can pick up an adequate microscope for less than you would pay for another lens for your 35mm camera, a new boron-fiber fly rod or even a used pair of downhill skis. I therefore urge you to look forward with eagerness rather than concern to your first crack at wood identification with a microscope.

THE WOODS IDENTIFIED
IN THIS BOOK

This book focuses on the woods most commonly encountered in North America, primarily those native to the United States and Canada. Where appropriate, similar European species are referred to, even though most cannot be separated from their North American counterparts. In addition, an attempt has been made in Chapter 12 to include at least a sampling of the more common imported tropical hardwoods.

A SUPRISINGLY DIVERSE
AUDIENCE

While researching this book, I was amazed at the diversity of interest in the subject of wood identification. It is no surprise that wood identification is a routine problem for wood-industry professionals, such as lumber producers, dealers and product manufacturers, as well as for carpenters and other professional tradesmen. The need for wood identification also commonly arises in related woodworking specialties, such as furniture repair, building restoration, cabinetry, shipbuilding and woodcarving, both at professional and hobby levels. Anyone who works closely with wood and has the proverbial "sawdust in the veins" seems to take pride in being knowledgeable about the material and in being able to identify the wood.

It is fascinating how many occupations, although unrelated directly to woodworking, nevertheless have a critical need for wood identification. Anthropologists, archaeologists, museum curators and conservators, forensic scientists, biologists and teachers in many disciplines all need to identify wood. Many who do not do the actual identification work are involved in collecting and processing samples, for which some understanding of the overall identification process is essential.

IF YOU WANT TO TAKE
WOOD IDENTIFICATION
A STEP FURTHER

The subject of wood identification can be extremely complex; this book is designed as an introduction and understandably has limits to its breadth and technical depth. Many readers will eventually become interested in additional wood species, in special techniques or in more rigorous confirmations, and will want to expand their expertise. The references at the end of the text are annotated to assist the reader in selecting suitable sources. Because many readers will progress from this book to others on the subject, an attempt has been made to use standard terminology throughout the text and glossary.

Most of the book focuses on the identification of common woods in typical situations. However, the appendices deal with special topics, including atypical woody materials such as bamboo and rattan, the identification of decayed wood and charcoal, wood-fiber identification and chemical methods of wood identification. The final appendix lists sources of information on wood and wood identification, as well as suppliers of wood samples, reference slides and equipment.

It is my hope that this book will serve as a useful, informative introduction and manual for wood identification, and that each reader will share in the intrigue, challenge and reward of wood identification that I have enjoyed over the years.

CHAPTER ONE

THE CLASSIFICATION AND NAMING OF WOODS

The practice of classifying and assigning names to living things is called **taxonomy**. It begins by placing all organisms in either the plant or animal **kingdom**. The plant kingdom is subdivided into major divisions or **phyla** (singular **phylum**). The division *Spermatophyta*, which includes all seed plants, is separated into two broad groups based on seed type. One group is the **gymnosperms**, which have exposed seeds; the other is the **angiosperms**, whose seeds are encapsulated. These groups are further divided into **orders**, **families**, **genera** (singular **genus**) and **species** (singular also **species**).

CLASSIFICATION OF EASTERN WHITE PINE
(*Pinus strobus*)

Kingdom: Plant
    Division (or phylum): Spermatophyta
        Subdivision (or class): Gymnospermae
            Order: Coniferales
                Family: Pinaceae
                    Genus: *Pinus*
                        Species: *strobus*

CLASSIFICATION OF THE SPERMATOPHYTES
Within the plant kingdom, the seed plants, or spermatophytes, include three major groups that yield woody material.

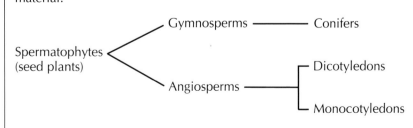

Spermatophytes (seed plants)
— Gymnosperms — Conifers — Softwood lumber (pine, fir, etc.)
— Angiosperms — Dicotyledons — Hardwood lumber (maple, mahogany, etc.)
— Monocotyledons — Woody monocots (bamboo, palm, etc.)

STEM FORMS OF SOFTWOODS, HARDWOODS AND MONOCOTS

Evergreen conifers such as pines, spruces and firs typically have an excurrent form (left). Deciduous hardwoods have a branching form (center) that is typical of maple, walnut and cherry. Woody monocots (right) are typified by bamboo and palms.

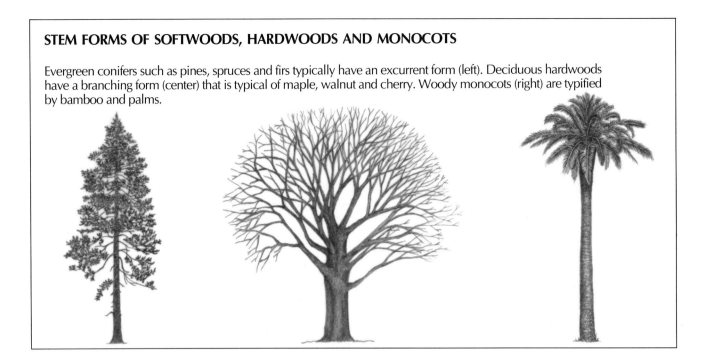

Gymnosperms and Angiosperms

Gymnosperms include all trees that yield the lumber we call **softwoods**. Softwoods fall within four families of the order *Coniferales,* more commonly called **conifers**. They are characterized by needlelike or scalelike foliage, which is usually evergreen. They have an **excurrent** form, which means that there is a dominant main stem (trunk) with lateral side branches. You have probably seen this form in pines, spruces, firs, hemlocks and cedars.

The conifers include the largest living plants — the sequoias — and the oldest plants — the bristlecone pines. Table 1.1 on pp. 4-6 lists the conifers that are included later in the softwood-identification chapter.

Angiosperms are separated into the **monocotyledons,** or **monocots,** and **dicotyledons,** or **dicots**. Monocots have one initial seed leaf, dicots have two. Dicots include such familiar plants as string beans and roses; they also include all tree-sized plants that yield hardwood lumber — oak, birch and mahogany, for example. Hardwood trees are typically **dendritic** or **deliquescent**, that is, characterized by branching or rebranching of the main stem.

Dicot trees usually have broad-leafed foliage, and in temperate zones they are mainly **deciduous,** losing their foliage during winter dormancy. Table 1.1 lists the dicot woods that are covered later in the hardwood-identification chapter.

Although we generally think of wood as coming from the conifers of the gymnosperms — softwoods — or the dicots of the angiosperms — hardwoods — monocots also produce important woody material. Most monocots, especially in temperate zones, are familiar to us as grasses or corn. However, particu-larly in warmer climates, monocots can attain large size and produce woody stems in vine or tree form. Although their cell structure and arrangement are distinctly different from those seen in hardwoods and softwoods, their overall size and the nature of their woody material make them functionally useful. Some of these monocots are bamboo, palm and rattan, which we see in a variety of products, including furniture, fishing rods, utensils and jewelry. The brief discussion of monocots in Appendix I on pp. 183-199 will help you recognize these woods when you run across them.

Species

A species comprises individual organisms that are similar in form and structure and that can interbreed to produce fertile offspring. For identification purposes we are not generally satisfied calling a tree by its generic name alone — birch, for example. Rather, we want a specific name: yellow birch, sweet birch, river birch, white birch, gray birch.

Scientific Names

The obvious objective in identifying an unknown sample of wood is to call it by its correct name. The fact that you have three categories of names to choose from is hardly a blessing. Scientific names, common names and trade names are used interchangeably even though common names and trade names are often inappropriate or misleading. A single common name can denote several species. A common name for a tree in a foreign country may differ from the name for the same tree in the United States, and, of course, a common name in a different language may be completely unrecognizable. Some trade names are deliberately designed to be deceptive for marketing advantage. Fortunately, because there is one, and only one, scientific name for each plant, scientific nomenclature helps avoid confusion.

The renowned Swedish botanist Carl Linnaeus (1707-1778) adopted a system of binomial (two-name) nomenclature that forms the basis for our present system of naming plants. Because the scientific names in this system are in Latinized form, they are italicized (in handwritten or typed work, they are underlined to indicate italics). The first name designates the genus (the generic name) and the second, the species (the specific name). The genus always begins with a capital letter, the species with a lowercase letter. For example, the scientific name for the tree species commonly called eastern white pine is *Pinus strobus*. Where the genus is repeated or clearly understood it is sometimes abbreviated: *P. strobus* for eastern white pine, *P. lambertiana* for sugar pine.

Generally, these names are proposed by botanists or taxonomists and then adopted by the scientific community. Thus, a complete species designation includes the name of the author in addition to genus and species. The names of well-known authors are abbreviated. For example, eastern white pine is formally written *Pinus strobus* L., the appended initial for Linnaeus. Rock elm, *Ulmus thomasii* Sarg., credits Charles Sprague Sargent. However, author designation is used only in more formal or highly technical writing; in more routine work it is now usually omitted.

Once accepted, scientific names are usually not changed. Occasionally, however, new scientific evidence calls for a name change. It takes time for this new name to become fully established, which often causes confusion. It is helpful in such cases to indicate either the former or proposed name in parentheses with an "equals" sign immediately following or beneath the name that is in current use. For example, yellow birch is *Betula alleghaniensis (= B. lutea)*; afrormosia is Pericopsis elata (= *Afrormosia elata)*. To determine which name is current, consult a regularly updated handbook such as the U.S. Department of Agriculture's *Checklist of United States Trees, Native and Naturalized* (see the Bibliography on pp. 212-215 for the full listing).

Binomial nomenclature is simple in principle, but the obvious challenge is getting used to the Latin names. But this task is not as imposing as it might initially appear. The common genera will be encountered repeatedly and will soon become familiar. The second name — the species — usually refers to the author or another person, the place the plant was discovered or a distinguishing feature, which also makes it easy to remember.

Common Names

In contrast with the precision of scientific names, common names can be confusing, inconsistent and misleading.

Common names (also called vernacular or local names) often note the area in which the wood is found, as in California laurel and Oregon myrtle. The tree's most obvious visual characteris-

tics are evident in designations like bigleaf maple, blue spruce, cottonwood, trembling aspen and gray birch. Fruits are noted in butternut, apple and black cherry; favored sites in river birch, swamp maple and mountain hemlock. The tree's properties are denoted by ironwood and rock maple, and its ultimate use by arrow wood, spoonwood and pencil cedar.

Common names are notoriously variable from region to region. It seems that every country has a redwood, a whitewood and an ironwood. A single species may have different names in different localities. In the United States *Platanus occidentalis* is sycamore; in England it is American plane. England's sycamore is *Acer pseudoplatanus*. We call *Pinus strobus* eastern white pine; in England it is yellow pine. The English call both Norway spruce *(Picea abies)* and Norway fir *(Abies alba)* "whitewood."

The examples go on. *Liriodendron tulipifera*, whose preferred common name is yellow-poplar, is also called tuliptree, tulip poplar, white-poplar or poplar; the preferred common name of the lumber from this tree is also yellow poplar (no hyphen), but it is known in various localities as whitewood, tulipwood, tulip poplar, hickory poplar, white poplar or simply poplar or popple. But poplar and popple correctly refer to a completely different wood — cottonwood or aspen in the genus *Populus*. Moreover, tulipwood is a South American rosewood—*Dalbergia*—and numerous species share the name whitewood.

Fortunately, most checklists designate a **principal** or **preferred common name** (also referred to as a standard or pilot name) for both tree and wood. In most cases this name uses two words, one word representing the genus, the other the species, as in black walnut, American elm and shagbark hickory. Less commonly a single term such as butternut, tamarack or hackberry has been adopted based on well-established traditional use.

Some common names are hyphenated or joined together as compound words, while others are not. Common names are compounded to make it clear that the two words, if taken separately, would be somewhat misleading. For example redcedar (compounded) is really a juniper, not a red cedar. Brief, simple compounds are written as a single word: redcedar, tanoak. But somewhat longer compound names are hyphenated as an aid in reading and pronunciation: yellow-poplar, white-cedar, Osage-orange.

Trade Names

Trade or commercial names are those used to designate lumber or other forms of wood in commerce. Some woods simply retain their common tree name — red alder, sycamore, yellow poplar, Sitka spruce, eastern white pine — or a shortened version — beech for American beech, cherry for black cherry. Place of origin occasionally creeps into the trade name. For example, bigleaf maple becomes Oregon maple, western white pine is sold as Idaho white pine.

It is quite common for more than one species to be included under a single lumber designation, either because the woods are functionally equivalent and can be used interchangeably, or because once cut into lumber they are very difficult or impossible to tell apart. Examples of such groups are western fir, southern yellow pine, red oak and African mahogany. The so-called hem-fir group often used in building now include western hemlock and firs; white woods designates a group which may include any of 13 western pines, hemlocks, firs and spruces.

Conversely, the wood of a single species may be segregated into more than one commercial category, resulting in more than one trade name. For example, the sapwood (outer wood) of sugar maple is sold as white maple, while wood of this species with the popular bird's-eye figure is sold as bird's-eye maple. The sapwood of sweetgum is sapgum; its heartwood (inner wood) is sometimes sold as figured gum or redgum. Baldcypress with a disease known as brown pocket rot is called pecky cypress.

Unfortunately, trade names have been invented that borrow names from more popular or desirable species to help market a lesser-known wood. This is probably the explanation for Philippine mahogany, used for species of *Shorea*, *Parashorea* and *Pentacme*. In truth, these are not mahoganies at all; the only genera usually accepted as genuine mahogany are *Khaya* and *Swietenia*.

To assure consistency and to avoid confusion, it is good practice to consult a standard or recognized authority. I relied on the *Checklist of United States Trees, Native and Naturalized* as my source for names of American species used in this book, and I recommend it as a guide to scientific names, principal common names and commercial names for lumber. Its guidelines governing scientific and common names are well worth reading. For foreign woods, I use British Standards publication BS 881 & 589 (a single booklet). More information on this and other helpful references can be found in the Bibiliography on pp. 212-215.

TABLE 1.1:
WOODS IDENTIFIED IN THIS BOOK

Scientific Name	Common Name
I. SOFTWOODS	
Abies	(true) fir
A. alba	Norway fir
A. amabilis	Pacific silver fir
A. balsamea	balsam fir
A. concolor	white fir
A. grandis	grand fir
A. lasiocarpa	subalpine fir
A. magnifica	California red fir
A. procera	noble fir
Chamaecyparis	white-cedar
C. lawsoniana	Port-Orford-cedar
C. nootkatensis	Alaska-cedar
C. thyoides	Atlantic white-cedar
Juniperus virginiana	eastern redcedar
Larix	larch
L. decidua	European larch
L. laricina	tamarack
L. occidentalis	western larch
Libocedrus decurrens	incense-cedar
Picea	spruce
P. abies	Norway spruce
P. engelmannii	Engelmann spruce
P. glauca	white spruce
P. mariana	black spruce
P. rubens	red spruce
P. sitchensis	Sitka spruce
Pinus	pine
P. albicaulis	whitebark pine
P. contorta	lodgepole pine
P. echinata	shortleaf pine
P. elliottii	slash pine
P. flexilis	limber pine
P. lambertiana	sugar pine
P. monticola	western white pine
P. palustris	longleaf pine
P. ponderosa	ponderosa pine
P. resinosa	red pine
P. rigida	pitch pine
P. strobus	eastern white pine
P. sylvestris	Scots pine
P. taeda	loblolly pine
Pseudotsuga menziesii	Douglas-fir
Sequoia sempervirens	redwood

Sequoiadendron giganteum	giant sequoia
Taxodium distichum	baldcypress
Taxus	yew
T. baccata	European yew
T. brevifolia	Pacific yew
Thuja	thuja
T. occidentalis	northern white-cedar
T. plicata	western redcedar
Tsuga	hemlock
T. canadensis	eastern hemlock
T. heterophylla	western hemlock

II. HARDWOODS

Acer	maple
A. campestre	field maple
A. macrophyllum	bigleaf maple
A. negundo	boxelder
A. nigrum	black maple
A. platanoides	Norway maple
A. pseudoplatanus	sycamore (British)
A. rubrum	red maple
A. saccharinum	silver maple
A. saccharum	sugar maple
Aesculus	buckeye, horsechestnut
A. glabra	Ohio buckeye
A. hippocastanum	horsechestnut
A. octandra	yellow buckeye
Alnus	alder
A. glutinosa	common alder
A. incana	grey alder
A. rubra	red alder
Betula	birch
B. alleghaniensis	yellow birch
B. lenta	sweet birch
B. papyrifera	paper birch
B. pendula	European (silver) birch
B. pubescens	European (white) birch
Carpinus caroliniana	American hornbeam
Carya	(true) hickory
C. glabra	pignut hickory
C. laciniosa	shellbark hickory
C. ovata	shagbark hickory
C. tomentosa	mockernut hickory
Carya	pecan hickory
C. aquatica	water hickory
C. cordiformis	bitternut hickory
C. illinoensis	pecan
C. myristiciformis	nutmeg hickory

Castanea	chestnut
C. dentata	American chestnut
C. sativa	sweet chestnut
Castanopsis chrysophylla	giant chinkapin
Catalpa	catalpa
C. bignonioides	southern catalpa
C. speciosa	northern catalpa
Celtis	hackberry, sugarberry
C. laevigata	sugarberry
C. occidentalis	hackberry
Cornus	dogwood
C. florida	flowering dogwood
C. nuttallii	Pacific dogwood
Diospyros virginiana	common persimmon
Fagus	beech
F. grandifolia	American beech
F. sylvatica	European beech
Fraxinus	ash
F. americana	white ash
F. excelsior	European ash
F. latifolia	Oregon ash
F. nigra	black ash
F. pennsylvanica	green ash
Gleditsia triacanthos	honeylocust
Gymnocladus dioicus	Kentucky coffeetree
Ilex	holly
I. aquifolium	English holly
I. opaca	American holly
Juglans	walnut, butternut
J. californica	southern California walnut
J. cinerea	butternut
J. hindsii	northern California walnut
J. nigra	black walnut
J. regia	European walnut
Liquidambar styraciflua	sweetgum
Liriodendron tulipifera	yellow-poplar
Lithocarpus densiflorus	tanoak
Maclura pomifera	Osage-orange

(continued on next page)

TABLE 1.1 (continued)

Scientific Name	Common Name
Magnolia	cucumbertree, magnolia
M. acuminata	cucumbertree
M. grandiflora	southern magnolia
Malus spp.	apple
Morus rubra	red mulberry
Nyssa sylvatica	black tupelo
Ostrya virginiana	eastern hophornbeam
Platanus	sycamore, plane
P. occidentalis	sycamore
P. acerifolia	London plane
Populus	aspen, cottonwood, poplar
P. alba	white poplar
P. balsamifera	balsam poplar
P. canescens	gray poplar
P. deltoides	eastern cottonwood
P. grandidentata	bigtooth aspen
P. heterophylla	swamp cottonwood
P. nigra	European black poplar
P. tremula	European aspen
P. tremuloides	quaking aspen
P. trichocarpa	black cottonwood
Prunus	cherry
P. avium	European cherry
P. serotina	black cherry
Pyrus spp.	pear
Quercus	oak
Q. alba	white oak
Q. bicolor	swamp white oak
Q. coccinea	scarlet oak
Q. falcata	southern red oak
Q. garryana	Oregon white oak
Q. kelloggii	California black oak
Q. lyrata	overcup oak
Q. macrocarpa	bur oak
Q. palustris	pin oak
Q. petraea	sessile oak
Q. prinus	chestnut oak
Q. robur	European oak
Q. rubra	northern red oak
Q. stellata	post oak
Q. velutina	black oak
Q. virginiana	live oak
Rhus typhina	staghorn sumac
Robinia pseudoacacia	black locust
Salix	willow
S. alba	white willow
S. nigra	black willow
Sassafras albidum	sassafras
Tilia	basswood, lime
T. americana	American basswood
T. heterophylla	white basswood
T. vulgaris	European lime
Ulmus	elm
U. alata	winged elm
U. americana	American elm
U. crassifolia	cedar elm
U. glabra	wych elm
U. procera	English elm
U. rubra	slippery elm
U. thomasii	rock elm
Umbellularia californica	California-laurel

III. TROPICAL HARDWOODS

Scientific Name	Common Name
Acacia koa	koa
Cedrela spp.	Spanish-cedar
Chlorophora excelsa	iroko
Chloroxylon swietenia	Ceylon satinwood
Cybistax donnell-smithii	primavera
Dalbergia fructescens var. tomentosa	Brazilian tulipwood
Dalbergia latifolia	Indian rosewood
Dalbergia stevensonii	Honduras rosewood
Dicorynia guianensis	basralocus
Diospyros spp.	ebony
Dyera costulata	jelutong
Entandrophragma cylindricum	sapele
Entandrophragma utile	utile
Gonystylus macrophyllum	ramin
Guaiacum spp.	lignumvitae
Khaya spp.	African mahogany
Lophira alata	ekki
Mansonia altissima	mansonia
Ochroma pyramidale	balsa
Peltogyne spp.	purpleheart
Pericopsis elata	afrormosia
Phoebe porosa	imbuia
Pterocarpus spp.	padauk
Shorea spp.	white lauan
Shorea spp.	yellow meranti
Swietenia spp.	Central American mahogany
Tectona grandis	teak
Terminalia superba	korina
Triplochiton scleroxylon	obeche
Turraeanthus africanus	avodire

CHAPTER TWO

THE STRUCTURE OF WOODY PLANTS: GROSS ANATOMICAL FEATURES

In evaluating any unidentified piece of wood, you will routinely find and examine the larger, more obvious anatomical features that are discussed in this chapter. We first look at those elements of tree form and growth that are common to both softwoods and hardwoods. Then we go to the opposite extreme and look at wood cells, since these compose the larger anatomical features.

Although some features can be recognized on any exposed wood surface, many are seen best on a specific plane of cut. Accordingly, the second part of this chapter discusses the three principal planes that we routinely expose on a sample in order to observe many important features.

Tree Form and Growth

Wood comes from higher-order, vascular (fluid-conducting), perennial plants. Woody plants are capable of secondary thickening, that is, of adding new yearly growth layers of cells onto the accumulated growth of previous years. Woody plants include trees, shrubs and vines, which are separated from each other on the basis of size and form.

Species that are taller than 20 ft. when mature and have a dominant single stem are called trees; those species that do not reach 20 ft. are shrubs. The stems of shrubs may be single or multiple, erect or trailing. The stems of woody vines are elongated and usually climb, supported by another plant or by a structure.

The trunk of a tree has limbs that divide into branches and then into twigs. As shown in the drawings on p. 8, if we slice crosswise through a trunk, limb or twig, and look down at the roughly circular surface, we find the **bark** — which is not considered wood — to the outside. To the inside of the bark is the **cambium,** the thin layer of living reproductive tissue separating the bark from the wood. The wood itself is characterized by **growth rings** arranged concentrically around the **pith.**

Wood Cells

The basic unit of the structure of plants is the cell. Thus, wood is a collection of various kinds of cells, which are produced by division in the cambium. The cells are typically elongated, consisting of an outer **cell wall** encompassing a **cell cavity.**

Cells produced by division in the cambium contain living material called **protoplasm** in their cell cavities. As the newly divided cells develop, they assume different sizes and shapes and perform various functions in the tree. Certain cells called **parenchyma** (puh-REN-kih-muh) remain alive for years, performing metabolic functions such as carbohydrate storage in the growing stem. Most wood cells, however, develop into cell types that lose their protoplasm within days after being created. As nonliving cells they function as conductive, or vascular tissue, and/or contribute mechanical support to the tree. These cells include **vessel elements, tracheids** and **fibers;** we will study them more closely later.

MAIN PARTS OF A TREE

Wood is the vascular supportive tissue of trees. Sapwood remains active in the conduction of sap and in food storage in living parenchyma cells; heartwood gives the tree mechanical support.

Crown

Bark
Cambium
Inactive heartwood
(Sap flow— upward in sapwood

Stem

Roots

PRINCIPAL FEATURES OF A TREE STEM

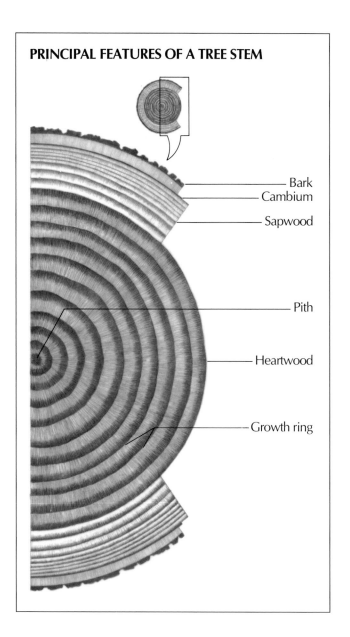

Bark
Cambium
Sapwood

Pith

Heartwood

Growth ring

The long axis of most wood cells parallels the tree stem — the cells are oriented vertically in the stem. This dominance of longitudinal cells is responsible for **grain direction,** the direction in which the wood most readily splits or cleaves. Where the longitudinal cells are almost perfectly parallel to the stem axis, the grain is described as straight; where cells are significantly nonparallel to the stem, some form of cross grain, such as **spiral grain,** may result.

Rays

Whereas most cells are elongated in a direction parallel to the stem, a small number of cells — generally accounting for less than 10% of the volume of the wood — are elongated perpendicularly to it. These cells form flattened ribbons of tissue called **rays** (see the drawing on the facing page). The wide side of the rays is oriented vertically, and the rays radiate outward from the pith and connect with the cambium. Consequently they cross the growth rings at right angles.

Rays vary considerably in size among wood species. At one extreme, they are a single cell wide and invisible to the naked eye; at the other, they are 30 to 40 cells wide and over an inch high. These rays are obvious without any magnification. As we shall see in the next chapter, all rays, regardless of size, are very important in wood identification.

LONGITUDINAL AND HORIZONTAL WOOD TISSUE

Wood consists mostly of longitudinal cells, which are elongated vertically along the stem and thereby determine the grain direction of the wood. The ray cells are elongated horizontally in the radial direction and form the rays.

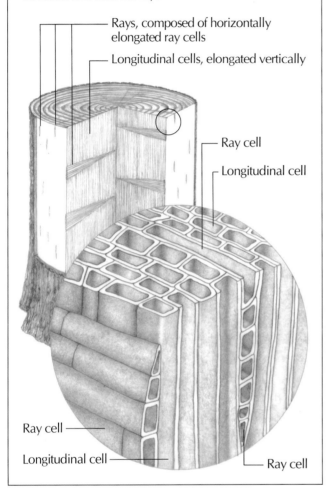

Rays, composed of horizontally elongated ray cells

Longitudinal cells, elongated vertically

Ray cell

Longitudinal cell

Ray cell

Longitudinal cell

Ray cell

HEARTWOOD AND SAPWOOD
Darker heartwood contrasts with lighter sapwood in tree stem, board and finished product.

Sapwood and Heartwood

In saplings (very young trees), the entire wood portion of the stem is involved in the upward conduction of sap. This wood is called **sapwood.** The conductive cells in sapwood have open cavities to accommodate the moving fluid. Parenchyma cells, which contain living material, are involved in food storage.

As the tree gets older and its trunk increases in girth, the tree no longer needs its entire trunk for the conduction of sap. In the central area of the trunk, that nearest the pith, parenchyma and conductive cells cease to function, save for the structural support they provide. As these cells shut down, the sapwood is transformed into **heartwood** (see the photos above right).

Residue of once-living protoplasts is occasionally found in heartwood parenchyma cells. Bubblelike structures may develop in what were formerly the conducting vessels. The cell walls may accumulate materials, called **extractives.** This gives heartwood its distinctive color. Living sapwood is typically a light, neutral shade of tan, yellow or near-white; any more pronounced color in the heartwood — the dark brown of walnut, the deep red of cedar, the black of ebony — is the result of extractives. However, some species — spruces, for example — do not have pigmented extractives, so the heartwood cannot be distinguished from sapwood on the basis of color.

It is important to remember that wood is always sapwood when it is formed. Once formed, however, no cells are added or lost in the sapwood area, nor does cell size or shape change. Yet, as mentioned, their contents may be modified slightly in the transition to heartwood.

Growth Rings

Tree growth — the result of reproductive activity in the cambium just under the bark — continues as long as the tree is alive to the extent that environmental conditions and tree vigor permit. In the kind of temperate climate that prevails across most of the United States, the characteristic annual life cycle includes a growing season and a dormant season. Cell formation in most trees follows this cycle, resulting in visible **growth layers** or **growth rings**. These increments are called **annual rings** when formed as the result of annual growth.

Ring width varies from species to species. It also varies within a species from year to year in response to more or less favorable conditions during the growing season.

Some growth rings are very obvious, others are nearly invisible. The distinctiveness of a growth ring in a particular species is determined by seasonal variation in cell diameter and cell-wall thickness and by the distribution of different kinds of cells within the wood. Variations in the growing environment of a particular species can have a pronounced effect upon the appearance of growth rings (see the photo on these pages).

Where there is a visible contrast between the first-formed and later-formed portions of a single growth ring, the first-formed portion is called **earlywood** and the remainder **latewood**. The terms **springwood** and **summerwood** are also used, but these terms are misleading because they wrongly suggest a consistent correlation with those seasons.

In certain species where earlywood and latewood are visible, the transition from one to the other is abrupt and obvious. But in other cases the transition is so gradual that we cannot point to a single place at which it occurs. In areas with more tropical climates, growth may be consistent year-round. As a result, some tropical species may not exhibit growth layers.

Grain, as we have seen, is used to denote the direction in which a wood splits. But the word "grain" can also be used to describe the visual contrast between earlywood and latewood when coupled with modifiers like "even" or "uneven." If a pronounced contrast exists, the wood is uneven-grained, as in southern yellow pine. If little contrast is evident, the wood is even-grained, as in basswood. Intermediate cases are indicated by adding further modifiers, as in "fairly even-grained" for a wood like eastern white pine or "moderately uneven-grained" for western fir, although the use of this terminology is somewhat subjective.

When working with an unidentified sample, it is extremely important to consider the evenness of grain and the transition of earlywood to latewood separately. Although it is common that an uneven-grained species has an abrupt earlywood/latewood transition, as in southern yellow pine, Douglas-fir, tamarack and white ash, an uneven-grained wood can also have a gradual transition, as in western hemlock. And a fairly even-grained wood like ponderosa pine, upon close examination, can show an abrupt earlywood/latewood transition.

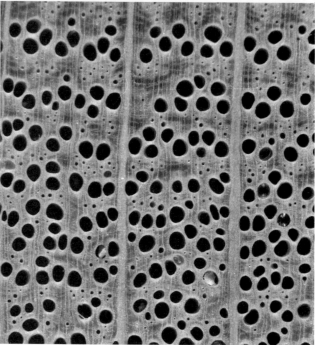

**EXTREMES OF GROWTH RATE
AS SEEN IN GROWTH RINGS**
The photographs in the top row are all of Douglas-fir (*Pseudotsuga menziesii*); those in the bottom row are of red oak (*Quercus rubra*). The center photograph in each row reflects average growth rate. Note how different extremely slow growth (left) or extremely fast growth (right) can appear in cross section. (15x)

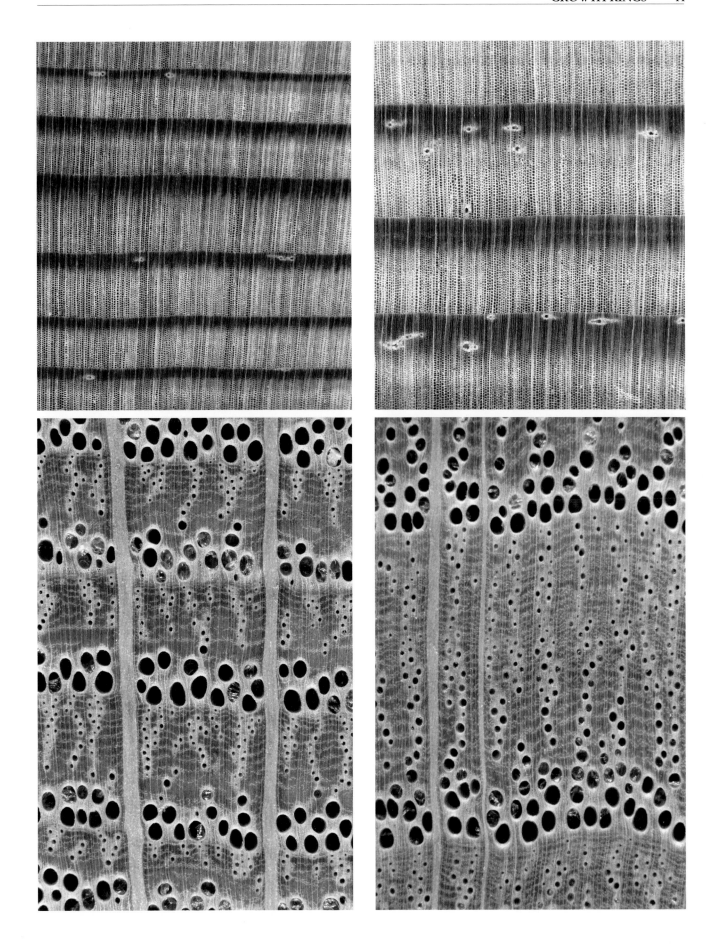

PRINCIPAL STRUCTURAL PLANES IN A STEM: X, R, T

The cross-sectional or transverse plane (X) is perpendicular to the stem axis. The radial plane (R) passes through the pith. The tangential plane (T) forms a tangent to the cylindrical plane of the growth rings; it is therefore a 'compromise', being most truly tangential where the plane forms a right angle with the rays.

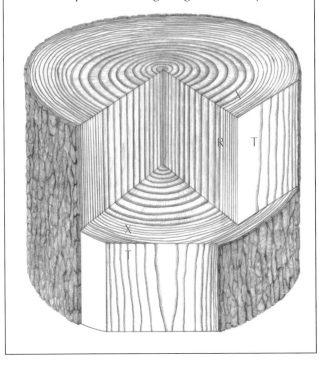

Viewing Features in a Sample

So far we have spoken about features as if they were always plainly visible. But that cannot be assumed. The features of a piece of wood can disappear or look dramatically different according to the surfaces exposed by cutting the wood. However, if we remember that cells are layered in a somewhat concentric cylindrical arrangement or that they radiate outward from a central axis, we realize that wood is actually a three-dimensional, triaxial material whose structure and cell orientation bear a close relationship to its grain direction, growth rings and rays (see the drawing above). We can then simplify our complex material somewhat by reducing our study to these three basic planes.

Sometimes we cut horizontally through the stem perpendicular to the grain direction. This is called a **transverse plane**, a **cross-sectional plane** or simply a **cross section**. All three terms are used interchangeably in this book and elsewhere.

Because this view allows us to see the concentric growth rings, we can evaluate growth-ring characteristics such as ring width, percentage of earlywood and latewood, evenness of grain and abruptness of transition from earlywood to latewood.

If the rays in the sample are large enough, they appear as lines — usually lighter than the background cell mass — crossing the growth rings at right angles.

Another view, the **radial plane**, is any plane whose surface parallels the stem and passes through the pith. If we take a short section of a log, stand it on end and cut it in half from top to bottom, we expose two radial planes.

Growth rings appear as nearly parallel lines in radial view. The radial plane essentially coincides with the plane of the rays. However, because of slight irregularities in the wood structure, exposure of the rays at the wood surface may be intermittent. The smallest rays are not visible at all or show up only as small patches or streaks that are more reflective than the surrounding area. Medium and larger rays, if present, usually appear as contrasting strips or patches.

In species such as maple and sycamore, rays show up as darker bands or stripes against a lighter background; in other species, such as black cherry or yellow-poplar, they are lighter than the background. The distinctive appearance of the rays on a radial surface is called **ray fleck** (see the photo at left on the facing page), which is an important identification feature, particularly in hardwoods. When cutting a sample in preparation for viewing, the changing appearance of ray fleck can be very helpful in locating the true radial plane.

The **tangential plane** or surface (see the photo at right on the facing page), like the radial plane, parallels the stem. The difference is that it does not pass through the pith and it is tangent to the cylindrical growth rings. We may also think of the tangential plane as being perpendicular to the rays — and therefore to the radial plane at that point.

In a small sample, growth-ring curvature is so slight as to be inconsequential. But on a gross scale — board surfaces, for example — the tangential plane cannot be accurately tangential over a large surface due to the growth-ring curvature. So we compromise and accept the surface of any outer slab from the length of a log as tangential.

In reality, growth rings are never perfectly round; their curvature varies from place to place. Therefore, in tangential view they usually appear as irregular U-shaped, V-shaped or oval markings. Rays are visible on tangential surfaces only if they are medium- to large-sized; they usually appear as dark lines along the grain. The length of the line indicates ray height and the breadth indicates width. In a species like sugar maple, rays are barely visible as hairlines up to $\frac{1}{32}$ in. long. In red oaks they are clearly visible as in lines up to 1 in., and in white oaks they may be several inches long.

Transverse, Radial and Tangential Sections

Remember that you'll often be working with small samples. If features are to show correctly in a smaller sample, it is critical to cut the wood accurately along the intended planes. The thin, transparent slices of wood tissue removed from these surfaces for microscopic study are termed **transverse**, **radial** and **tangential sections**. For convenience, both planes and sections are often designated by the letters X (for **cross**-sectional), R and T.

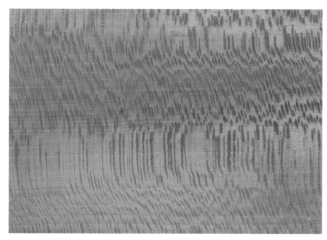

RAY FLECK
Striking ray fleck is apparent on the radial surface of this sycamore board.

I cannot overemphasize the importance of being able to locate these three planes, since further analysis usually hinges on it. Accurate orientation of planes is especially important in sectioning for microscopic examination; otherwise features will not appear as they should. To prepare a sample with one or more principal planes, first figure out the grain direction. This is sometimes indicated by large visible cells, by rays on an approximate tangential surface or by growth rings on a radial surface. Or you can split the wood. Since wood splits longitudinally, cutting a plane perpendicular to the grain direction and perpendicular to the plane of the split creates a transverse surface.

Once a transverse surface has been produced, the growth rings and rays apparent on it will assist in locating the orientation of radial and tangential planes. I strongly suggest plenty of practice on small pieces of scrap wood, whittling each as necessary to expose the X, R and T surfaces.

Summary

In examining any piece of wood, you should first study it to determine its relationship to the stem from which it came. Decide whether any of the exposed surfaces are true transverse, radial or tangential planes. If not, cut the piece to expose these surfaces (this technique is further described in Chapter 7). Then take note of the following gross anatomical features:
1. Sapwood and heartwood
2. Growth rings
   a. Width
   b. Percentage of earlywood and latewood
   c. Evenness of grain
   d. Transition from earlywood to latewood
3. Rays
   a. Visibility on transverse and tangential surfaces
   b. Ray fleck on radial surface

GROWTH RINGS AND RAYS AS SEEN IN THE TANGENTIAL PLANE
Growth rings and rays are clearly visible without magnification on this tangential surface of red oak.

CHAPTER THREE

THE STRUCTURE OF WOODY PLANTS: SOFTWOODS

Gymnosperms evolved on earth before angiosperms, and conifers retain a relatively primitive cell structure compared with the more specialized and complex anatomy of hardwoods. Because coniferous anatomy is simpler, softwoods will be discussed first. Their simplicity does, however, mean that there are fewer differences to distinguish one softwood from another, particularly at a macroscopic level. This contrasts with hardwoods, whose greater diversity of cell type, cell size and other features often makes identification possible even without a microscope. Thus, in the identification chapters, hardwoods are presented first because their greater anatomical complexity actually makes them easier to differentiate.

To accommodate readers who do not wish to use a microscope, this chapter first describes what can be seen without one. But, as mentioned, the primitiveness of softwood anatomy makes the microscope very important in finding features that separate one softwood from the next. You should, at the very least, read through the descriptions of the microscopic features and study the photographs taken at microscopic level.

An Introduction to the Softwoods: Eastern White Pine

We begin by examining a typical softwood, eastern white pine, which illustrates most of the features that are explained in detail in the pages that follow.

The oblique view of a cube of eastern white pine (facing page) was photographed with a scanning electron microscope. The circular insets show thin transverse, radial and tangential sections of the cube as they appear magnified 75 times under an ordinary microscope with transmitted light.

The wood tissue is composed mainly of **tracheids** (A), which are arranged in radial fashion and are fairly uniform in tangential diameter. The denser latewood (B) is the result of the tracheids' smaller radial diameter and flattened shape as well as their slightly thicker walls. In the transverse view we see a gradual transition from earlywood to latewood. Rays appear as narrow lines extending across the growth rings at right angles. The larger, rounded openings are **resin canals** (C), which are surrounded by thin-walled **epithelial cells** (D).

In tangential view, the elongated form of the tracheids (E) is evident; their length is 75 to 100 times their diameter. We see scattered rays, which appear as vertical tiers of small, rounded cells (F); we are actually looking into the ends of them. An occasional **fusiform ray** (G) is enlarged in the central position and has a resin canal (H) running through it horizontally.

The radial view also shows the elongated shape of the tra-

Center photo by Wilfred Côté

PRINCIPAL SURFACES OF EASTERN WHITE PINE

Cross-sectional (X), radial (R) and tangential surfaces of eastern white pine. The cube in the center has been magnified 25 times with an electron microscope. The large circular insets show typical hand-cut sections of these three surfaces magnified 75 times with a standard microscope. The small inset at the bottom shows the radial surface magnified 200 times. The letters correspond to features described in the text.

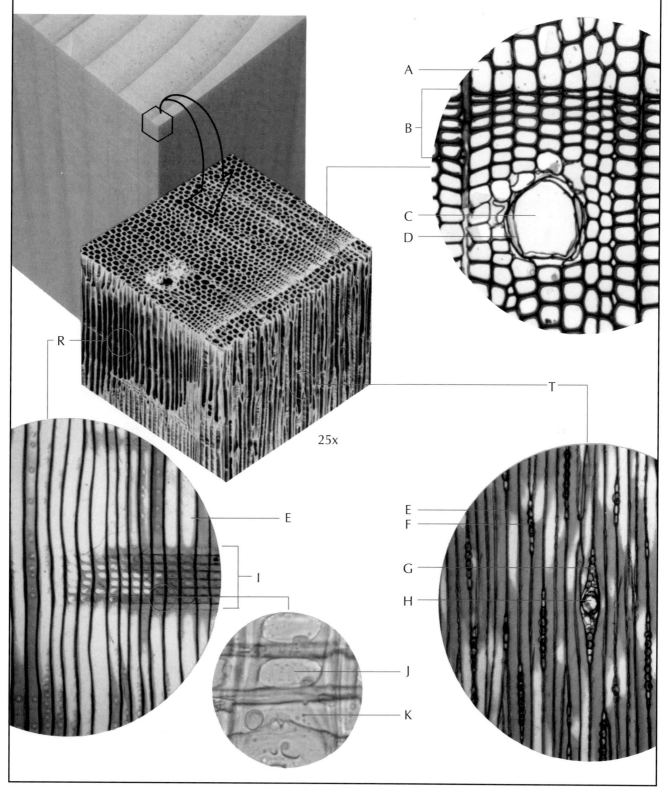

CONIFER CELL TYPES

This diagram shows the relative sizes and shapes of longitudinal cells and ray cells in conifers.

LONGITUDINAL CELLS

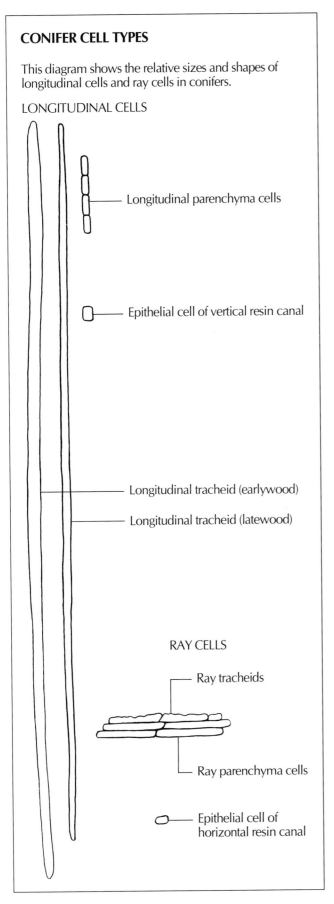

Longitudinal parenchyma cells

Epithelial cell of vertical resin canal

Longitudinal tracheid (earlywood)

Longitudinal tracheid (latewood)

RAY CELLS

Ray tracheids

Ray parenchyma cells

Epithelial cell of
horizontal resin canal

cheids. We see a ray (I) composed of seven rows of ray cells — both ray tracheids and ray parenchyma — that has been split by the sectioning. Where the ray parenchyma cells cross the longitudinal tracheids, rounded windowlike pitting occurs (J). The upper and lower rows of ray cells are ray tracheids (K).

These characteristics of eastern white pine are typical, in a general way, of most conifers. However, there are some specific points of variation as well as additional features in other conifers, and these are what provide a basis for species separation.

Tracheids and Texture

Although the general shape and proportions of longitudinal tracheids among conifers are so similar that they are of little value in identification, the actual dimensions can be valuable. Tracheid length is very difficult to measure in solid wood or in sections, so it is not often used in identification. But tracheid diameter is very informative since it has a bearing on the texture of softwoods (see the photos on the facing page).

The radial diameter varies with the tracheids' position in the growth ring, but the tangential diameter remains fairly constant and is therefore a useful identification feature. Among coniferous woods, tracheid diameter provides a basis for judging the coarseness or fineness of the wood's texture. The larger the tracheid diameter, the coarser the texture. Table 3.1 lists typical tracheid diameters for various conifers, and enables you to correlate the texture of a sample with a certain group of softwoods.

For example, eastern redcedar, with its small-diameter tracheids, is fine-textured. But redwood, at the opposite extreme, is coarse-textured. Douglas-fir and western redcedar fall in the middle. With a little practice, you will learn to estimate coarse, medium and fine textures. It is useful to have a set of labeled wood samples to help calibrate your eye. Sources for such sets are listed in Appendix V.

When transverse surfaces are viewed with a hand lens, individual tracheids can usually be seen clearly in coarse-textured species; they are virtually invisible in fine-textured woods. Just how "invisible" they are is subjective — it depends on your eyesight. Most people can see tracheids clearly in baldcypress, just barely in white pine and not at all in redcedar. How well you can see and judge texture with a hand lens also depends on how well you cut the transverse surface (this is discussed in Chapter 7). A microscope fitted with measuring scales can also be used to determine texture by measuring tracheid diameters directly.

TEXTURE IN CONIFERS

In conifers, 'texture' is a measure of relative tracheid diameter. In redwood (left), a coarse-textured conifer, individual tracheids are distinct when viewed with a hand lens. In a medium-textured wood, such as western redcedar (center), tracheids are barely discernible, and in fine-textured woods, such as eastern redcedar (right), individual tracheids cannot be seen clearly.

TABLE 3.1: DIAMETERS OF LONGITUDINAL TRACHEIDS IN VARIOUS SOFTWOOD SPECIES

Scientific Name	Common Name	Tracheid dia.*	Scientific Name	Common Name	Tracheid dia.
Abies	(true) fir		*Pinus*	pine	
A. amabilis	Pacific silver fir	40-60	P. albicaulis	whitebark pine	20-35
A. balsamea	balsam fir	30-50	P. contorta	lodgepole pine	35-55
A. concolor	white fir	35-50	P. echinata	shortleaf pine	35-60
A. lasiocarpa	subalpine fir	35-45	P. elliottii	slash pine	40-60
A. magnifica	California red fir	35-45	P. lambertiana	sugar pine	40-65
A. procera	noble fir	40-60	P. monticola	western white pine	35-60
			P. palustris	longleaf pine	35-60
Chamaecyparis	white-cedar		P. ponderosa	ponderosa pine	35-60
C. lawsoniana	Port-Orford-cedar	35-50	P. resinosa	red pine	30-45
C. nootkatensis	Alaska-cedar	25-40	P. strobus	eastern white pine	25-45
C. thyoides	Atlantic white-cedar	25-40	P. taeda	loblolly pine	35-60
Juniperus virginiana	eastern redcedar	20-35	*Pseudotsuga menziesii*	Douglas-fir	35-55
Larix	larch		*Sequoia sempervirens*	redwood	50-80
L. laricina	tamarack	30-45			
L. occidentalis	western larch	40-60	*Sequoiadendron giganteum*	giant sequoia	45-60
Libocedrus decurrens	incense-cedar	35-50	*Taxodium distichum*	baldcypress	45-70
Picea	spruce		*Taxus brevifolia*	Pacific yew	15-25
P. engelmannii	Engelmann spruce	25-35			
P. glauca	white spruce	25-35	*Thuja*	thuja	
P. mariana	black spruce	25-30	T. occidentalis	northern white-cedar	20-35
P. rubens	red spruce	25-35	T. plicata	western redcedar	30-45
P. sitchensis	Sitka spruce	35-55			
			Tsuga	hemlock	
			T. canadensis	eastern hemlock	30-45
			T. heterophylla	western hemlock	30-50

* Approximate range in micrometers
(1 micrometer = 1μm = 0.001 mm)

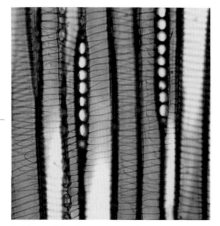 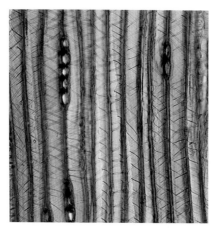 

SPIRAL THICKENINGS
In Douglas-fir (left), which has resin canals, spiral thickenings are nearly perpendicular to the tracheid axis. In yew (right), which lacks resin canals, spiral thickenings form a less regular pattern and a steeper spiral. (200x)

SPIRAL CHECKING
The spiral checking in the longitudinal tracheids of this eastern white pine sample is associated with reaction wood. (200x)

Spiral Thickenings

Because tracheids are conductive cells, they are devoid of contents and usually have relatively smooth inner walls. In a very few conifer species, however, their walls have **spiral thickenings** (see the photos above). These thickenings are visible only with a microscope, appearing in radial or tangential section as fine-lined, helically arranged ridges.

This feature is found in only two conifers, so its presence is important in identifying them. Spiral thickenings in conjunction with resin canals indicate Douglas-fir; spiral thickenings in the absence of resin canals indicate yew.

Spiral Checking

Spiral checking (see the photo above right), also seen in some tracheids, refers to the minute separations or checks that may appear in tracheid walls. They are seen microscopically as helical lines. Because spiral checking occurs among a variety of conifers, it is not particularly useful as an identification feature; it is mentioned here to avoid confusion with spiral thickenings.

Spiral checking can result from severe seasoning stresses and is most common in conifers with thick-walled tracheids; it is also associated with an abnormality known as reaction wood, which is described in Chapter 6. Spiral checks follow the cellulose orientation in the tracheid walls and are oriented at an angle of 45° or less with the tracheid axis. Note that the spiral thickenings in the photos above are more nearly perpendicular to the tracheid axis than the spiral checking in the photo above right — this will help you tell the two apart.

Pits and Pit Pairs

Voids in the secondary walls of cells are called **pits**. A pit in one cell occurs opposite a pit in an adjacent cell, forming a **pit pair**. The pits of a pair are separated by a sievelike **pit membrane**, which is the remnant of the original primary walls of the cells before the secondary wall was added. Pit pairs are the principal avenues through which fluids pass from cell to cell. When seen under a microscope, the face view of these pits (and sometimes the section view) can be extremely telling in identification.

If the pit looks like a plug removed from the cell wall, it is called a **simple pit**. Pits found in the walls of longitudinal or ray parenchyma cells in conifers are usually simple pits. Pits in tracheids have more elaborate, domelike shapes and are called **bordered pits**.

Bordered pits are abundant along the radial walls of coniferous tracheids. They look like circular targets or doughnuts when viewed radially. Since they are always present, their occurrence is not helpful for identification. But the number of these pits across a radial wall of earlywood tracheids can be very meaningful (see the photos on the facing page). For example, in larch, pits are commonly paired; in spruce, they are usually single; and in redwood and baldcypress, there may be up to four pits.

SIMPLE PITS

In this pair of simple pits of two adjacent cells (A), each simple pit is a more or less uniform-diameter void in the walls. In face view (B), we see the pit in outline. In sectional view (C), the pit membrane may be visible.

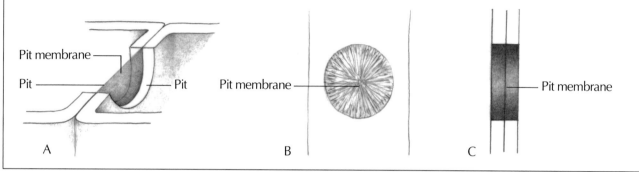

BORDERED PITS

These bordered pit pairs typically form between two coniferous tracheids (A). In face view (B), they appear as doughnutlike structures. In sectional view (C), the domed borders are apparent. A half-bordered pit pair (D) is found where a bordered pit joins with a simple pit.

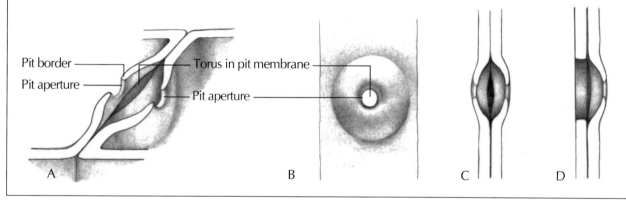

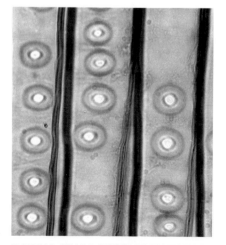
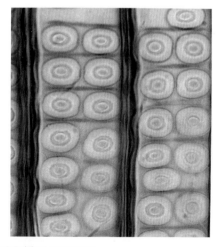
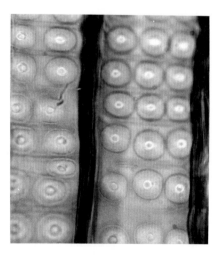

RADIAL WALL PITTING IN EARLYWOOD TRACHEIDS
The number of bordered pits across the radial wall of earlywood tracheids has some diagnostic value. For example, in spruce (left) there is rarely more than one bordered pit, whereas in larch (center), pits are commonly paired. In redwood (right), as many as four bordered pits may occur. (450x)

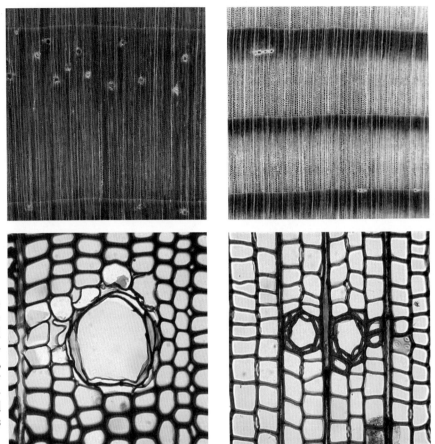

RESIN CANALS
In pines (top row, left [10x]), resin canals are usually larger, numerous and quite evenly distributed, occurring singly; the epithelial cells (bottom row, left [200x]) are thin-walled. In Douglas-fir (top row, right [10x]), spruces and larches, resin canals are usually smaller, fewer in number and erratically distributed, occurring singly or in tangential groups of four to several; the epithelial cells (bottom row, right [200x]) are thick-walled.

Resin Canals

Resin canals are tubular passages in wood that are actually intercellular spaces surrounded by special parenchyma cells called **epithelial cells**. The epithelial cells exude **pitch** or **resin** into the canals. The canals apparently serve a protective function for the tree by exuding pitch to seal off wounds caused by mechanical damage or boring insects. Resin canals occur in all species of four genera within the family *Pinaceae:* pines (*Pinus* spp.), spruces (*Picea* spp.), larches (*Larix* spp.) and Douglas-fir (*Pseudotsuga menziesii)*, as shown in the photos above.

The presence of resin canals provides an initial basis for separating the pines, spruces, larches and Douglas-fir from the balance of the conifers. In some species, resin canals are readily visible with the naked eye, but the smallest resin canals in other species can be seen only with hand-lens magnification.

In pines, resin canals are relatively large and numerous. They are usually present in all growth rings and are evenly distributed throughout the rings. They tend to occur singly, and, as can be seen microscopically, they have thin-walled epithelial cells. These cells sometimes bulge out and seem to block or fill the resin canal. Such protrusions are called tylosoids. Thin-walled epithelial cells are easily damaged during sectioning of a sample and therefore may not be clearly defined when viewed microscopically.

In spruces, larches and Douglas-fir, the resin canals are smaller, fewer and less uniformly distributed. They may appear to be absent from some growth rings. They commonly occur in groups of two to several, arranged tangentially to one another. As seen with a microscope, the epithelial cells are thick-walled and usually distinct.

The appearance of resin canals is variable, and, consequently, so is one's ability to locate them. Some caution is in order here. Their distinctness depends not only on size, but also on the condition of the pitch or resin within them. When a resin canal is empty, it appears as a minute pinhole in transverse view. In some instances, however, the canal is filled with yellowish or amber resin in a liquid state. In other instances, the resin may permeate surrounding tracheids or migrate onto the adjacent wood surface and produce dark spots far larger than the resin-canal diameter. This is easy to see but can be very misleading.

In still other cases, the resin forms sugary crystals in the canal and appears as white or yellow specks. I have the greatest difficulty finding resin canals in certain pieces of eastern spruce, where even with a hand lens the canals often appear only as scattered white specks, barely distinguishable from cutting debris and dust particles. One trick that helps in deciding whether or not a doubtful speck is a resin canal is to slice a new section from the same area (the method for doing this is de-

P   P   C   P   P   C   Strand of parenchyma cells

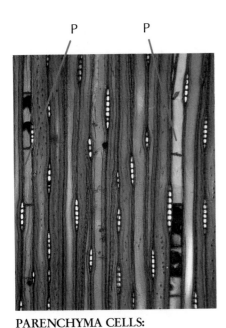

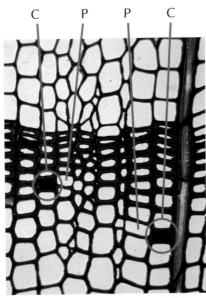

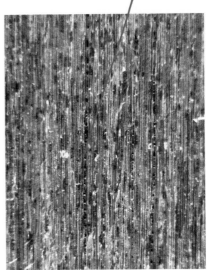

PARENCHYMA CELLS: LONGITUDINAL VIEW
Longitudinal parenchyma cells are of approximately the same cross-sectional dimensions as the surrounding longitudinal tracheids, but are short and occur in strands. Cross walls clearly separate them, and darkened contents also help make them visible. (50x)

PARENCHYMA CELLS: CROSS-SECTIONAL VIEW
In cross section, longitudinal parenchyma cells can be distinguished by their thin walls (P) and, in some cases, by their dark cell contents (C). (100x)

PARENCHYMA CELLS: LONGITUDINAL VIEW WITH HAND LENS
When longitudinal parenchyma cells are abundant, as in redwood, the strands of cells are just barely visible with a hand lens on split longitudinal surfaces. They appear as minute dashed lines due to the dark contents in the cells.

scribed in Chapter 7). If the speck disappears, it was a dust particle. But if it is a resin canal, it will still be in the same location after the next cut.

Longitudinal resin canals show up to varying degrees on radial and tangential surfaces. Their clarity depends on how closely the plane of the cut coincides with the axis of the canal, the size of the canal and the color and condition of the canal's contents. In species like soft pines, particularly sugar pine, they appear as distinct dark-brown lines; in spruce, they are either invisible or show only as fine, faint yellow lines.

Smaller transverse resin canals located within certain rays are also found in those species that have normal longitudinal resin canals. Transverse resin canals are discussed below under rays.

Finally, it should be noted that resin canals occasionally form as a response to environmental stress or injury in species that normally do not have them. These are called **traumatic resin canals**. They appear in transverse view as a single continuous line extending for a considerable distance along a growth ring, and are usually absent elsewhere. They are very small, barely larger than the tracheid diameter. Fortunately, they are rare and should cause little confusion since they are easily recognized and so different from normal resin canals.

Longitudinal Parenchyma

Although almost all longitudinal cells in conifers are tracheids, a few cells in certain species are longitudinal parenchyma. These cells have about the same cross-sectional dimensions as tracheids, and therefore are not individually distinct to the unaided eye. Viewed microscopically in longitudinal section, parenchyma cells are relatively short and occur in vertical series or strands — hence the term **strand parenchyma**.

In longitudinal (either radial or tangential) section (see the photo above left) parenchyma cells are easily distinguished microscopically because of their cross walls. In cross section (above center) they are about the same size as tracheids, but are always thin-walled, whereas tracheids can be thin- to thick-walled.

In sapwood, the parenchyma cells are alive and have cell contents. When the transition to heartwood occurs and the cells die, they often retain enough contents to fill the cell cavity partially or completely. The contents may also appear as scattered globules. The contents range in color from yellow to orange, red or amber, or they can be opaque.

Microscopically, contents may cause parenchyma cells to appear as darkened cells. With a hand lens, they are seen in coarse-textured woods like redwood and baldcypress (above right) as fine, dark, broken lines on longitudinal surfaces. These lines are a series of the same darkened individual cells that are visible through the microscope.

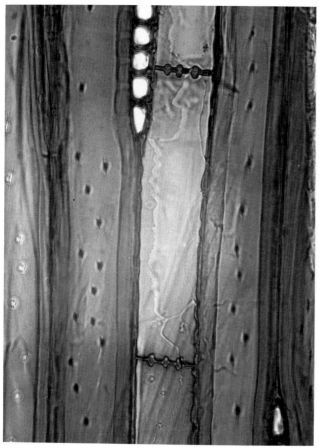

NODULES
Nodules appear in association with simple pitting on the end walls of longitudinal parenchyma in tangential view in some species, such as the baldcypress shown here. (300x)

PARENCHYMA CELLS: CLASSIFICATION OF ARRANGEMENT

Parenchyma arrangement is described according to the placement of the parenchyma within the growth ring, as shown in these cross-sectional views. Parenchyma at the growth-ring boundary (A) is termed marginal or terminal parenchyma. If it occurs singly within the ray (B), it is called diffuse parenchyma, and if it is concentrated in a tangential arrangement (C), it is referred to as metatracheal or zonate parenchyma.

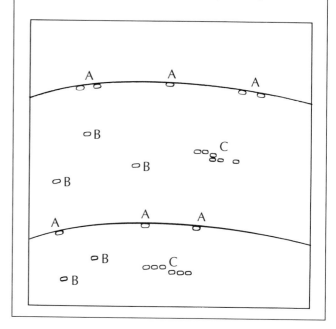

Simply noting the presence or absence of longitudinal parenchyma cells assists in identification; sometimes estimating their number is of value. For example, they are absent in the genera *Pinus* and *Picea*, but they are always abundant in *Sequoia* and *Taxodium*. In other genera they occur to varying degrees, consistently in some, sporadically in others.

Noting their position and distribution within the growth ring as seen in cross section is also very helpful, as shown in drawing above. We use the term **diffuse** to describe single cells that are scattered throughout the growth ring. **Terminal** or **marginal parenchyma** denotes cells that are located at the growth ring boundary. And **zonate parenchyma** are cells that are numerous enough in one area of the growth ring to be obvious.

The abutting horizontal walls of adjacent longitudinal parenchyma cells within a strand sometimes show pronounced simple pit pairs. Where multiple pairs occur, as viewed microscopically in tangential section, the remaining walls may appear as bumps in the cross wall. These are known as **nodular end walls,** or nodules (see the photo above).

Nodules are useful in identification. Their presence in baldcypress, for example, distinguishes it from redwood, whose

longitudinal parenchyma cross walls lack nodular thickenings. Determining microscopically whether nodules are present or absent can be very difficult where dark contents are present in the parenchyma cells; end walls show up much more clearly in cells whose cavities are devoid of contents.

Rays

Rays are important identification features in conifers when viewed microscopically, since magnification produces valuable information on how many cells high or wide a ray is, what type or types of cells compose the ray and what sort of pitting or sculpting is evident on the cell walls.

Seriation refers to the width of a ray as measured in cells (see the top photos on the facing page). Viewed tangentially, most rays in any conifer are **uniseriate**; that is, they consist of a single vertical tier of cells. A few species — redwood, for example — commonly have rays that are two cells wide, at least at some points along the ray. These are **biseriate** rays. Both uniseriate and biseriate rays are considered narrow rays.

The maximum height of the rays, as defined by a cell count in

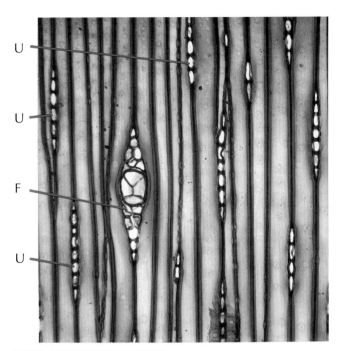

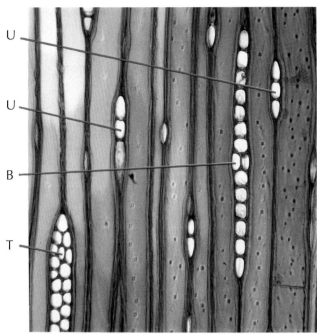

RAYS: TANGENTIAL VIEW
Conifers have narrow rays that are mostly uniseriate (U), that is, one cell wide, as seen in these tangential views. But occasionally the rays are biseriate (B) or triseriate (T). Species that have resin canals have fusiform rays (F), specialized rays that contain horizontal resin canals. (150x)

tangential section, also helps identify the wood. Species that have normal longitudinal resin canals also have horizontal resin canals; these small canals occur in the middle of specialized rays called **fusiform rays** (F in the photo above left).

As with vertical resin canals, the epithelial cells surrounding horizontal resin canals are either thin-walled, as in pines, or thick-walled, as in spruces, larches and Douglas-fir.

On a transverse surface as viewed with a hand lens, a sectioned horizontal canal in a fusiform ray will stand out as a bright yellow or white line (photo at right). Fusiform rays are often similarly apparent on radial surfaces, as long as the cuts are precisely radial. Generally these rays are invisible with a hand lens on tangential surfaces, although in soft pines, where they are larger, they may appear as minute dark specks.

Fusiform rays are extremely valuable as microscopic identification features. For example, a sliver of wood might not show any longitudinal resin canals with the eye or hand lens. But finding a fusiform ray on a tangential section with the microscope would indicate that the wood was one of the four genera of *Pinaceae* that do have normal longitudinal resin canals. Evaluating the thickness of the walls in epithelial cells would immediately place the unknown sample into either the pine or the spruce-larch-Douglas-fir group.

Most narrow rays contain two types of cells, **ray tracheids** and **ray parenchyma**. Typically, the ray parenchyma is in the central portion of the ray and the ray tracheids are in one or more rows along the upper and lower margins of the ray. But ray tracheids may also occur in the central portion of the ray.

Uniseriate ray   Horizontal resin canal

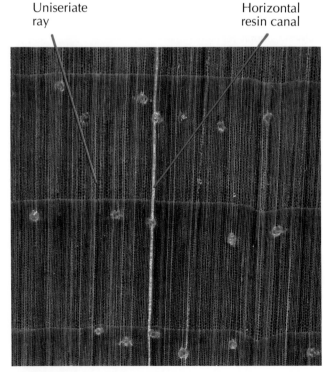

HORIZONTAL RESIN CANALS
A fusiform ray in this cross-sectional view of eastern white pine has been sectioned through its horizontal resin canal. The fusiform ray shows as a conspicuous radial line, distinct from the barely visible uniseriate rays. (15x)

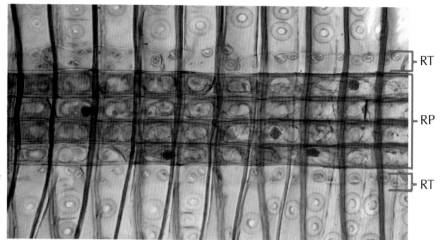

HETEROCELLULAR RAYS

Heterocellular or heterogeneous rays are rays that are composed of both ray tracheids (RT) and ray parenchyma (RP), as seen in this radial view of eastern white pine. (250x)

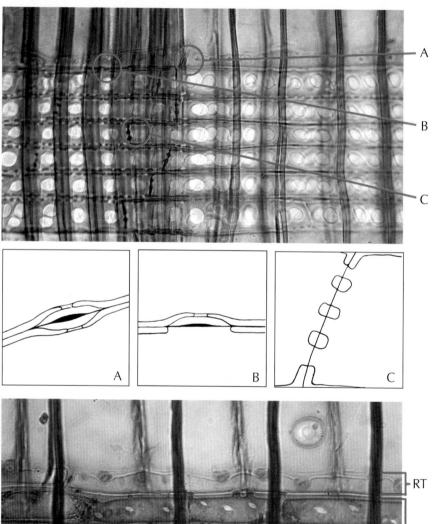

RAY CELL PITTING

Ray tracheids can be distinguished by the bordered pit pairs (A) that separate adjacent cells. Half-bordered pit pairs (B) occur between ray tracheids and ray parenchyma. Simple pit pairs (C) are found between ray parenchyma cells. (250x)

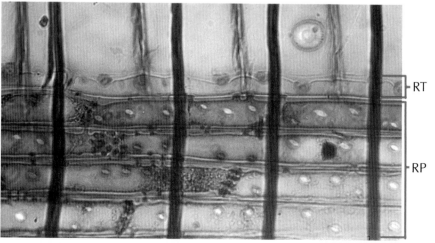

NARROW RAY TRACHEIDS IN HEMLOCK

In hemlock, ray tracheids (RT) may be difficult to see, sometimes appearing only faintly compared with ray parenchyma cells (RP). (550x)

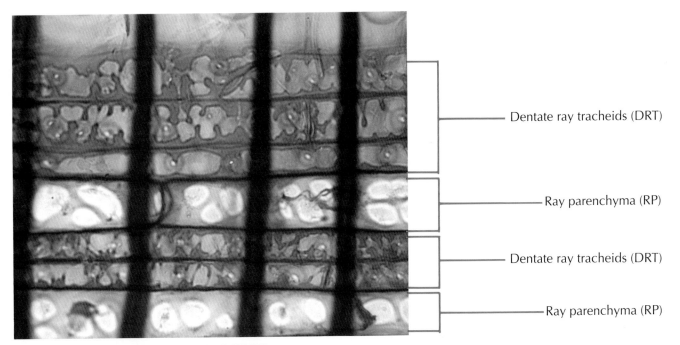

Dentate ray tracheids (DRT)

Ray parenchyma (RP)

Dentate ray tracheids (DRT)

Ray parenchyma (RP)

DENTATE RAY TRACHEIDS IN HARD PINES
In hard pines, ray tracheids have jagged inner walls and are called dentate ray tracheids (DRT). Ray parenchyma (RP) are also visible. (600x).

A ray with both types of cells is **heterocellular** or **heterogeneous** (facing page, top photo). A ray composed of either ray tracheids or ray parenchyma is **homocellular** or **homogeneous**. Some species, such as true firs and cedars, have only homogeneous parenchyma rays. Alaska yellow cedar has homogeneous rays composed of either all ray parenchyma or all ray tracheids.

Ray tracheids are distinguished by their bordered pits (facing page, center photo). In radial view, where rays cross longitudinal tracheids, small, rounded bordered pits can be seen. At abutting end walls, bordered pit pairs can be seen in section. Ray tracheids at the ray margin may be exceptionally narrow, irregular along the outer edge or very faint (facing page, bottom photo).

In most species ray tracheids are smooth-walled. In the hard pines, however, ray tracheids have irregular walls. In radial view these tracheids have toothlike projections and are consequently called **dentate ray tracheids** (photo above).

Ray parenchyma cells have simple pits. Viewed radially, these pits are sometimes evident along horizontal walls or on end walls. Where multiple simple pits occur on the end walls of ray parenchyma cells, the cell wall remnants may appear knobby and are called nodules.

An important identification feature in conifers is the pitting that occurs between ray parenchyma cells and earlywood longitudinal tracheids. As viewed radially at higher magnifications, the common wall between these cells is delineated above and below by the horizontal walls of a ray parenchyma cell, and along the sides by the lateral walls of the longitudinal tracheid. This area of pitting is called a **cross field** or **ray crossing**.

CROSS FIELD

The common wall joining a ray parenchyma cell and an earlywood longitudinal tracheid, as seen in radial view, is called a cross field or ray crossing.

Longitudinal tracheid

Cross field

Ray parenchyma cell

TYPES OF CROSS-FIELD PITTING

This chart summarizes the major classifications of cross-field pitting types. There is considerable intergrading between taxodioid and cupressoid pitting and, to some extent, between piceoid and cupressoid pitting. The photos show radial views.

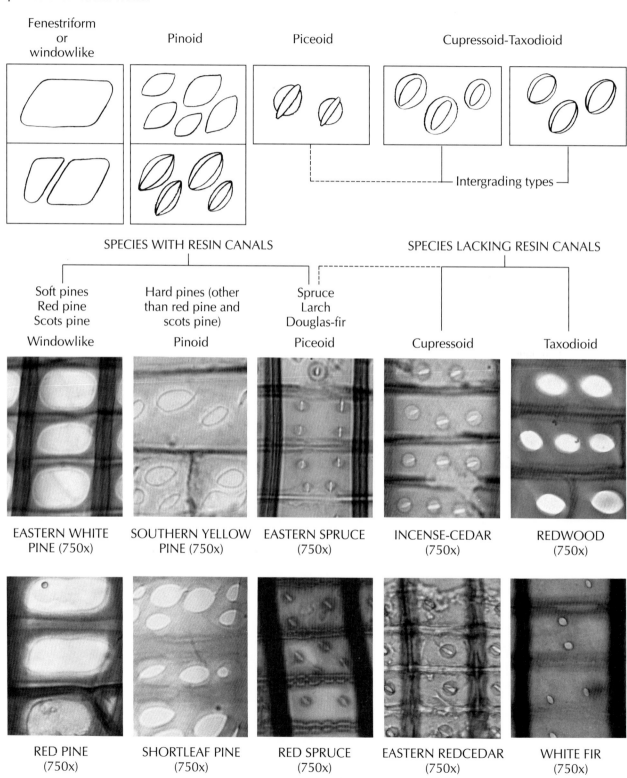

| Fenestriform or windowlike | Pinoid | Piceoid | Cupressoid-Taxodioid |

TABLE 3.2: SUMMARY CHECKLIST OF ANATOMICAL FEATURES FOR SOFTWOOD IDENTIFICATION

Sapwood/Heartwood:
Color/visual contrast
Width of sapwood

Growth Rings:
Width (rate of growth)
Evenness of grain
Abruptness of earlywood-to-latewood transition

Tracheids:
Diameter/texture
Radial wall pitting

Resin Canals:
Present/absent
Size
Number/distribution
Epithelial cell wall thickness

Longitudinal Parenchyma:
Present/absent
Abundance
Plain/nodular end walls
Contents
Location:
Terminal or marginal
Diffuse
Metatracheal or zonate

Rays:
Height
Seriation
Homocellular or heterocellular
Fusiform rays:
Present/absent
Epithelial cell walls: thick or thin
Ray tracheids:
Smooth or dentate
Ray parenchyma:
Cross-field pitting: windowlike, pinoid, piceoid, cupressoid or taxodioid
End walls: smooth or nodular
Contents

Latewood tracheids are narrow in radial diameter and therefore form narrow cross fields. To evaluate the cross-field pitting types discussed below, cross fields formed at earlywood tracheids should be examined. The drawing on the facing page shows the various types of cross-field pitting found in coniferous woods. As you will see, cross-field pitting can be an extremely important feature for identifying softwoods.

Generally speaking, **fenestriform** or **windowlike pitting** is seen in soft pines and red pine, **pinoid pitting** in hard pines, **piceoid** or **piciform pitting** in spruces, larch, and Douglas-fir, **cupressoid pitting** in hemlock and cedars (except *Thuja*) and **taxodioid** pitting in redwood, firs, *Thuja* and cypress. Among the last three types, pitting is not consistent for every sample of certain species.

Fenestriform and pinoid pitting are relatively easy to discern. Pinoid pitting is made even easier to recognize by the fact that it occurs only when dentate ray tracheids are also present. But the other three types—piceoid, cupressoid and taxodioid—are smaller and more difficult to see, and they intergrade with one another. You should consider them in conjunction with other features. For example, piciform pitting is associated with three genera that also have resin canals; cupressoid and taxodioid pitting is associated with species that usually have fairly abundant longitudinal parenchyma. Cupressoid and taxodioid pitting are the most confusing.

Summary

Although the softwood features that you observe with your eye and hand lens will help suggest some of the possibilities for identification, there are only a few species that can be identified consistently without microscopic examination. Any serious involvement with softwood identification requires microscopic examination. And, as you shall see later in the softwood identification section, even microscopy will fail to separate all species of certain softwood genera.

The checklist at left summarizes what we have just covered on softwood identification.

CHAPTER FOUR

THE STRUCTURE OF WOODY PLANTS: HARDWOODS

Among coniferous woods, similarity in structure and appearance makes identification of most species uncertain without the use of a microscope. In the case of most hardwoods, however, the extreme variability in cell type and arrangement usually produces unique combinations of features that can be easily recognized with a 10x ("ten-power") hand lens. The secret to success in hand-lens identification of hardwoods does not come through memorization of visual designs and patterns, but through an understanding of these visual features in terms of the cells or cell masses that produce them.

In the discussion of each of the major cell types that follows, the features that are visible to the eye or with a hand lens are explored first in some detail, since macroscopic features alone suffice to identify many woods. Within each wood group, the discussion then moves on to cover microscopic features. These features are most reliable in the identification of the more difficult woods, and they also provide a positive check as well as an alternative approach in the identification of pieces that cannot be routinely examined on a cross section.

The Four Basic Cell Types

In the initial examination of a cross-sectional surface of a hardwood with the unaided eye and hand lens, the features that are visible are usually associated with the following:

Vessels (or "pores");
Fibers;
Ray cells;
Longitudinal parenchyma and tracheids.

Pores are the transverse openings of vessel elements. In hardwoods, these are the largest-diameter openings you see on a transverse plane. The smallest pores are barely visible with a 10x hand lens, but the prominent pores in some species are obvious to the unaided eye. Pores usually appear as dark holes in the wood. Sometimes they are filled with various kinds of material, but they nonetheless remain recognizable as individual cells.

In the other three cell types, cells are too small to be seen individually with a hand lens. However, we can observe them en masse, and the differences among these cell groups usually enable us to tell them apart.

Fibers are among the smallest-diameter cells and have the thickest walls. On a transverse surface, masses of fibers take on a dense appearance, often forming a dark background or ground mass against which pores and lighter-colored masses of parenchyma and ray cells contrast.

Ray cells form rays, one to many cells wide. Some rays are

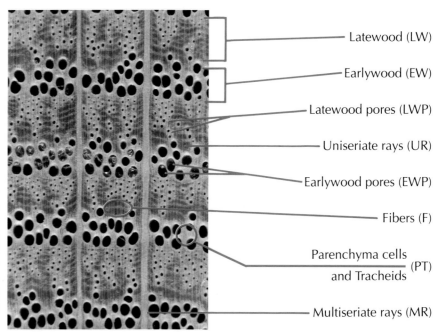

Latewood (LW)

Earlywood (EW)

Latewood pores (LWP)

Uniseriate rays (UR)

Earlywood pores (EWP)

Fibers (F)

Parenchyma cells
and Tracheids (PT)

Multiseriate rays (MR)

INTRODUCTORY VIEW OF HARDWOOD ANATOMY: RED OAK This is a cross-sectional surface of northern red oak as seen with a hand lens. Within each growth ring, the first-formed earlywood is clearly defined by a zone of large pores, followed by latewood with its smaller pores. Other visible features include large multiseriate rays, narrow uniseriate rays, lighter-colored masses of mixed parenchyma cells and tracheids and darker masses of fibers.

barely visible with a hand lens, others are clear to the unaided eye. They appear as fairly straight radial lines across a transverse surface and are usually of a lighter shade than the ground mass. In woods that lack obvious growth rings, locating rays can be an important aid in establishing a radial surface and the tangential surface perpendicular to it.

Longitudinal parenchyma cells are found in varying numbers among hardwoods. In species like black cherry, they are absent entirely; in sweetgum, they are present but so sparse that they cannot be seen. In many woods, however, they are visible en masse in transverse section as lighter-colored lines or areas. **Tracheids** are less frequent among hardwoods. In cross section they are virtually indistinguishable from parenchyma because both are thin-walled and about the same diameter. Moreover, the two are usually intermixed in the vicinity of pores. As seen with the hand lens, such combined masses of tracheids and parenchyma are described as "lighter-colored tissue," or, less accurately, simply as parenchyma.

An Introduction to the Hardwoods: Red Oak

Eastern white pine was used to introduce softwood anatomy in Chapter 3. Here we will examine northern red oak *(Quercus rubra)* to provide an overview of hardwood features (see the photo above).

Within each growth ring, the earlywood is evident as a distinct band of larger pores that are surrounded by lighter-colored tissue. This lighter-colored tissue is actually a mixture of parenchyma and tracheids. In the latewood, smaller pores occur in more or less radial lines, surrounded by lighter-colored parenchyma and tracheids. The rays are of two distinct sizes. There are conspicuously larger rays, many cells wide, that are obvious

at a glance. There are also numerous fine, uniseriate rays that are barely visible with a hand lens. In the latewood, tangential bands of parenchyma appear as continuous or broken lighter lines across the darker masses of fiber.

The combination of features just described is unique to the red oak group (which is discussed more fully in the hardwood-identification section, Chapter 10) and would be sufficient to identify an actual red oak sample. Other hardwood species show variations of the same basic features.

Choice of Viewing Surface and Instrument

When a generously sized cross section is available for examination, enough of the pore, parenchyma and ray characteristics are usually visible with a hand lens to permit identification. But sometimes, as when examining thin veneers, an adequate cross section is not available. In such cases we turn to the longitudinal surfaces. Fortunately, many additional features are apparent here. For example, ray height — the dimension of the ray along the grain — can be measured on a tangential surface. And ray fleck — the exposed ray surfaces — can be seen in radial view. Consequently, this chapter also reviews gross and macroscopic (nonmicroscopic) features on longitudinal surfaces.

Although hardwoods can frequently be identified based on macroscopic features alone, there are instances in which a microscope is mandatory. Fragments are sometimes so thin or tiny that they preclude macroscopic identification. In other cases, the macroscopic features are similar between two species, and only a microscopic examination will separate them accurately. Hence, some of the more important microscopic features are reviewed on the following pages. To give you an idea of how much more you can see with a microscope, compare the hand-

TRANSVERSE, RADIAL AND TANGENTIAL SURFACES OF RED OAK
The circled insets show, from left to right, typical handcut thin sections from the cross-sectional (X),
radial (R) and tangential (T) surfaces as viewed for identification under a standard microscope.

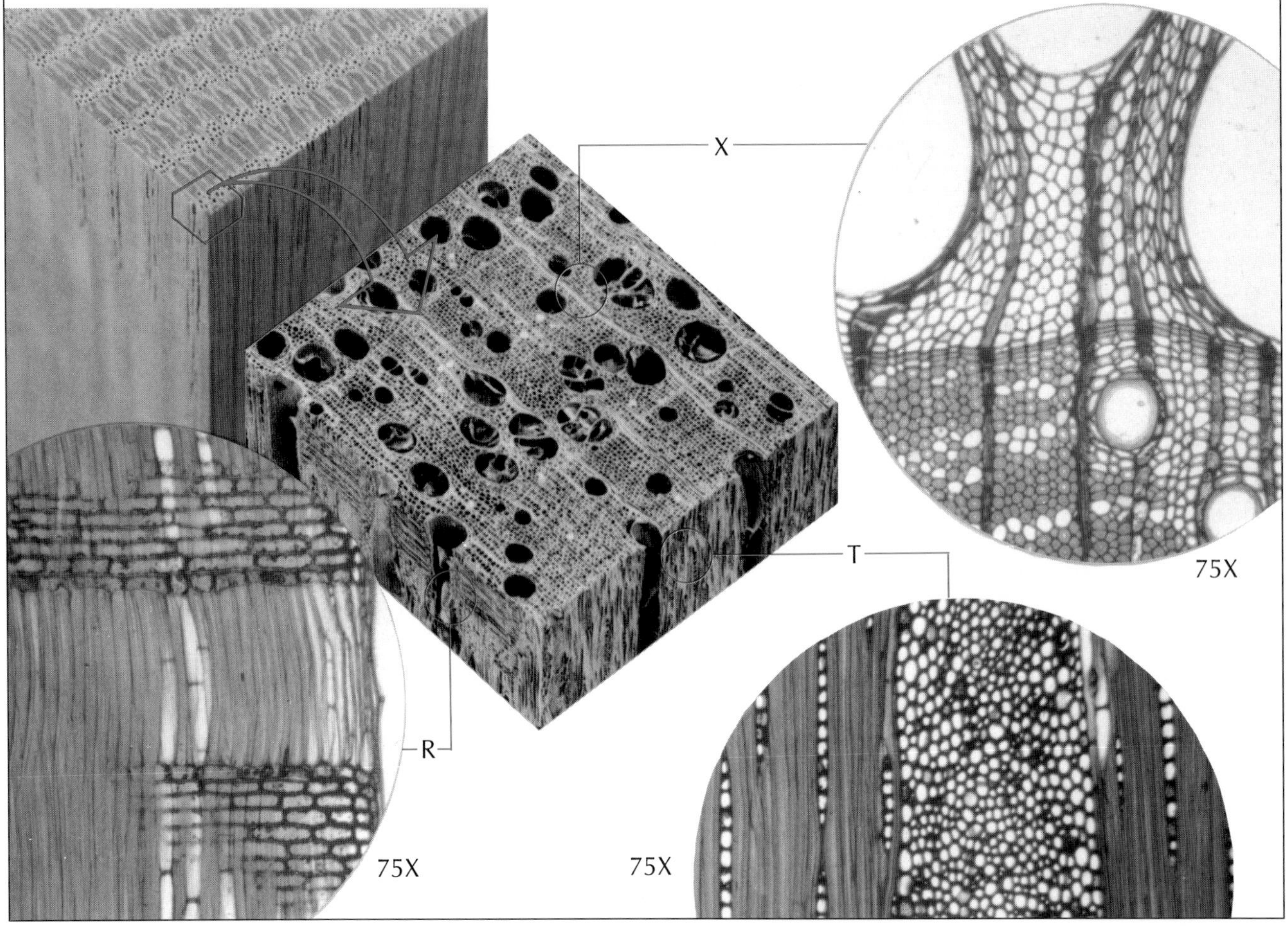

Center photo by Wilfred Côté

X

75X

T

R

75X   75X

lens view with the microscopic view of the same red oak sample (above). Although red oak can be identified readily with the hand lens, here we see that it is anything but simple in structure.

Growth Rings

As mentioned in Chapter 2, hardwoods that grow in the temperate zone undergo annual growth/dormant cycles that produce wood in annual growth rings. In some species these growth rings may be clearly evident due to a concentration of large earlywood pores; in others the rings may be very faint because of insignificant variation in cell characteristics. The trained eye can usually recognize even the more indistinct growth rings on a cleanly cut cross-sectional surface.

It should also be noted, however, that in tropical regions growth may be continuous, so it is possible for wood to form without typical rings. However, growth may be interrupted at intervals, although not necessarily on an annual or other cycle.

These interruptions can result from variation in rainfall or from other causes, and the consequent growth rings can be either periodic or sporadic. For example, in Central American mahogany — the traditional and familiar furniture wood — visible growth rings are formed periodically due to a distinct fine line of parenchyma called marginal parenchyma.

In some other woods, visible growth rings may result from cyclic variation in pigmentation. However, these zones of pigmentation may not coincide with the concentric anatomical arrangement of the wood.

At this point it is appropriate to examine the various cell types found in hardwoods. Each cell type is discussed separately in terms of shape, size, arrangement within the wood and features. As you will see, each cell type contains a number of features that are useful in identification. The drawing at left on the facing page characterizes and compares these basic cell types.

HARDWOOD CELL TYPES

This diagram shows the relative sizes and shapes of typical cell types found among hardwoods.

Longitudinal cell types

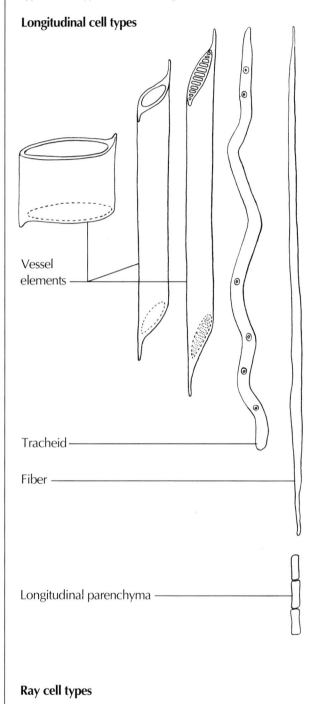

Vessel elements

Tracheid

Fiber

Longitudinal parenchyma

Ray cell types

Upright cells

Procumbent cells

Vessel Elements

Vessel elements are the largest cells in hardwoods (drawing below). With their thin walls and spacious cell cavities, they are well adapted to sap conduction. They always occur end to end in vertical series, and they are unique in that each end wall has an opening called a perforation. This perforation is matched to a similar perforation in the adjacent vessel element, so the vertical series of elements forms a continuous conductive passageway called a **vessel**.

When a vessel is split lengthwise, as often happens when you section a longitudinal surface, it appears as a minute groove called a **vessel line**. Depending on vessel diameter and the presence or absence of contents, vessel lines may have a conspicuous and unique appearance. This makes them important identification features.

When cut transversely, the opening of a vessel on a cross-sectional surface is called a pore. Pores vary in size according to vessel diameter from a minimum of 50 to 60 micrometers, as in red gum, to over 300 micrometers, as in oaks. Among hard-

RELATIVE HARDWOOD CELL DIAMETER AND WALL THICKNESS

This diagrammatic transverse view of hardwood longitudinal cell types gives an approximate comparison of cell diameter and cell-wall thickness.

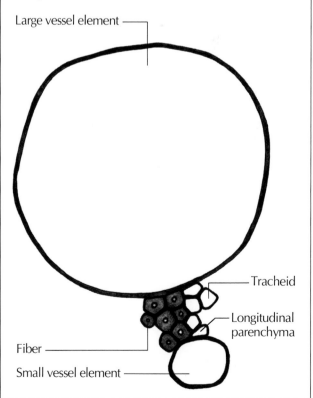

Large vessel element

Tracheid

Longitudinal parenchyma

Fiber

Small vessel element

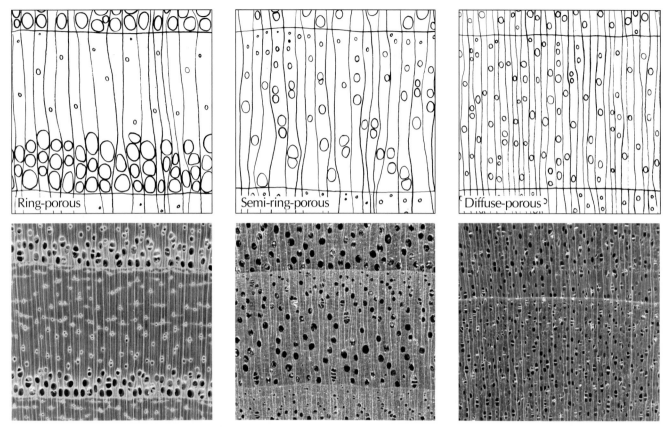

CLASSIFICATION OF RING POROSITY
Hardwoods are classified as ring-porous, semi-ring-porous (or semi-diffuse-porous) or diffuse-porous, based on pore size and distribution within a growth ring as viewed with a hand lens in cross section.

woods, texture designations are based on pore size and range from fine-textured for the smallest pores to coarse-textured for the largest.

One of the first steps in identifying a hardwood sample is to classify it according to pore size and distribution within a growth ring (photos and drawings above). If the pores are uniform in size and evenly distributed, the wood is **diffuse-porous**. Maples (*Acer* spp.) are diffuse-porous woods with fine texture; American mahoganies (*Swietenia* spp.) are diffuse-porous but coarse-textured.

A **ring-porous** wood is one that has a distinct concentration of larger pores in the earlywood. The earlywood may consist of a single row of these pores, as in American elm, or may be two to several pores wide, as in white ash or sassafras.

In ring-porous species, the zone of large pores identifies the earlywood portion of the ring. The transition from earlywood to latewood may be sharply defined by an abrupt change in pore size, or it may be poorly defined due to gradually decreasing pore diameter.

If a wood has a concentration of earlywood pores at the growth-ring boundary but the pores are about the same size as others throughout the growth ring, the wood is considered diffuse-porous, not ring-porous. A good example is

black cherry, where a distinct line of pores is usually conspicuous at the very first portion of the growth ring, tempting us to classify it as ring-porous.

Woods whose pore diameters vary gradually from large at the earlywood edge to much smaller at the latewood edge, with no clear separation of earlywood and latewood zones, are described as **semi-ring-porous** or **semi-diffuse-porous**. Black walnut, butternut and persimmon are semi-ring-porous woods.

Although woods of a given species usually fall into one of the above categories, some species range from one classification to the next. True hickories, for example, are typically ring-porous, but pecan hickories vary from ring-porous to semi-ring-porous. Cottonwood and willow are typically diffuse-porous, but some species show enough variation in pore size to be considered semi-diffuse. It should also be noted that tropical woods are typically diffuse-porous. An exception is teak, which can be either, although it is most often ring-porous.

Pores are further described according to their position relative to one another in cross section. This configuration is known as pore arrangement.

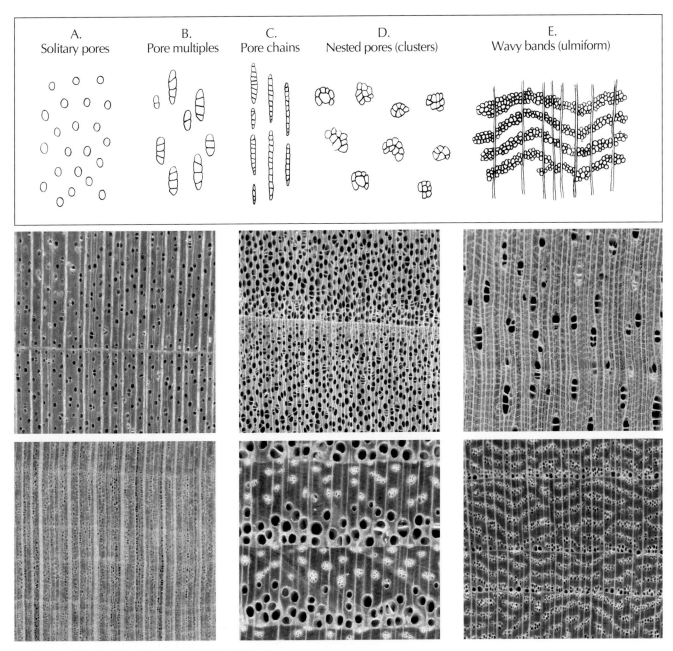

PORE ARRANGEMENTS IN HARDWOODS
The drawing above illustrates various pore arrangements, which are described according to the relationship of pores to one another within a growth ring as they appear on a cross section. Examples of each type are shown in the corresponding photographs. Top row: (A) maple, (B) cottonwood and jelutong; bottom row: (C) holly, (D) Kentucky coffeetree, (E) American elm. (15x)

In hardwoods, pores are arranged in various ways (see the drawing above). A **solitary pore** is a single pore that does not touch any others (e.g., as in maple, top row, left). A **pore multiple** occurs where two or more pores come in contact with one another (cottonwood, top row, center). Multiples are most commonly arranged in radial groups, which are more precisely called **radial multiples**, as are seen, for example, in jelutong (top row, right). When tightly crowded, the multiple may appear as a single large oval pore, elongated in a radial direction with one or more divisions. In multiples where larger numbers of pores contact one another both radially and tangentially, as in Kentucky coffeetree (bottom row, center), the term **nested pores** or **pore cluster** is used. Radially arranged series of multiples or closely arranged solitary pores are called **pore chains**, as seen in holly (bottom row, left). Pores arranged in irregular concentric bands that are more or less tangential are called **wavy bands** (bottom row, right), also referred to as **ulmiform** because this arrangement is a distinctive feature of all elms (*Ulmus* spp.).

On longitudinal surfaces, vessel lines may hint at pore arrangement , as shown in the photos below. For example, on the tangential surface of ash, the earlywood vessel lines are crowded and in multiple layers, as opposed to the predominantly single rows of more widely spaced pores in hickory. In woods such as elm and hackberry, the wavy bands of latewood pores are revealed as distinctive zigzag patterns of vessel lines in the latewood portions of tangential board surfaces.

Not all vessels have contents, but where contents are present, they can be valuable in identification. **Tyloses** (see the photos on the facing page) are bubblelike structures that form in the cell cavities of vessels of some species.

Tyloses may be absent or sparse (as in most red oaks), variable (as in chestnut and ash), abundant (as in most white oaks) or densely packed (as in black locust). In some species, including black locust, tyloses have a characteristic iridescence or sparkle.

VESSEL LINES ON LONGITUDINAL SURFACES

Vessel lines on longitudinal surfaces can be diagnostic. In ash (top left), the earlywood is comprised of a multiple layer of large pores, whereas in hickory (top right), earlywood is clearly a single row of the largest pores. In elm (bottom left) and hackberry (bottom right), the ulmiform latewood pores show as jagged patterns of short vessel lines in the latewood areas.

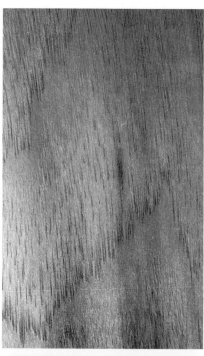

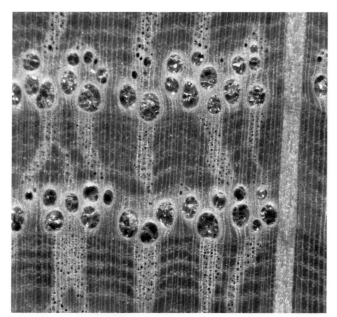

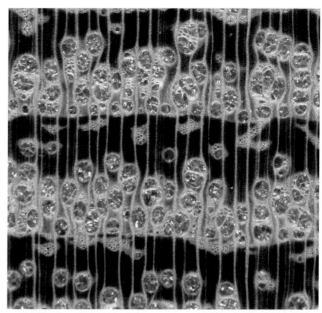

TYLOSES IN VESSEL ELEMENTS
Abundant tyloses can be seen in the vessel elements in these cross-sectional views of white oak (left) and black locust (right). (20x)

Other significant vessel contents include whitish or chalky mineral deposits and reddish or brown gummy deposits.

MICROSCOPIC DETAIL

The double end wall joining two consecutive vessel elements is called a perforation plate because it has a perforation through it.

There are two basic types of perforation plates, simple and scalariform. A **simple perforation** denotes a single opening, usually occupying almost the entire plate. The remaining wall at the edge of the plate is called the **perforation rim**. Simple perforations are found in many species, including oaks, ashes, maples and basswood. A **scalariform perforation** consists of parallel openings, the remaining **bars** lending a ladderlike or gratelike appearance to the plate. Bars range in number from few, as in yellow-poplar, to many, as in sweetgum. In some species, such as sassafras, both types of perforation plates occur. Beech has a transitional type of perforation plate with a simple opening in the center and scalariform areas above and below. Irregular or distorted bars in scalariform plates are called **reticulate** or **foraminate perforations**.

Of the two types of perforations (as seen in the drawing at right), scalariform is easier to detect because it has a distinctive appearance. Even where the scalariform plate is damaged or partially cut away, the barlike pattern is still recognizable. Simple perforations are more difficult to confirm. It helps to remember that in large vessels where the plates are simple, they are nearly perpendicular to the vessel. In smaller vessels, the plate is usually at an angle, sometimes forming a very long sloping end wall. Often the orientation of the plate approaches the radial plane. Therefore, a radial section is more apt to show a

SIMPLE AND SCALARIFORM PERFORATION PLATES

The composite end wall formed where adjoining end walls of two vessel elements have matching openings, or perforations, is called a perforation plate. A perforation plate is simple if it has a single large opening, scalariform if it has multiple, slotlike openings.

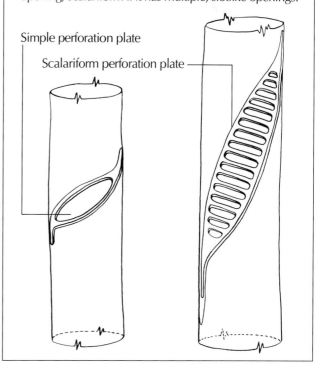

Simple perforation plate

Scalariform perforation plate

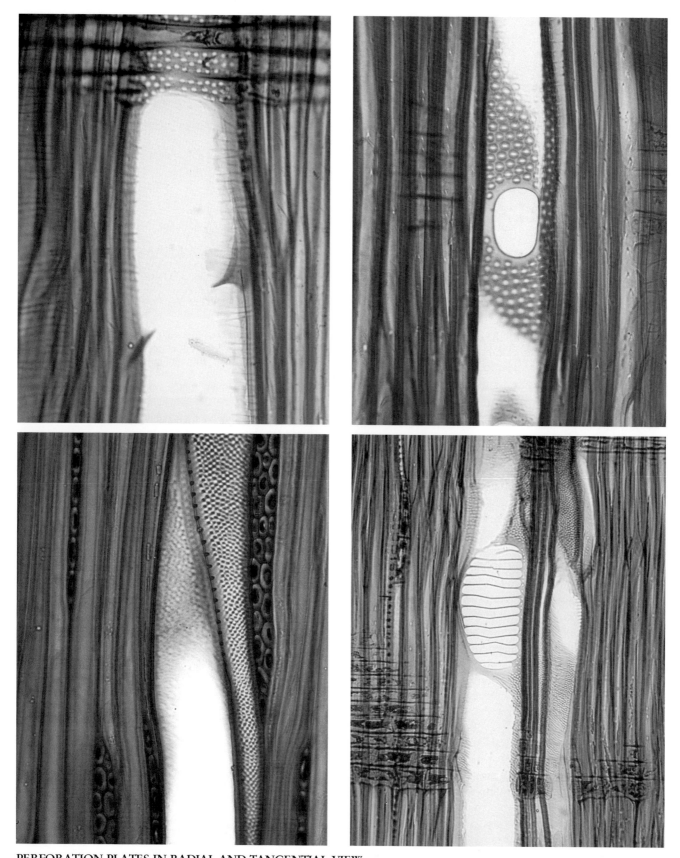

PERFORATION PLATES IN RADIAL AND TANGENTIAL VIEW
These micrographs show simple perforation plates in maple (viewed tangentially at top left and radially at top right) and scalariform perforation plates in yellow birch (viewed tangentially at bottom left and radially at bottom right). (350x)

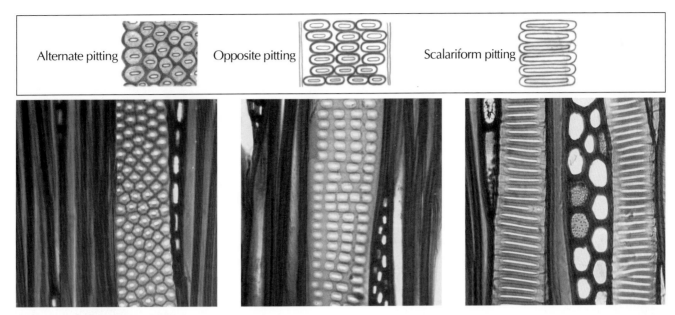

Alternate pitting   Opposite pitting   Scalariform pitting

INTERVESSEL PITTING TYPES

Intervessel pitting occurs along the common lateral walls joining two vessels. Pitting is classified according to shape and arrangement into three broad types, as indicated in the micrographs of aspen (alternate pitting), left, yellow-poplar (opposite pitting), center, and magnolia (scalariform pitting), right. (275x)

perforation plate in face view (see the photos on the facing page). You should consider this when cutting sections to observe perforation plates.

Where two vessels are in lateral contact, the fairly wide common wall joining them is usually heavily pitted. The characteristic appearance of this **intervessel pitting** can also be very helpful in identification.

Alternate pitting designates pits that are crowded together, resulting in an irregular or diagonal arrangement. **Opposite pitting** refers to short rectangular or slightly rounded pits in a distinct horizontal arrangement. Elongated barlike pits in parallel, ladderlike arrangement are **scalariform,** as in magnolia.

The various types of pitting may intergrade. In sweetgum, for example, pitting intergrades between scalariform and opposite, and the pits range in shape from short and rectangular or oval to linear.

When inspecting a sample for intervessel pitting, first carefully study the cross-sectional view to determine the general orientation of pore multiples. This helps you decide which longitudinal plane is best for sectioning. If the pores are predominantly in radial multiples, a tangential section is more apt to include an intervessel wall.

Characteristic pitting sometimes occurs between vessel elements and other cell types, especially between ray cells and vessel elements, as in willow and aspen.

Because spiral thickenings are present in the vessel elements of only some species, they are also useful in identification. For example, hophornbeam has thickenings, but birch does not. In sweetgum and black tupelo, spiral thickenings are restricted to the ligules (elongated tips) of the vessel elements. The reticulate

(netlike) thickenings that occur sporadically in the latewood vessels of black walnut (*Juglans nigra*) are referred to as **gash pits** or **gashlike pitting** (see the photo below). This feature, when present, provides a positive means of separating black walnut from butternut (*J. cinerea*) and English walnut (*J. regia*), which lack gash pits.

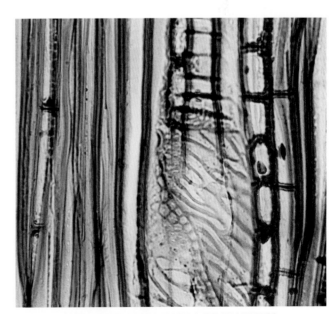

GASH-PIT THICKENINGS IN BLACK WALNUT

Reticulate thickenings, often referred to as 'gash pits,' on a latewood vessel wall in black walnut. (300x)

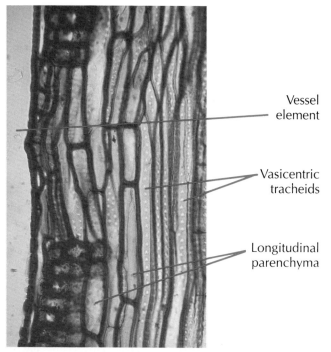

VASICENTRIC TRACHEIDS AMONG LONGITUDINAL PARENCHYMA
Vasicentric tracheids with abundant bordered pits are easily distinguished from the strands of longitudinal parenchyma cells as seen in this radial view of northern red oak. (200x)

Vessel element

Vasicentric tracheids

Longitudinal parenchyma

Vasicentric Tracheids

Tracheids are functionally intermediate between vessel elements and fibers. They are imperforate — closed at the ends — but fairly thin-walled and usually have numerous pits on the side walls. As the term vasicentric implies, they occur near vessel elements. Vasicentric tracheids are usually intermixed with longitudinal parenchyma (photo at left). The two form areas of "lighter-colored tissue" as viewed on cross sections with a hand lens.

MICROSCOPIC DETAIL
Vasicentric tracheids only appear separate and distinct microscopically. In longitudinal section they show up as small-diameter, thin-walled, abundantly pitted cells. In this view they are readily distinguishable from longitudinal parenchyma. Parenchyma cells are short, as indicated by the frequent cross walls.

Tracheids are present in oaks and chestnut in the vicinity of the large earlywood pores. In ash, cells resembling vasicentric tracheids also occur among the earlywood vessels. Since ash, oak and chestnut can be identified readily on the basis of macroscopic features, these microscopic clues are useful primarily when only small slivers or fragments are available for identification. Tracheids intergrade somewhat with fibers, which are discussed in the next section.

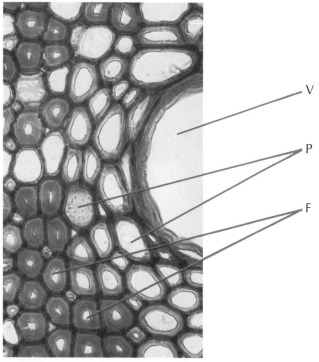

FIBERS IN OAK
In this cross-sectional view of latewood in red oak, the small-diameter fibers (F) are easily distinguished from the thinner-walled parenchyma cells (P). Note the thick wall of the vessel (V) in the pore at right. (200x)

V

P

F

PARENCHYMA ARRANGEMENTS

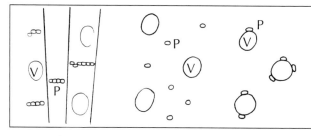

Apotracheal diffuse-in-aggregates

Apotracheal diffuse

Paratracheal scanty

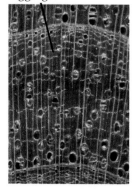

(Apotracheal diffuse and paratracheal scanty parenchyma are not visible with hand-lens magnification.)

Butternut (10x)

Fibers

Fibers are typically long, slender, straight cells whose ends taper to a point. They are among the smallest-diameter cells. Because of their small size and thick walls, they appear as darker areas when seen en masse in cross section. Without higher magnification they simply form a background for pores, rays and parenchyma. A microscope must be used to observe greater detail.

MICROSCOPIC DETAIL

With a microscope, there is a clear contrast between the thick-walled fibers and the thinner-walled parenchyma and tracheids (see the photo at bottom left on the facing page).

Note, however, that in lower-density woods such as aspen and balsa, the fibers can be fairly thin-walled. In cross section they are usually somewhat rounded in outline. But in a few species, sweetgum for example, the fibers are distinctly angular, which can be helpful in identification.

Fibers can be separated into two intergrading types, **libriform fibers** and **fiber tracheids**. Fiber tracheids have bordered pits; libriform fibers have simple pits. However, these differences are difficult to distinguish with an elementary microscope. Therefore, this book will refer to fibers collectively; it will not distinguish between the two types.

Parenchyma

Parenchyma arrangement is described according to the relationship between parenchyma cells and pores as viewed in cross section (see the photos below). If the parenchyma cells make direct contact with the pores, they are **paratracheal parenchyma**. If they are separated from the pores by rays or fibers, they are **apotracheal parenchyma**.

Single, isolated cells of apotracheal parenchyma are called **diffuse parenchyma**; they cannot be seen without a microscope. Apotracheal parenchyma occurring in short tangential lines is described as **diffuse-in-aggregates parenchyma**. In woods like basswood and birch, this parenchyma is visible only with a microscope, but in other woods like black walnut and butternut, diffuse-in-aggregates parenchyma can be seen with a hand lens.

Single cells of paratracheal parenchyma are called **scanty paratracheal parenchyma**, or simply, **scanty parenchyma**. If parenchyma cells are numerous enough to form a complete or partially complete sheath around a pore, they are called **vasicentric parenchyma**. As with scanty parenchyma, a broken layer of vasicentric parenchyma might not be visible with a hand lens. But a complete layer of parenchyma will usually be apparent at that level, particularly where the layer is several cells thick, as in the latewood of white ash.

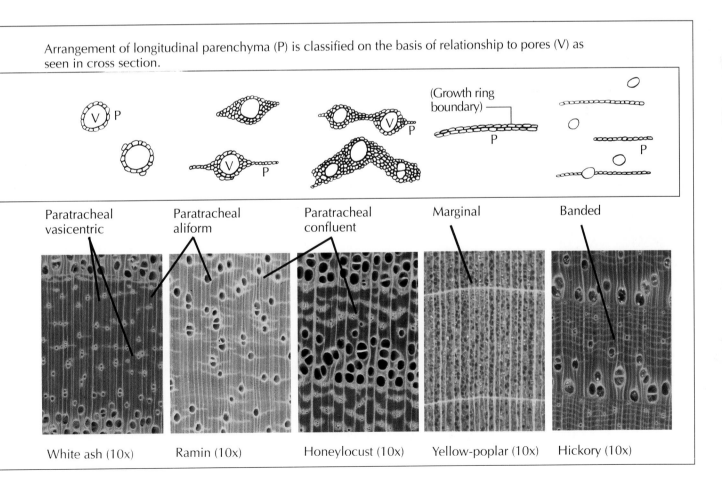

Arrangement of longitudinal parenchyma (P) is classified on the basis of relationship to pores (V) as seen in cross section.

(Growth ring boundary)

Paratracheal vasicentric	Paratracheal aliform	Paratracheal confluent	Marginal	Banded
White ash (10x)	Ramin (10x)	Honeylocust (10x)	Yellow-poplar (10x)	Hickory (10x)

Vasicentric parenchyma that occurs in a well-developed layer with additional cells forming tangential winglike extensions on either side is called **aliform parenchyma**. The aliform parenchyma of ramin, a tropical wood, is a good example. When parenchyma forms a continuous tangential or diagonal zone connecting two or more pores, it is called **confluent parenchyma**. The term **zonate parenchyma** is often used when cells form well-developed continuous bands.

Banded parenchyma describes the distinct tangential lines of cells that usually occur with fairly regular spacing, as in hickory. The lines may intergrade with shorter lines of diffuse-in-aggregates parenchyma, as in black walnut. In species with banded parenchyma, some of the lines appear to make contact with the pores while others do not; consequently no attempt is made to classify banded parenchyma as either apotracheal or paratracheal. When lines of banded parenchyma join with ray lines of approximately the same width to form a meshlike or netlike pattern—as in hickory and persimmon—the term **reticulate parenchyma** is used.

Parenchyma occurring at the growth-ring boundary is called **marginal parenchyma**. The terms **initial** and **terminal** parenchyma are also used. Single cells are not discernible with a hand lens, but in species such as yellow-poplar and American mahogany the growth ring is clearly delineated by a line of marginal parenchyma several cells thick.

MICROSCOPIC DETAIL

Microscopically, longitudinal parenchyma cells are seen to occur in strands of several to many cells — hence, the term **strand parenchyma**. In some species, crystals or gummy deposits are found in parenchyma cells, and this is important in identification. For example, crystals occur in black walnut (see the photo below), but not in butternut or English walnut.

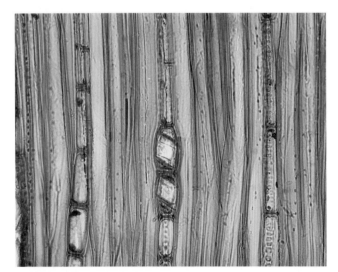

CRYSTALS IN BLACK WALNUT
This micrograph shows crystals in the longitudinal parenchyma cells of black walnut. (250x)

Rays

Ray size in hardwoods varies widely among species or even within a single species. Thus, the relative size and appearance of rays on all three principal surfaces are important identification characteristics. Measurements of particular significance are height along the grain and tangential width. Ray length, however, is of no value, since rays are continuous along a radial plane once they have formed.

The narrowest rays are only one cell, or 10 to 15 microns (about 0.01mm or 1/2000 in.) wide. They are invisible to the unaided eye. The widest rays may be 40 or more cells or 300 microns (about 1/64 in.) wide; they are distinctly visible. It should be noted, however, that it is not only size that determines visibility, but also contrast with adjacent tissue. In oak, even one-cell-wide rays can be seen on a cross-sectional surface with a hand lens against the contrasting background mass of dense fibers.

Actual ray size is difficult to judge. It is often useful to compare ray width to pore diameter (see the photos at the top of the facing page). In maples, for example, the widest rays are approximately the same width as the largest pores; in birch, maximum pore diameter is much larger than the widest rays. Comparing ray width to parenchyma band width can also be informative.

The number or frequency of rays over a given area and their relative spacing are also important. One method of counting is to measure the number of rays that cross the growth-ring boundary over a fixed distance along the boundary — over a distance of one millimeter, for example. Another method is to estimate how many pore diameters would fit into the space between two rays an average distance apart.

You can also attempt to estimate what percentage of the cross-sectional surface is occupied by ray tissue. Obviously, this is a less precise, more subjective estimate, but it can be useful if you are reasonably consistent in your criteria for judging from sample to sample. Another subjective observation I make when examining rays is whether they are straight or appear to weave as they pass the pores.

When rays swell at the growth-ring boundary they are called **noded rays** (see the photos at the bottom of the facing page). These are seen in yellow-poplar, beech and sycamore.

A tangential surface cuts through the rays and shows them in cross section. They may appear as fine vertical lines, sharply pointed at the top and bottom. On this surface, they range from invisible to clearly distinct, depending on size, shape and contrast with background longitudinal tissue. The tiny rays of chestnut cannot be seen in this view at all, even with a hand lens. In black cherry rays are perceptible with a hand lens but not with the unaided eye. In maple they are visible with the naked eye and distinct with the lens. And in beech, sycamore and oak (see the photos on p. 42) they are easily seen without any magnification.

In a ring-porous wood like oak, the rays show up much more distinctly when the wood is sectioned through the latewood

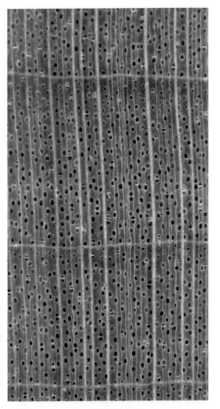
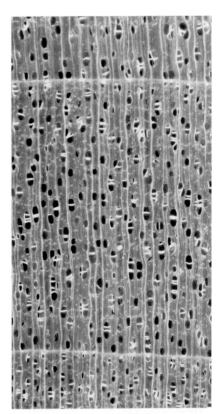

RAY WIDTH/PORE DIAMETER
COMPARISON
Comparing ray width to pore diameter
quickly distinguishes maple (near right) in
which the largest rays (R) are about the
same width as the largest pores (P), from
birch (far right) in which the largest rays
are much narrower than the largest
pores. (20x)

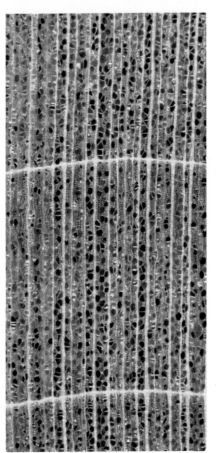
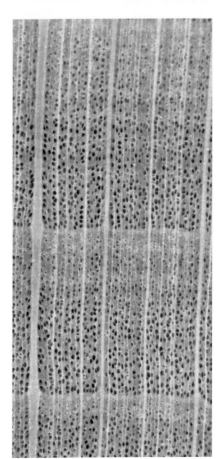

NODED RAYS
Noded rays appear to flare or become
swollen as they cross the growth-ring
boundary, as shown in the cross-sectional
views of yellow-poplar (near right) and
beech (far right). (20x)

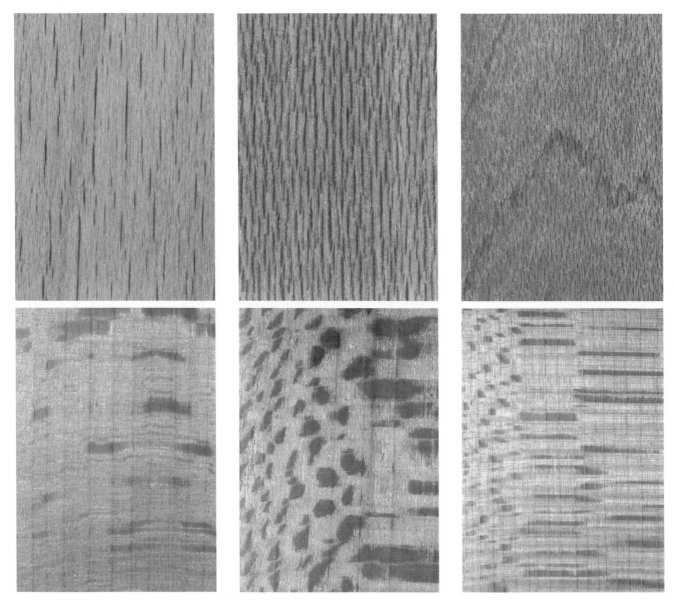

RAY HEIGHT AND RAY FLECK IN THREE HARDWOODS
These pairs of macrographs compare ray heights as seen tangentially (top row) and ray fleck as seen radially (bottom row) in American beech (left), sycamore (center) and maple (right).

rather than through the earlywood. The dense fibers make a uniform background against which the outline of the rays is distinctly seen.

Ray height (see the photo at right) is best evaluated in tangential view. It varies from imperceptibly small to several inches high, as in white oaks.

Aggregate rays are large rays that have longitudinal cells running irregularly through them. When the sample is viewed in tangential section, it appears as though there are multiple rays that are closely arranged but separated by longitudinal cells. These rays are usually found in addition to smaller rays in the same species; the aggregate rays are relatively few in number and may be sporadic in distribution. Red alder and American

RAY HEIGHT IN WHITE OAKS
Ray height in white oaks usually exceeds 1¼ in.

AGGREGATE RAYS IN ALDER
The large, scattered aggregate rays in red alder are conspicuous on the tangential surface.

hornbeam are two species that show aggregate rays (see the photo above).

In some species the rays are of fairly even height and are spaced at even horizontal levels throughout the tree. As a result, they form even horizontal rows when the wood is sectioned tangentially. These are called **storied rays** and produce **ripple marks** (see the photos at right).

Ripple marks appear to the unaided eye as horizontal striations of lighter and darker wood ¼₄ to ⅓₂ in. apart. They are rare in domestic species — persimmon is one of the few that has ripple marks — but fairly common in tropical woods, notably in Central American mahogany (although not as a constant feature), rosewood and ebony. It should be noted that longitudinal cells can also be storied. Ripple marks can result from these alone, or they can result from a combination of longitudinal cells and storied rays.

When wood is sectioned radially, many rays are split and exposed at the surface. The resulting patches of ray tissue — to the extent that they contrast with the background longitudinal tissue — are called **ray fleck** (see the photos on p. 44).

The prominence of ray fleck (see the photos on the facing page) depends on the size of the rays and how closely the cut approximates a true radial plane. In yellow-poplar, ray fleck appears lighter against a darker background; in maple and

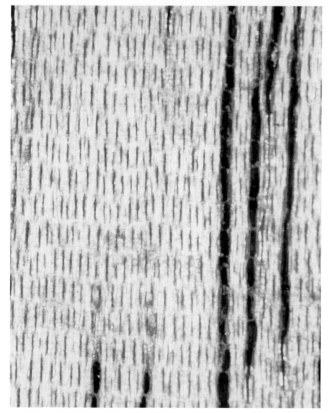

RIPPLE MARKS AND STORIED RAYS
The ripple marks commonly seen on tangential surfaces of American mahogany (top) result from storied rays, which can be seen more clearly at higher magnification (bottom). (20x)

RAY FLECK
Ray fleck on radial surfaces of cherry (top) and oak (above).

sycamore, the rays are darker against a lighter background. The contrast can also vary between sapwood and heartwood in the same species. In black cherry, for example, the ray fleck appears darker than the background in the sapwood, but lighter than the background in the heartwood.

The term **silver grain** is sometimes used to describe the glossy ray fleck in oak. In veneer and lumber manufacturing, white oak is often riftcut (cut at an angle to the rays) to produce a uniform, narrow and vertical ray fleck called **needlepoint grain** or **comb grain.** Lacewood, a well-known tropical hardwood, is also cut at an angle to the radial plane; its uniformly sized and spaced rays form its familiar repetitive, lacelike surface appearance.

With a little experience, you will recognize the characteristic ray fleck, where present, of various species. Remember that ray fleck is particularly dependent on the plane of the cut and may not appear on a particular sample. But on a spindle or turning, look for the radial sides where ray fleck will likely appear.

MICROSCOPIC DETAIL

Viewed microscopically, rays have several important identification characteristics.

On a tangential surface, we first note the ray width. A ray that consists of a single vertical series of cells — that is, a ray that is only one cell wide — is a **uniseriate ray**. If it is two cells wide, it is a **biseriate ray,** and if it is three or more cells wide it is **multiseriate.** To be more precise, you can attach specific numbers to the designation: 5-seriate, 8-seriate, etc.

Ray "seriation" can be extremely valuable. For example, willows *(Salix* spp.), aspens and cottonwood (both of which are *Populus* spp.) and buckeyes *(Aesculus* spp.) have uniseriate rays. But sweetgum, basswood, yellow-poplar and magnolia have multiseriate rays. In soft maples the rays are generally no more than 5-seriate, but in hard maples they can be up to 8-seriate.

Ray height is not as broadly useful for identification due to its variability within species, but occasionally it is definitive. Pacific dogwood, for example, has rays up to about 40 cells high, whereas the rays of flowering dogwood may reach 80 cells in height.

Although rays in coniferous woods may contain ray tracheids as well as ray parenchyma, hardwood rays consist entirely of ray parenchyma cells. However, there are two types of ray parenchyma cells, classified on the basis of overall shape. In radial view, **procumbent ray cells** are elongated horizontally; **upright ray cells** are either squarish or vertically oriented (photo at bottom on the facing page).

A hardwood ray that contains both upright and procumbent ray cells is called a **heterocellular** or **heterogeneous** ray (note that these terms have a different meaning when used in reference to conifers). The upright cells are usually located in one or more rows along the margins of the rays; the main body of the ray, especially if multiseriate, is usually composed of procumbent cells. If only one type of cell is present

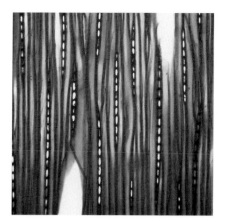

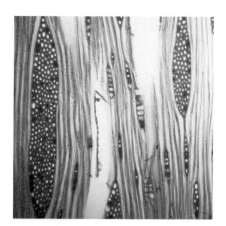

RAY SERIATION
Ray seriation, easily determined from tangential sections at relatively low microscopic power, is an extremely useful identification feature. Shown here are uniseriate rays in aspen (left), biseriate rays in sweetgum (center) and multiseriate rays in sugar maple (right). (100x)

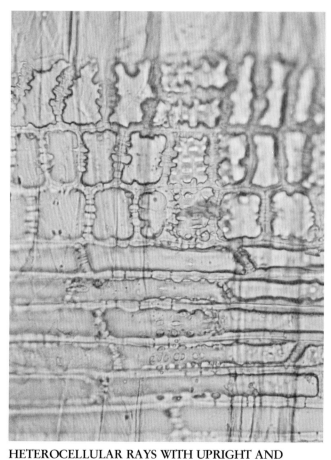

in the ray, the ray is **homocellular** or **homogeneous**. Most homogeneous rays have procumbent cells; homogeneous rays with upright cells are less common.

Ray composition is sometimes helpful in separating woods. In cottonwoods and aspens, for example, the rays are homocellular; in willows they are heterocellular.

As in the case of longitudinal parenchyma, ray parenchyma cells of some species contain crystals that help in identification.

HETEROCELLULAR RAYS WITH UPRIGHT AND PROCUMBENT CELLS
A heterocellular ray, shown here in yellow-poplar, has both upright and procumbent cells. (400x)

CHAPTER FIVE

PHYSICAL AND CHEMICAL PROPERTIES

In the identification of wood, the physical and chemical properties of a sample are important supplements to its anatomical features. In a few cases these properties provide an immediate answer.

Such physical characteristics as color, luster, odor and hardness can be detected simply by observing and handling a piece of wood. But other characteristics, especially chemical properties, reveal themselves only through the indirect process of testing. Some of the more sophisticated chemical analyses require complicated procedures, but there are also a number of relatively simple tests that are useful in wood identification.

Color

Color is probably the first thing we perceive when we look at a piece of wood. As discussed earlier, the distinctive colors associated with many woods result from extractives deposited during heartwood formation. Basic wood substance — cellulose and lignin — has little color of its own, so any distinctive color is associated with heartwood.

In evaluating color in a sample, it is important to consider whether it is heartwood or sapwood. A dark color always indicates heartwood, but a light color can be either heartwood or sapwood. Sapwood is typically pale, ranging from near-white to pastel shades of tan, yellow and cream, although it sometimes contains a tinge of blue-grey or red. Heartwood can be light-colored too. But even in woods with light-colored heart-

wood, a subtle color difference may help to distinguish sapwood from heartwood.

Many common woods are known for their distinctive heartwood colors: the chocolate brown of black walnut, the deep purple-red of junipers, the black of ebony, the near-white of holly, the brilliant red-orange of padauk (vermilion) wood.

In some species, color is reasonably constant — for example, the soft medium-brown of chestnut. But in other species shades vary. Black cherry ranges from light cinnamon brown to deep reddish brown. A species may even exhibit different colors. In yellow-poplar, whose sapwood is a flat creamy white, the heartwood is generally a characteristic green, but specific areas can be reddish, brown-to-black or streaked in combinations of these colors.

Some woods undergo considerable color change with age or exposure. Freshly cut mahogany is usually light brown with a pink cast. With age, however, it turns deep reddish brown. Purpleheart is more brown than purple when first cut; eventually it turns a rich purple at the surface. Just the opposite occurs in eastern redcedar — the deep purple of the fresh wood ages to brown.

Color may be deposited in uneven patterns, giving the wood a characteristic appearance. This is called pigmented figure (discussed in Chapter 6).

Even where color is perfectly obvious and consistent, matching that color to a verbal description in an identification book can be difficult. Terms such as "light reddish brown" and "deep olive brown" mean different things to different people. It is valuable to have a reference collection of known woods to compare with written descriptions of wood colors.

Luster

Luster is a measure of light reflection. It varies among species and it also varies with grain direction in a single sample. Pronounced luster — or the dullness where it is lacking — may be helpful in identification. Among conifers, spruce is noted for its lustrous longitudinal surfaces. This helps to separate it from white pines or true firs, with which it is often confused.

Within a species, radial planes often show greater luster than tangential planes due to the collective contribution of ray surfaces. The ribbon or stripe figure apparent on a radial surface of wood with interlocked grain results in large part from differences in luster associated with variable grain direction. Likewise, the visual effect of blister figure, quilted figure, fiddleback figure, and many other distinctive wood patterns discussed in the next chapter result in large measure from variable luster.

Tactile Properties

Virtually all woods can be machined and finished to a reasonably smooth surface. However, freshly machined surfaces of certain woods have a unique or characteristic feel. The denser, harder woods usually have a smooth, sleek-feeling surface, whereas low-density woods usually have softer, drier-feeling surfaces.

The texture of the wood or the evenness or unevenness of grain may contribute to the tactile quality, and extractives and resins can also play a role. For example, the surface of baldcypress is noticeably tacky, whereas redwood feels dry; tamarack sometimes feels waxy. (By contrast, I well remember a dense wood called manwood *[Minquartia guianensis]* whose dry, planed surfaces were so slippery and slick it was impossible to stack many pieces — they simply slid off one another. In this case, an extremely low coefficient of friction is a noteworthy property.)

Odor

Odor is an important identification feature in woods where it is both distinctive and constant. Among conifers, this includes Douglas-fir, incense-cedar, Alaska yellow-cedar and the junipers. The familiar fragrance of pine is shared among many species. Among hardwoods, teak and sassafras have distinctive odor. Although the ability to detect odor varies with the individual, many find the subtler odors of basswood, baldcypress, catalpa and other woods distinctive enough to be informative.

In some species, the green wood has an odor that virtually disappears as the wood dries. This is true of the sour odor of cottonwood. Generally speaking, odor in dry wood is more pronounced in recently cut surfaces and is usually enhanced by moistening the cut surface. Where odor results from extractives, heartwood will probably have a stronger odor than sapwood.

However, where odor results from resin, as in pines, the sapwood may be quite fragrant, especially in the new wood.

Odor can be transferred among samples. Many a student has been dismayed to find that an entire wood collection packed in a tightly closed box has acquired the odor of the one eastern redcedar sample.

Density and Specific Gravity

Density is expressed as weight per unit of volume — pounds per cubic foot in the English system, grams per cubic centimeter in the metric system. Water has a density of 62.4 lb./ft.^{3} or, metrically, 1 g/cm^3. **Specific gravity** is relative density, calculated as the ratio of the density of a substance — in this case, wood — to the density of water.

To avoid the effect of the weight of water in the wood, we use oven-dry weight for standard measurements of density and specific gravity. Since the volume of wood varies as the wood shrinks and dries, volume is usually measured at a specified moisture content, such as 12% ("air-dry").

As an example, if a given sample of wood has an oven-dry weight of 37.44 lb./ft.^{3} — 0.6 g/cm^3 — it is six-tenths (0.6) as heavy as water and thus has a specific gravity of 0.6. Woods denser than water have a specific gravity greater than 1.0 and sink; woods that are less dense have a specific gravity less than 1.0 and, if dry, will float.

Specific gravity varies somewhat within a species. Among temperate woods, specific gravities of most individual samples will fall within ±0.05 of the published species average. For this reason, specific gravity is useful in identification. For example, butternut and black walnut are quite similar in many ways, but butternut has an average specific gravity of 0.38, whereas black walnut's is 0.55. Tropical species may be more variable in density. Mahogany, for example, varies from as low as 0.40 to as high as 0.83.

Table 5.1 on pp. 49-51 gives specific gravities for the woods covered in this book. It should help narrow the range of possibilities for a given sample. It also conveys a sense of the range of specific gravity among woods and indicates the relative position of any particular species among other species.

If you have a collection of some or all of the woods listed in Table 5.1 machined to exactly the same dimensions, you can get a sense of the range of specific gravity simply by picking them up. Moreover, an unknown wood, machined to these same dimensions, can be weighed and compared to the known species or simply lifted to get an approximate comparison. (For this comparison to be meaningful, the woods should be of about the same moisture content, which can be accomplished by allowing them to equalize in the same atmosphere.)

If you wish to measure specific gravity, the method described on p. 49 will give a fairly close value. It applies only to those woods whose specific gravity is less than 1 — in other words, to woods that float. This accounts for most of the species in this book.

FLOTATION METHOD FOR SPECIFIC GRAVITY: MATERIALS

In using the flotation method to determine specific gravity, float the stick of wood endwise to determine its depth of immersion. Use a slender vessel if available (left); otherwise, fashion a guide from a single strand of wire as shown (right). Be sure the piece of wood can move freely within the wire ring guides before noting the waterline.

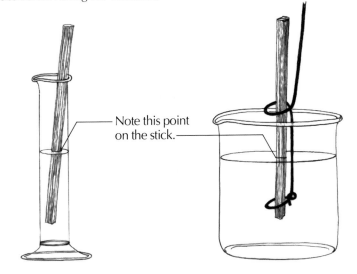

Note this point on the stick.

FLOTATION METHOD FOR SPECIFIC GRAVITY: CALCULATION

When using the flotation method, calculate the ratio of the immersed length (L') to the total length (L) of the sample. For example, a stick of maple is $7\frac{1}{2}$ in. long (L) and is assumed to be at 8% moisture content. When sealed and floated, it has an immersed length (L') of $5\frac{1}{16}$ in. The ratio is $5\frac{1}{16}$ in. / $7\frac{1}{2}$ in. = 5.0625 in. / 7.5 in. = 0.675.

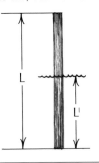

CHART FOR DETERMINING SPECIFIC GRAVITY/DENSITY

Find the calculated ratio of immersed length to total length (A = 0.675 in the example given in the illustration above right) on the bottom axis, then move directly upward to the intersection with the appropriate moisture content line (B). Now move horizontally left to find the specific gravity of 0.625 (C). The equivalent density is 39 lb./ft.^{3}

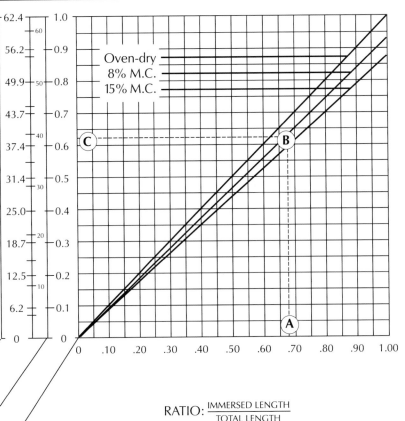

DENSITY (lb./ft.^{3})
SPECIFIC GRAVITY

RATIO: $\dfrac{\text{IMMERSED LENGTH}}{\text{TOTAL LENGTH}}$

Measuring Specific Gravity

Step 1: Estimate the moisture content of the wood:

a: If possible, dry the wood in an oven for 24 hours at 212°F. This allows you to assume a 0% moisture content. Otherwise;

b: If it has been stored outdoors in a dry location, assume a 15% moisture content;

c: If it has been stored indoors, assume an 8% moisture content.

Step 2: Cut a strip from the sample.

Step 3: Coat the strip with shellac if it is fine-textured, or with paraffin, which works for fine-textured and coarse-textured woods. A sample can be coated by quickly dipping it into melted paraffin, allowing the paraffin to drain off and then placing it on a paper towel to cool. Be careful heating paraffin. Use a flameless heat source and keep the heat low, just above the melting point.

Step 4: Float the strip in a tall cylinder or hold it endwise with a wire guide (drawing , facing page, top left).

Step 5: Mark the depth at which the piece was immersed while floating. Measure the immersed portion (L¹), then measure the total length of the strip (L), as explained in the drawing, facing page, top right.

Step 6: Calculate the ratio of the immersed depth to the length of the strip (L¹/L)

Step 7: Using the chart on the facing page,

a: Locate the calculated ratio from Step 6 on the lower scale;

b: Find the intersection of this ratio with the appropriate moisture content line;

c: Move horizontally to the specific gravity/density scale at the left and read the value.

For example, because a piece of wood has been stored indoors, you estimate its moisture content at 8%. You remove a strip from one of its edges and coat it with shellac. Once dry, you place the strip in a cylinder filled partly with water. The strip measures 212mm in length and sinks to a depth of 143mm. The ratio of the immersed depth of the strip to the total length of the strip is 143/212mm, or .675. Using the chart, you find .675 on the lower scale, then move upward to the 8% moisture line. You follow this line to the left and find that the specific gravity, corrected for the 8% moisture content, is about .625 .

Hardness

Hardness of wood is closely related to specific gravity: the higher its specific gravity, the harder the wood. Special equipment is needed to measure hardness precisely, but you can obtain a surprisingly informative reading simply by running your thumbnail across the grain on a longitudinal surface. For example, the difference in hardness between butternut and black walnut can usually be detected with the thumbnail test. Through trials on known woods of various densities, you will soon learn how deeply your thumbnail indents woods of different hardness. The secret is

TABLE 5.1: SPECIFIC GRAVITIES OF WOODS DISCUSSED IN THIS BOOK

Scientific Name	Common Name	Average Specific Gravity*
I. SOFTWOODS		
Abies	(true) fir	
A. alba	Norway fir	.43
A. amabilis	Pacific silver fir	.38
A. balsamea	balsam fir	.36
A. concolor	white fir	.37
A. grandis	grand fir	.40
A. lasiocarpa	subalpine fir	.34
A. magnifica	California red fir	.39
A. procera	noble fir	.38
Chamaecyparis	white-cedar	
C. lawsoniana	Port-Orford-cedar	.42
C. nootkatensis	Alaska-cedar	.44
C. thyoides	Atlantic white-cedar	.32
Juniperus virginiana	eastern redcedar	.47
Larix	larch	
L. decidua	European larch	.53
L. laricina	tamarack	.53
L. occidentalis	western larch	.51
Libocedrus decurrens	incense-cedar	.37
Picea	spruce	
P. abies	Norway spruce	.42
P. engelmannii	Engelmann spruce	.34
P. glauca	white spruce	.40
P. mariana	black spruce	.40
P. rubens	red spruce	.41
P. sitchensis	Sitka spruce	.40
Pinus	pine	
P. albicaulis	whitebark pine	—
P. contorta	lodgepole pine	.41
P. echinata	shortleaf pine	.51
P. elliottii	slash pine	.61
P. flexilis	limber pine	—
P. lambertiana	sugar pine	.36
P. monticola	western white pine	.38
P. palustris	longleaf pine	.58
P. ponderosa	ponderosa pine	.40
P. resinosa	red pine	.44
P. rigida	pitch pine	.52
P. strobus	eastern white pine	.35
P. sylvestris	Scots pine	.46
P. taeda	loblolly pine	.51
Pseudotsuga menziesii	Douglas-fir	.48

*oven-dry weight; volume at 12% moisture content

(continued on next page)

Sequoia sempervirens	redwood	.40
Sequoiadendron giganteum	giant sequoia	—
Taxodium distichum	baldcypress	.46
Taxus	yew	
T. baccata	European yew	.64
T. brevifolia	Pacific yew	.62
Thuja	thuja	
T. occidentalis	northern white-cedar	.31
T. plicata	western redcedar	.33
Tsuga	hemlock	
T. canadensis	eastern hemlock	.40
T. heterophylla	western hemlock	.44

II. HARDWOODS

Acer	maple	
A. campestre	field maple	—
A. macrophyllum	bigleaf maple	.48
A. negundo	boxelder	.46
A. nigrum	black maple	.57
A. platanoides	Norway maple	—
A. pseudoplatanus	sycamore (British)	.56
A. rubrum	red maple	.54
A. saccharinum	silver maple	.47
A. saccharum	sugar maple	.63
Aesculus	buckeye, horsechestnut	
A. glabra	Ohio buckeye	—
A. hippocastanum	horsechestnut	.45
A. octandra	yellow buckeye	.36
Alnus	alder	
A. glutinosa	common alder	—
A. incana	grey alder	.47
A. rubra	red alder	.41
Betula	birch	
B. alleghaniensis	yellow birch	.62
B. lenta	sweet birch	.65
B. papyrifera	paper birch	.55
B. pendula	European (silver) birch	.59
B. pubescens	European (white) birch	.59
Carpinus caroliniana	American hornbeam	.70
Carya spp.	(true) hickory	
C. glabra	pignut hickory	.75
C. laciniosa	shellbark hickory	.69
C. ovata	shagbark hickory	.72
C. tomentosa	mockernut hickory	.72
Carya	pecan hickory	
C. aquatica	water hickory	.62
C. cordiformis	bitternut hickory	.66
C. illinoensis	pecan	.66
C. myristiciformis	nutmeg hickory	.60
Castanea	chestnut	

C. dentata	American chestnut	.43
C. sativa	sweet chestnut	—
Castanopsis chrysophylla	giant chinkapin	.46
Catalpa	catalpa	
C. bignonioides	southern catalpa	—
C. speciosa	northern catalpa	.41
Celtis	hackberry, sugarberry	
C. laevigata	sugarberry	.51
C. occidentalis	hackberry	.53
Cornus	dogwood	
C. florida	flowering dogwood	.73
C. nuttallii	Pacific dogwood	—
Diospyros virginiana	common persimmon	.74
Fagus	beech	
F. grandifolia	American beech	.64
F. sylvatica	European beech	.64
Fraxinus	ash	
F. americana	white ash	.60
F. excelsior	European ash	.61
F. latifolia	Oregon ash	.55
F. nigra	black ash	.49
F. pennsylvanica	green ash	.56
Gleditsia triacanthos	honeylocust	.66
Gymnocladus dioicus	Kentucky coffeetree	.60
Ilex	holly	
I. aquifolium	English holly	.68
I. opaca	American holly	.56
Juglans	walnut, butternut	
J. californica	southern California walnut	—
J. cinerea	butternut	.38
J. hindsii	northern California walnut	—
J. nigra	black walnut	.55
J. regia	European walnut	.56
Liquidambar styraciflua	sweetgum	.52
Liriodendron tulipifera	yellow-poplar	.42
Lithocarpus densiflorus	tanoak	.65
Maclura pomifera	Osage-orange	.81
Magnolia	cucumbertree, magnolia	
M. acuminata	cucumbertree	.48
M. grandiflora	southern magnolia	.50
Malus spp.	apple	.62
Morus rubra	red mulberry	.66

Nyssa sylvatica	black tupelo	.50
Ostrya virginiana	eastern hophornbeam	.70
Platanus	sycamore, plane	
P. acerifolia	London plane	.55
P. occidentalis	sycamore	.49
Populus	aspen, cottonwood, poplar	
P. alba	white poplar	—
P. balsamifera	balsam poplar	.33
P. canescens	grey poplar	.40
P. deltoides	eastern cottonwood	.40
P. grandidentata	bigtooth aspen	.39
P. heterophylla	swamp cottonwood	—
P. nigra	European black poplar	—
P. tremula	European aspen	—
P. tremuloides	quaking aspen	.38
P. trichocarpa	black cottonwood	.35
Prunus	cherry	
P. avium	European cherry	—
P. serotina	black cherry	.50
Pyrus spp.	pear	.73
Quercus	oak	
Q. alba	white oak	.68
Q. bicolor	swamp white oak	.72
Q. coccinea	scarlet oak	.67
Q. falcata	southern red oak	.59
Q. garryana	Oregon white oak	.72
Q. kelloggii	California black oak	.57
Q. lyrata	overcup oak	.63
Q. macrocarpa	bur oak	.64
Q. palustris	pin oak	.63
Q. petraea	sessile oak	.62
Q. prinus	chestnut oak	.66
Q. robur	European oak	.62
Q. rubra	northern red oak	.63
Q. stellata	post oak	.67
Q. velutina	black oak	.61
Q. virginiana	live oak	.88
Rhus typhina	staghorn sumac	.47
Robinia pseudoacacia	black locust	.69
Salix	willow	
S. alba	white willow	.40
S. nigra	black willow	.39
Sassafras albidum	sassafras	.45
Tilia	basswood, lime	
T. americana	American basswood	.37
T. heterophylla	white basswood	—
T. vulgaris	European lime	.48
Ulmus	elm	
U. alata	winged elm	.66
U. americana	American elm	.50
U. crassifolia	cedar elm	.64

U. glabra	wych elm	.60
U. procera	English elm	.49
U. rubra	slippery elm	.53
U. thomasii	rock elm	.63
Umbellularia californica	California-laurel	.55

III. TROPICAL HARDWOODS**

Acacia koa	koa	.60
Cedrela spp.	Spanish-cedar	.44
Chlorophora excelsa	iroko	.62
Chloroxylon swietenia	Ceylon satinwood	.87
Cybistax donnell-smithii	primavera	.41
Dalbergia fructescens var. tomentosa	Brazilian tulipwood	.86
Dalbergia latifolia	Indian rosewood	.76
Dalbergia stevensonii	Honduras rosewood	.90
Dicorynia guianensis	basralocus	.72
Diospyros spp.	ebony	.90
Dyera costulata	jelutong	.40
Entandrophragma cylindricum	sapele	.60
Entandrophragma utile	utile	.59
Gonystylus macrophyllum	ramin	.59
Guaiacum spp.	lignumvitae	1.14
Khaya spp.	African mahogany	.63
Lophira alata	ekki	1.00
Mansonia altissima	mansonia	.59
Ochroma pyramidale	balsa	.16
Peltogyne spp.	purpleheart	.83
Pericopsis elata	afrormosia	.62
Phoebe porosa	imbuia	.57
Pterocarpus spp.	padauk	.67
Shorea spp.	white lauan	.46
Shorea spp.	yellow meranti	.50
Swietenia spp.	Central American mahogany	.58
Tectona grandis	teak	.57
Terminalia superba	korina	.49
Triplochiton scleroxylon	obeche	.34
Turraeanthus africanus	avodire	.52

**Note: In tropical hardwoods, specific gravity can vary widely among samples of a given species or species group. For example, reported specific gravity values for Spanish-cedar range from 0.33 to 0.67; for Central American mahogany, from 0.43 to 0.75. Values listed are reported weighted averages or means of extreme values.

learning to use approximately the same thumbnail pressure each time.

You will soon find yourself estimating specific gravity and relative hardness during the routine process of cutting a sample with a knife or razor blade in preparation for an examination of its cell structure. The greater the resistance, the higher the specific gravity and hardness.

Fluorescence

Certain woods emit a glow or **fluorescence** when exposed to a source of ultraviolet radiation, which is also sometimes referred to as "black light." In a few cases, woods that fluoresce can be quickly and easily separated from similar woods that do not.

There are two basic kinds of black lights; it is important to know the difference between them because one is very dangerous to the eyes and should not be used. The commercially available black light used for various visual effects emits ultraviolet radiation in the range of 3,200 to 3,800 angstrom units. (One angstrom unit equals approximately 4 billionths of an inch.) It is called a long-wave or near-ultraviolet light, and is not harmful to the eyes. Many of these lamps also emit non-ultraviolet light in the visible range at wavelengths up to 4,500 angstrom units, which accounts for the purple glow associated with them. Although the inexpensive black lights used for entertainment and poster illumination do show the fluorescent properties of wood, the more expensive long-wave black lights of this same general type filter out more of the visible light and produce a more vivid effect.

A second type of black light that emits radiation at wavelengths between 2,900 and 3,200 angstrom units is called a short-wave or far-ultraviolet light. This light is used to kill bacteria and is *very dangerous* to the eyes. Sterilization units and other sources of short-wave ultraviolet light should never be used to test for fluorescence.

The term fluorescence was coined in 1852 by Sir George Stokes, who first observed the phenomenon in the mineral fluorite. Fluorescence occurs when invisible light energy is absorbed by chemicals that transform the energy and emit it at wavelengths visible to the eye. The visible spectrum ranges from wavelengths of about 3,800 angstrom units (violet) to about 8,000 angstrom units (red).

Among the woods that show a response to ultraviolet light, fluorescence is barely recognizable in some but strikingly brilliant in others. The fluorescent response of wood decreases on exposed surfaces; the decrease is apparently associated with the familiar darkening or aging effect. When testing for fluorescence, the surface should be scraped or shaved down to fresh wood to obtain the best color response.

The North American species in this book that exhibit marked fluorescence are listed in Table 5.2 on the facing page. The colors indicated are typical for each species, but variations in hue and brilliance occur among individual samples of a given wood. Countless other species from around the world

also show fluorescent properties. Those species comprise some 78 genera, 45 in the family *Leguminosae* alone.

As can be seen from the table, yellow is the predominant fluorescent color; it is also the most brilliant, as exemplified by black locust. Fluorescence in this wood quickly separates it from a look-alike, Osage-orange. Other common woods with brilliant yellow fluorescence are honeylocust, Kentucky coffeetree and acacia. Barberry, a shrub whose wood is lemon yellow even under normal light, has perhaps the most brilliant yellow fluorescence. One of the most interesting woods is staghorn sumac, whose sapwood has a pale lavender-blue fluorescence. In the heartwood, each growth ring repeats a striking yellow, yellow-green, lavender-blue sequence. Other species show less spectacular but nonetheless interesting colors. Yucca and holly have a soft blue-grey fluorescence. Purpleheart emits a dim coppery glow.

Fluorescence is sometimes seen in specific anatomical features, including resin canals, oil cells, vessel contents, bark, fungal stains and pigment streaks. The brilliant yellow fluorescence that usually occurs at the margins of fungus-stained areas in aspen helps identify that wood.

Chemical Tests

Most chemical tests that are used to separate woods detect differences in the extractives developed in the heartwood.

The simplest test involves soaking shavings of the wood in water and watching to see if any color disperses into solution. For example, when Osage-orange shavings are immersed in lukewarm water, they liberate a yellow extractive. But soaking the shavings of black locust, with which Osage-orange is often confused, causes no significant color change in the water.

A more sophisticated kind of test—which I include mainly to give an idea of the types of tests that can be performed—involves the application of a chemical reagent to the surface of a sample to see if a certain color reaction occurs. For example, in separating subalpine fir (*Abies lasiocarpa*) from Pacific silver fir (*A. amabilis*), Ehrlich's reagent (composed of 1 gram of p-dimethyl amino benzaldehyde in 23 milliliters of ethyl alcohol containing 2 milliliters of concentrated hydrochloric acid) is brushed across a freshly exposed edge-grain heartwood surface. A purple color is a positive reaction that indicates subalpine fir; no color or only a pale greenish discoloration indicates Pacific silver fir. When a saturated water solution of ferrous sulfate is applied to red maple, a deep blue-black color develops; when the same solution is applied to sugar maple, a greenish color results.

The chemical analysis of extractives involves even more complex procedures, but it can provide invaluable separations of very similar woods. For example, analysis of essential oils distinguishes Jeffrey pine, which contains 90-95% normal heptane, from ponderosa pine, which contains pinene and limonene.

Appendix IV on p. 196 lists additional examples of woods that can be separated based on chemical composition and the

TABLE 5.2: PARTIAL LIST OF NORTH AMERICAN WOODS EXHIBITING NOTEWORTHY FLUORESCENCE UNDER ULTRAVIOLET LIGHT

Scientific Name	Common Name	Color of Fluorescence
Acacia greggii	Gregg catclaw	Deep yellow
Annona glabra	Pond-apple	Dull yellow
Asimina triloba	Pawpaw	Faint yellow-green
Berberis thunbergi	Japanese barberry	Bright yellow
Cercidium floridum	Blue paloverde	Yellow-green
Cercidium microphyllum	Yellow paloverde	Yellow
Cercis canadensis	Eastern redbud	Bright yellow
Cladrastis kentukea	Yellowwood	Pale yellow-light blue
Cotinus obovatus	American smoketree	Deep yellow
Gleditsia aquatica	Waterlocust	Pale yellow
Gleditsia triacanthos	Honeylocust	Bright yellow
Gymnocladus dioicus	Kentucky coffeetree	Bright deep yellow
Ilex verticillata	Common winterberry	Light blue
Magnolia virginiana	Sweetbay	Faint pale yellow
Mangifera indica	Mango	Pale orange
Piscidia piscipula	Florida fishpoison-tree	Dull yellow
Rhus copallina	Shining sumac	Bright yellow
Rhus glabra	Smooth sumac	Bright yellow
Rhus integrifolia	Lemonade sumac	Bright yellow
Rhus ovata	Sugar sumac	Yellow
Rhus typhina	Staghorn sumac	Bright yellow or greenish yellow to pale blue
Robinia neomexicana	New Mexico locust	Bright yellow
Robinia pseudoacacia	Black locust	Bright yellow
Robinia viscosa	Clammy locust	Bright yellow
Torreya taxifolia	Florida torreya	Dull yellow
Yucca brevifolia	Joshua-tree	Yellowish grey

tests used to separate them. It also lists sources that describe the tests in greater detail.

As it turns out, most species that can be separated by chemical tests can be separated just as readily by physical and anatomical features. Examples are the black locust/Osage-orange and red maple/sugar maple separations already mentioned. Naturally, the most valuable chemical tests are those that allow separations that cannot be achieved any other way. Examples are subalpine fir/Pacific silver fir and Jeffrey pine/ponderosa pine.

Chemical identification is often the only method that can be used when the form of the wood — sawdust, for example — precludes anatomical analysis. In addition, there are a few situations where anatomical identification methods can be used but simple chemical color reaction tests are faster and more convenient. An example is the sorting of mixed species at a sawmill log deck.

To date, relatively few woods have been found to contain unique chemicals whose detection provides an instant means of separating that wood from all others. For the present, physical properties and anatomy remain the principal bases for identification. But the field of **chemotaxonomy**, the classification of plants based on their chemical composition, is expanding rapidly. As more is learned about the chemistry of extractives of various woods, the range of woods that can be identified through chemical analysis will surely grow.

CHAPTER SIX

IRREGULARITIES AND SPECIAL FIGURES

Thus far we have considered wood from normal trees with round trunks, uniform growth rings and other typical anatomical characteristics. But there are many departures from the norm. Some, such as the bird's-eye figure in maple, are inherent to the species; others, such as pith flecks in paper birch and other species, are caused by external factors. Either way, they interest us because they are often limited to a few or even a single species, which helps in the separation process.

It is tempting to call unusual features abnormalities or defects. But the strong negative connotation of these terms is inappropriate when the "abnormal" wood — walnut burl, for example — often has a commercial value many times that of "normal" wood. Thus, I prefer the word "irregularities."

Some irregularities — the blister figure of bigleaf maple or the bee's-wing mottle of Ceylon satinwood, for example — are the hallmarks of their respective species. Thus, these irregularities are very useful in identification. But other irregularities can occur in any species — decayed wood and reaction wood, for example. Since these irregularities are nonspecific, they are of little value in identification. Because they present special identification problems, we must also learn to recognize these irregularities.

Reaction Wood

Reaction wood is an abnormal condition found in stems and limbs that are not parallel to the pull of gravity. Reaction wood formation is associated with the tree's physiological mechanism for redirecting stem growth to the vertical in leaning trees, resulting in bowing in the living stem. Boards or pieces from a log with noticeable sweep usually contain reaction wood.

The gross, macroscopic and microscopic features of reaction wood are often drastically different from those of normal wood and may complicate identification. Consequently, it is important to recognize and understand the specific cellular differences of reaction wood and to avoid it whenever possible.

Reaction wood has different characteristics in softwoods and hardwoods. In softwoods, it forms mainly toward the underside of the leaning stem. Because the pull of gravity puts the lower side of the leaning stem in compression, the reaction wood in conifers is called **compression wood.** Several features aid in its detection. The portion of the growth ring containing reaction wood is usually wider than normal. This results in an eccentrically shaped stem with the pith offset toward the upper side. Abnormal tracheids in reaction-wood growth rings usually form latewood that is wider than normal.

This wider, denser latewood makes woods like eastern white pine appear uneven-grained. But in woods like southern yellow pine that are notably uneven-grained in normal wood, the latewood in reaction wood is duller and tends to even out the contrast.

Compression wood is denser but brittler than normal wood. Where it fails, fractured surfaces are abrupt and clean, without the splintering seen in normal wood. Compression wood has abnormally high shrinkage along the grain. Since reaction wood formation is uneven, the longitudinal shrinkage is also uneven

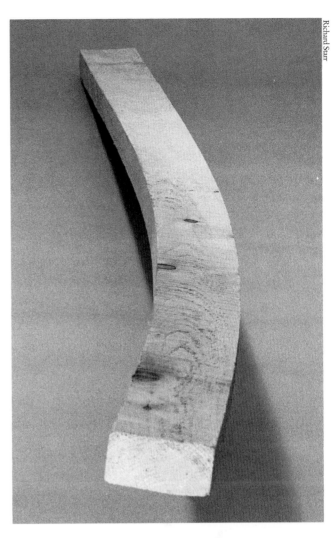

Richard Starr

REACTION WOOD
The sweep in the trunk of the leaning hemlock tree on the left is an indication that compression wood will be found within. The uneven and abnormally higher longitudinal shrinkage of reaction wood is a major cause of the crook in the lumber at right.

ABNORMALLY WIDE LATEWOOD
In sections of leaning stems of eastern hemlock (left) and Norway spruce (right), reaction wood is evident in areas where growth rings widen and appear to have an abnormally high percentage of latewood.

and leads to longitudinal warping. It may cause a crook, bow or twist in the lumber. Excessive shrinkage may result in localized cross breaks.

Microscopically, compression wood is characterized by tracheids that are rounded rather than angular in cross section, with many intercellular spaces (photos below). Tracheid walls are thicker than normal and may show deep spiral checking in cross or longitudinal section.

These abnormalities in tracheid characteristics can complicate the appearance of microscopic features routinely used in identification. For example, the thicker walls with spiral checks may affect the appearance of cross-field pitting. Also, among conifers, eastern redcedar is the only species whose normal wood has intercellular spaces among the longitudinal tracheids — but *without* spiral checks — as a normal feature. Therefore, intercellular spaces that occur together with spiral checking should be interpreted as probable reaction wood.

In hardwood trees, reaction wood forms mainly on the upper side of the leaning stem. Because this side is in tension, the reaction wood is called **tension wood** (shown in the photo at the top of the facing page). In hardwoods, the pith is less likely to be off-center, and the tension wood may develop irregularly around the stem. Tension wood is often hard to detect. Sometimes it looks silvery, other times it is dull and lifeless, and in some cases there is little if any visual difference between it and normal wood.

Crookedness or sweep in a hardwood log signals the possible presence of tension wood. Another indication may be seen during cutting, planing or sanding. Because the fiber structure may not sever cleanly, a fuzzy or woolly surface often results; subsequent staining produces a blotchy effect. As with compression wood, longitudinal shrinkage is greater than in normal wood. It is also irregular, resulting in warping and machining problems.

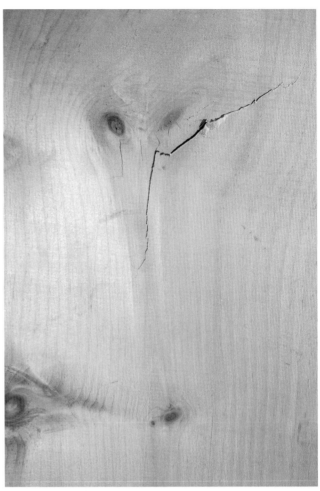

CROSS BREAKS
Cross breaks within a board due to excessive longitudinal shrinkage are usually traceable to reaction wood.

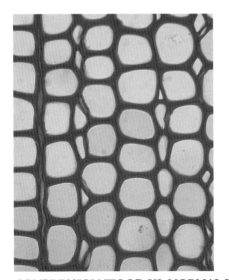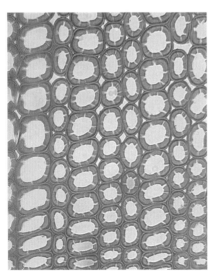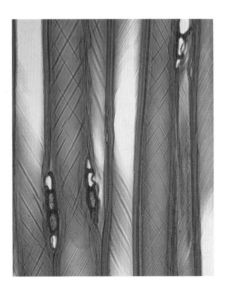

COMPRESSION WOOD VS. NORMAL WOOD
Compared to normal tracheids in eastern white pine (left), reaction-wood tracheids (center) are rounded, with intercellular spaces. The cell walls are thick, with spiral checks that are also evident in tangential view (right). (250x)

TENSION WOOD
Reaction or tension wood appears as silvery areas in this aspen cross section.

Microscopically, tension-wood fibers show the greatest abnormality. Tension-wood fibers typically develop an inner gelatinous layer, so called because of its appearance. Fibers modified in this way are called gelatinous fibers. They are usually more abundant in the earlywood portion of the growth ring. Vessel elements associated with tension wood may be slightly smaller in diameter than normal, but otherwise, microscopic features such as perforation plates, intervessel pitting and spiral thickenings can be evaluated correctly.

Juvenile Wood

Most of the discussion in this book focuses on the features of mature stem wood. But wood formed adjacent to the pith and limb wood can be substantially different and sometimes present identification problems.

Juvenile wood refers to atypical wood formed around the pith during the first few years of growth. It is especially pronounced in trees with unusually fast initial growth. In plantation-grown conifers, which grow quickly due to lack of competition, juvenile-wood formation often continues for 15 or

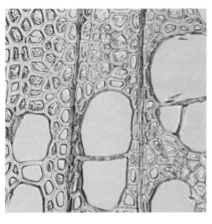
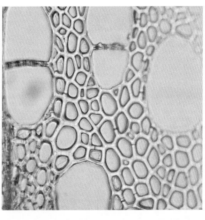

GELATINOUS FIBERS
Normal aspen fibers are shown at near left. By comparison, gelatinous fibers in tension wood have abnormally thick walls whose inner gelatinous layer often separates when sectioned, as shown at far left. (250x)

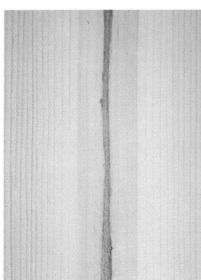

JUVENILE WOOD
Wide growth rings nearest the pith (far left) are juvenile wood. Needle scars (left) also indicate juvenile wood.

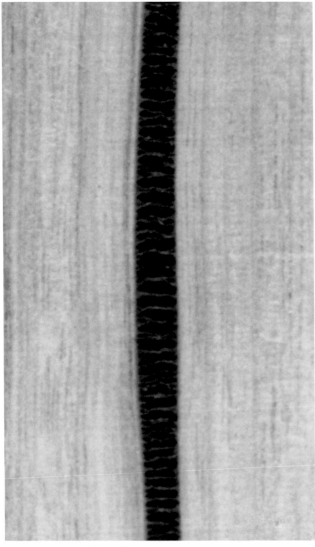

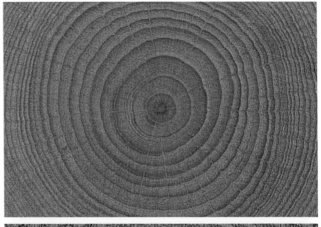

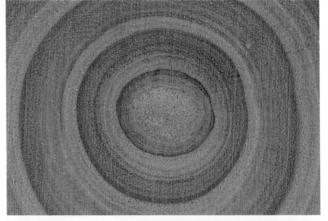

PITH CHARACTERISTICS
The longitudinal surface of a butternut board (above) shows that the pith is chambered. Cross sections at right compare the tiny pith of beech (top photo) with the medium-sized pith of catalpa (second from top) and the large pith of sumac (third from top). At the bottom is the five-pointed star-shaped pith of oak.

more years, forming a core several inches in diameter. In other cases, however, juvenile wood is either restricted to a few rings adjacent to the pith or is virtually nonexistent.

The transition from juvenile wood to mature wood is gradual in some cases, abrupt in others. Juvenile wood may be lower than normal in density, especially in conifers. In some cases there are pronounced differences in cell size, the relative number or arrangement of cells or their microscopic appearance. Wood from the first few growth rings should be compared to mature wood to get a sense of how different juvenile wood can be. In an unidentified piece or sample, mature wood should be examined whenever possible.

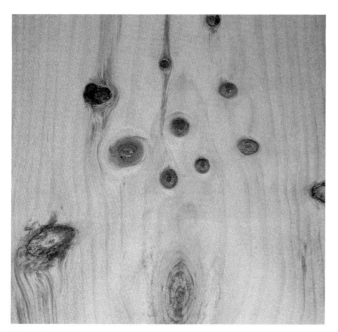

PIN KNOTS
Small pin knots in spruce.

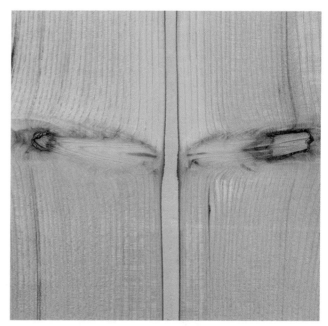

OPPOSITE BRANCHING
The paired spike knots in this wood reveal opposite branching. Because the wood is ring-porous, we can therefore narrow its identity to either ash or catalpa. The light heartwood color suggests ash, which it is.

Pith

Now that stem wood is more completely utilized for commercial purposes, products more often contain center-cut pieces in which the pith is visible; thus, the pith is deserving of greater attention as an identification feature.

The size, shape and cellular characteristics of pith can be quite distinctive, as shown in the photos on the facing page. In birch the pith is small and inconspicuous; in sumac it is pencil-sized. In conifers, the pith is usually round and quite small with only occasional exceptions. For example, among the southern yellow pines the pith of most species is less than ⅛ in. in diameter; in longleaf pine, however, the pith near the base of the tree may have a diameter of ¼ in. to ⅜ in.

Pith diameter is often suggested by twig characteristics. The slender twigs of grey birch indicate a correspondingly slender pith. The finger-thick twigs of staghorn sumac hint at its pencil-thick pith. Pith diameter also varies according to the vigor of the growing tips. Dendrology texts and tree identification manuals are good sources of information on twig and pith dimensions.

In cross section, the pith is not always round. In ash, basswood and maple it is typically elliptical; in alder, beech and birch it is triangular. In oaks it forms a five-pointed star, and in teak it is square.

Pith varies in color from nearly white to dark brown or almost black. It may be solid, or porous and spongy. Some pith has chambers, with the hollow areas separated by cross walls, and some pith is completely empty.

Limbs

Limbs — and the knots that occur where they intersect the main stem — are present in every tree. Thus they are seldom of value in identification. But there are exceptions. The knot tissue of aspens exhibits fluorescence. Spruce lumber cut from the inner bole (tree stem) may have numerous small intergrown or encased pin knots, as shown in the photo above left. This is so because spruces typically retain many small branches along the lower portions of the stem.

Most hardwood species have alternate branching. However, certain hardwoods have opposite branching. Where opposite branching is revealed in the wood, it may assist in identification. Among ring-porous hardwoods, only ashes and catalpas have opposite branching (see the photo above right). Among diffuse-porous and semi-diffuse-porous hardwoods, opposite branching is characteristic of maples, buckeyes and dogwoods.

The structural inconsistencies of limb wood can also complicate identification. In the knot portion of the limb where the limb is embedded in the main stem, the wood is extremely abnormal. It has a much higher-than-normal density, an abnormal resin content and an unfamiliar cell structure.

To summarize, then, in attempting to identify wood, you should first determine whether juvenile wood, reaction wood, knots or limb wood are present. Whenever possible, avoid these areas and choose instead the more regular structure of mature stem wood.

PITCH POCKETS
Pitch pockets exposed on a radial surface of a spruce board.

Pitch and Pitch Pockets

Although resin in conifers is usually confined to resin canals, injury to the tree may cause it to infiltrate other cell structures. These accumulations are referred to as **pitch**. If they occur in a well-defined longitudinal pattern they are called pitch streaks.

Pitch pockets are ellipsoidal voids in the wood that are up to 3 in. in diameter and flattened in the plane of the growth ring. They usually contain liquid or solidified pitch. On radial or cross-sectional surfaces the pockets are lens-shaped. Because pitch pockets generally occur in species that normally have resin canals, their presence is of little value in identification other than to signal that the wood is probably pine, spruce, larch or Douglas-fir. Conifers that do not normally have resin canals produce pitch pockets only under abnormal circumstances.

Decay and Stain

Wood-inhabiting fungi flourish under the right moisture, oxygen and temperature conditions. Mold is restricted to the wood surface, but stain and decay fungi penetrate the cell structure. Because stain fungi depend on the carbohydrate residue of parenchyma cells as food, they are restricted to sapwood. These organisms do not break down the cell walls, so they do not weaken the wood structure; rather their microscopic threadlike hyphae pass from cell to cell through the pit structures. They typically impart a grey or blue-grey color to the wood they inhabit. Consequently, the terms sap stain and blue stain are applied to them. Because the sapwood of virtually every species is susceptible to sap stain, the presence of stain fungi is not diagnostic.

Decay fungi actually break down and consume cell-wall material, and eventually destroy or rot the wood. The rot is broadly categorized as brown **rot**, white rot or soft rot. Because the many species of fungi are for the most part indistinguishable

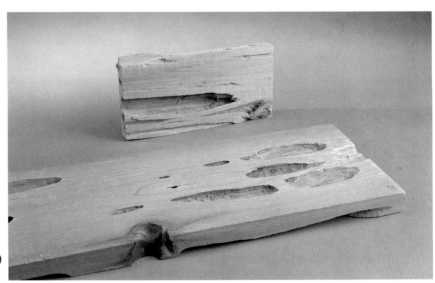

POCKET ROTS
Pecky dry rot in incense-cedar (at bottom) and in 'pecky' cypress.

TABLE 6.1: COMPARATIVE DECAY RESISTANCE OF HEARTWOOD OF VARIOUS WOODS

EXTREMELY RESISTANT	RESISTANT	MODERATELY RESISTANT	SLIGHTLY RESISTANT OR NON-RESISTANT	
Temperate:	**Temperate:**	**Temperate:**	**Temperate**	
Black locust	Baldcypress (old	Baldcypress	Alders	Hickories
Osage-orange	growth)	(young growth)	American	Hollies
Red mulberry	Black cherry	Douglas-fir	hornbeam	Horsechestnut
Yews	Black walnut	Eastern white pine	Ashes	Magnolia
	Catalpa	European cherry	Aspens	Maples
	Cedars	European walnut	Basswood	Oaks (most
	Chestnuts	Honeylocust	Beeches	of the red
	Junipers	Longleaf pine	Birches	oak group)
	Oaks (most of the	Slash pine	Buckeyes	Pines (other
	white oak group)	Western larch	Butternut	species)
	Redwood		Cottonwoods	Poplars
	Sassafras		Elms	Spruces
			Firs (true firs)	Sweetgum
			Hackberry	Willows
			Hemlocks	Yellow-poplar
Tropical:	**Tropical:**	**Tropical:**	**Tropical:**	
Afrormosia	Central American	African mahogany	Avodire	
Basralocus	mahogany	Primavera	Balsa	
Ceylon satinwood	Imbuia	Sapele	Jelutong	
Ebonies	Spanish-cedar	Utile	Limba	
Ekki		Yellow meranti	Obeche	
Iroko			Ramin	
Lignum vitae			White lauans	
Mansonia				
Padauks				
Purpleheart				
Rosewoods				
Teak				

with the untrained eye, and because they may inhabit so many different wood species, they are seldom useful in identification.

The decay produced by certain white rots is sometimes characterized by **zone lines**, which give the wood a somewhat marbled appearance. Wood infected in this way is called **spalted wood**. Because spalted wood is found in many species, it is of limited use in identification. But there are a few recognizable species-restricted forms of decay. For example, baldcypress and incense-cedar are susceptible to a distinct form of rot called **peckiness**, also known as **pecky dry rot** or **pocket rot** (see the bottom photo on the facing page).

On the surfaces of boards, peckiness occurs as distinct regions of advanced decay surrounded by apparently sound, firm wood. In cypress the pockets are commonly pencil-sized or larger. They may be empty or contain yellow-brown or brown residue. In incense-cedar the pockets are typically finger-sized, up to a few inches long, and usually contain dark brown crum-

bly wood. This wood is separated at intervals into cubical blocks by cross breaks. Pocket rot sometimes occurs in redwood, western redcedar and other western conifers, although it is less common in these species.

Food supply is another important requirement for fungal growth. Many species of fungi thrive on the wood substance found in sapwood. The extractives in heartwood, however, are often toxic or repellent to fungi, making it resistant to decay. Heartwood decay occurs only as the toxic extractives are leached from the surface.

A familiarity with the decay resistance of various species can be very useful in field identification problems (see Table 6.1 above). For example, let us assume that a fence post set in moderately damp soil has lasted for 25 years. It is observed to be a ring-porous hardwood, and its dense heartwood is still relatively sound. This rules out ash and hickory because they have little or no decay resistance. Catalpa and chestnut can also be ex-

Gum defects

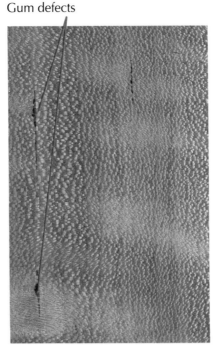

PITH FLECKS
Insect-caused pith flecks are conspicuous on this tangential surface of a paper birch board.

GUM DEFECTS
Insect-caused gum defects are a familiar characteristic in black cherry.

cluded because they are much less dense. More likely possibilities include white oak, black locust and Osage-orange, three woods that are commonly used for fence posts and that have both high density and superior decay resistance.

Insect Damage

Insects also cause damage. We often hear the term "wormy chestnut," which refers to the fact that old chestnut is commonly attacked by beetles. Boring insects of all sorts attack countless species of wood, so insect holes per se are of little value in wood identification. Other types of insect damage, however, can be informative.

Pith flecks are initiated by the larvae of tiny flies that burrow downward in the tree stem along the cambial layer during the growing season. These tunnels eventually fill with scar tissue. When the wood is machined, the tunnels appear as dark brown oval or oblong spots in cross section and as intermittent streaks on longitudinal surfaces. The similarity in appearance between these irregularities and the pith of some woods accounts for the terms pith flecks and medullary spots.

Pith flecks can help distinguish among species that cannot reliably be separated on cell structure alone. For example, they are usually abundant in grey birch *(Betula populifolia)*, white birch and river birch *(Betula nigra)*, but sparse or absent in yellow birch. They are abundant in red, silver and black maple, but infrequent in sugar and bigleaf maple. They are also seen sporadi-

cally in other hardwoods, including aspen, basswood, willow and cherry.

Cambium-mining insects are suspected of causing the traumatic gum canals found in cherry. These so-called gum defects are useful in identification.

Figure

Figure is broadly defined as any distinctive marking on a longitudinal wood surface. Wood without any visual characteristics is the exception rather than the rule. The normal figure commonly associated with a particular species results from its inherent cell and growth-ring characteristics. But there can be infinite variability of figure within a species because tree-stem form is never geometrically perfect.

Growth rings are seldom perfectly circular, and this irregularity is responsible for the V-shaped markings, ellipses and hyperbolas on typical flatsawn (tangential) surfaces, as well as for the parallel lines seen on radial surfaces (see the top photo on the facing page).

How pronounced this figure is depends on the evenness of the grain. Uneven-grained conifers like Douglas-fir and southern yellow pine have pronounced figure. So do ring-porous hardwoods like white ash and chestnut. Sometimes the visual contrast of one cell type against a different background causes figure, as with the wavy bands of latewood pores in elm and hackberry or the pronounced ray fleck of sycamore, oak and cherry.

FIGURE   63

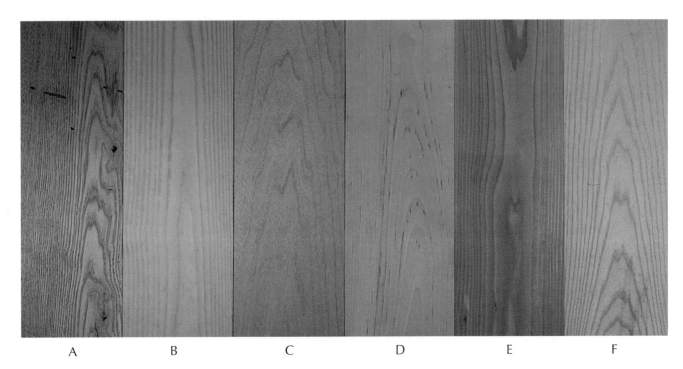

A   B   C   D   E   F

GROWTH-RING FIGURE
Flatsawn figure is characterized by oval, U-shaped and V-shaped patterns. The degree of contrast between earlywood and latewood determines how distinct the figure is. Shown above are chestnut (A), pitch pine (B), butternut (C), paper birch (D), tamarack (E) and white ash (F).

SPECIAL FIGURE
The word figure sometimes has a more specific meaning that denotes wood surfaces with a distinctive or decorative appearance. The term **figured veneer**, for example, implies more than just the common patterns caused by wood structure. Special figure may result from uneven heartwood pigmentation, irregular ring formation, deviation in cell and grain direction or any combination of these. The examples of figure discussed below have been selected from the broad array of figures for their value in identifying woods.

Pigment figure, which results from uneven pigmentation of the heartwood, can be quite striking. The normal figure of a wood already present may be even further accented if the variability of shade or hue coincides with the anatomical differences between earlywood and latewood. In staghorn sumac, for example, the larger earlywood pore area shows up as a flat light brown, not unlike the color of butternut or chestnut; sometimes it has a distinct orange tinge along the growth-ring boundary. As these earlywood pores make the gradual transition to fibrous latewood, the color changes to a distinct green or yellow-green, and finally back to a brown at the outer margin of the growth ring.

In some woods, contrasting pigment occurs in concentric bands. However, these bands are not necessarily concentric with the growth rings, crossing over the rings in some places. The banded color pattern is less regular in rosewood and is quite irregular in koa or figured red gum (photo at right).

PIGMENT FIGURE
This photograph shows several pieces of pigment-figured red gum (sweetgum) that have been glued together to create an interesting visual effect.

Although tree stems are typically round with some random irregularity, some species have rings with flutes at regular intervals around their entire circumference (photo above). In flatsawn surfaces, the usual U- and V-shaped pattern becomes a wavy, irregular figure (photo at right). This is a variable trait in a number of species, but it is consistent enough in some to be useful in identification. As shown in the photos on this page, butternut is quite consistent in producing fluted rings. The flutes apparently coincide with plates of the outer bark.

Black walnut occasionally shows fluting, as does basswood. But basswood is so fine-textured and even-grained that the figure is usually not apparent. The rings of American hornbeam (photo at top left on the facing page) are deeply contoured, with ridges and valleys of varying depth and spacing around the circumference of the stem.

Indented rings show vertical, groove-shaped depressions in the growth rings that may extend vertically in the stem for several inches. Dimples are prevalent in lodgepole pine, and are also common in ponderosa pine and Sitka spruce. They have been reported to occur less commonly in red spruce and tamarack.

Bird's-eye figure consists of localized conical elevations of the growth ring in which the grain swirls. Seen on a tangential surface, each sectioned swirl suggests a tiny bird's eye.

The term bird's-eye maple is essentially synonymous with bird's-eye figure because the irregularity occurs most commonly in the hard maples, in which it produces an even display of small, uniformly distributed bird's eyes. However, a similar figure is occasionally seen in other maples, birches and ashes. It should be noted that similar but larger swirls of grain are sometimes referred to as "curly" figure, but this term is more aptly applied to wood with wavy grain, as discussed on p. 66.

FLUTED GROWTH RINGS
Fluted growth rings can be recognized in cross-sectional (above) and tangential (below) surfaces of butternut.

Richard Starr

FIGURE 65

FURROWED GROWTH RINGS
Wavy or furrowed growth rings in American hornbeam.

DIMPLES
Dimples, which are depressions in the growth rings, are obvious on this split tangential surface of lodgepole pine.

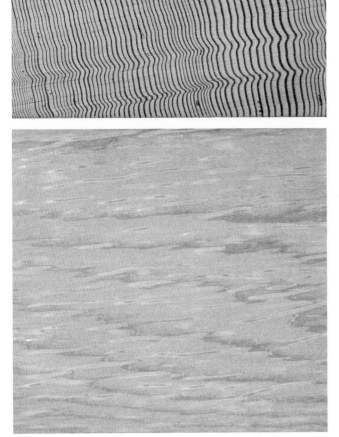

INDENTED GROWTH RINGS
Indented growth rings (top) produce a pattern of streaks called 'bear scratches' (above), as illustrated in this tangential surface of Sitka spruce.

BIRD'S-EYE FIGURE
Bird's-eye figure in maple.

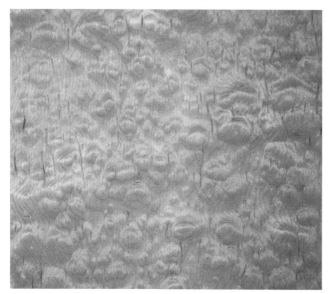

BLISTER FIGURE
Blister figure in bigleaf maple.

QUILTED FIGURE
Quilted figure in bigleaf maple.

CURLY FIGURE
Curly figure is most pronounced when wavy grain is cut radially, as in these pieces of maple.

Blister figure describes any series of domelike bulges in the plane of the growth ring. It occurs sporadically among so many woods — including birch, maple, yellow-poplar and mahogany — that it has diagnostic value only when it is visually unique. In yellow poplar, blisters may be striking because of the sapwood/heartwood contrast. Blister figure is perhaps best known and most beautifully displayed in bigleaf maple.

Light reflection, more than anatomical layering, produces blister figure. In bigleaf maple, where the bulges are elongated transversely and are closely crowded, a **quilted figure** results.

Curly figure is the result of **wavy grain**, a distortion that produces a horizontal washboard effect on a split radial surface. When machined to a flat plane, a cross-barred effect is produced

as a result of variable light reflection from cells intersecting the surface at changing angles.

In maple this same figure is called tiger maple or **fiddleback;** the latter term derives from the fact that this is a preferred wood for violin backs. In traditional American furniture, curly soft maple is more commonly found than curly hard maple. But the figure also occurs in many other species, including birch, ash, walnut and mahogany, so it is only of secondary value in identification.

Ribbon or **stripe figure** appears when wood with interlocked grain is surfaced radially. **Interlocked grain** is caused by alternating left-hand and right-hand spiral grain. The successive ribbons or stripes are the result of light reflecting from longitudinal

FIGURE   67

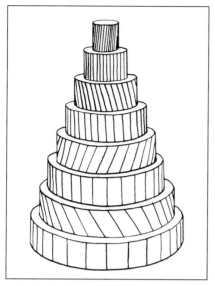  

INTERLOCKED GRAIN AND RIBBON FIGURE
Turning down a log in successive steps reveals a cyclically reversing spiral grain (left). Split radial surfaces (center) show interlocking grain. Smooth radial surfaces through interlocked grain produce ribbon or stripe figure (right).

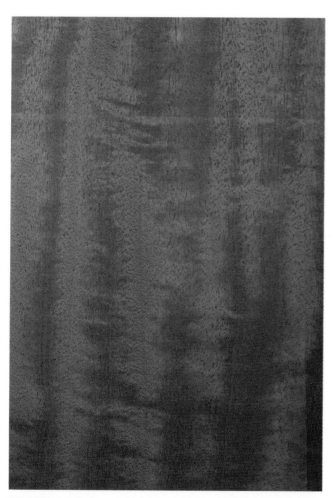 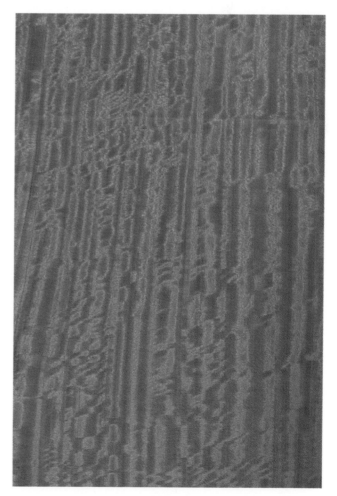

BROKEN-STRIPE AND BEE'S-WING MOTTLED FIGURES
Wavy grain and interlocked grain can occur in varied combinations. Examples include broken stripe figure in mahogany on the left and bee's-wing mottled figure in Ceylon satinwood on the right.

CROTCH FIGURE
Crotch figure in black walnut.

BURL
Veneer sliced from a walnut burl.

cells as well as the varying angles at which the vessels are severed. The cutting forms long vessel lines where the grain is parallel to the surface, and lines that shorten to little more than vessel pore diameter where the spiral grain intersects the plane of the cut at the maximum angle. Irregular interlocked grain is sometimes called **roey grain**.

Interlocked grain is seen in many species. Only where it is routinely present, however, is it useful in identification, and then only as a secondary feature. Among domestic woods, American elm and tupelo fall into this category.

There are several variations on the ribbon figure, especially when combined with curly figure, including **broken-stripe** and **bee's-wing** mottled.

Occasionally, the appearance of figure, as learned from known samples or even from photographs, may be characteristic enough to aid in identification (as shown in the photos on p. 67). As one good example, the classic bee's-wing mottled figure and golden-brown color of Ceylon satinwood make a recognizable combination that enables tentative identification.

Crotch figure appears when the crotch of a tree is cut in the plane common to the piths of the main stem and the branches. However, the feather crotch of black walnut and the **moonshine crotch** of mahogany are visually distinctive (see the photo at top left).

Burls are large, knoblike projections formed on the trunks or limbs of trees; their precise cause is not known. The abnormal wood tissue within them is extremely disoriented and typically contains numerous bud formations. Burls develop in virtually every species; only in a few instances are the surfaces of burl wood distinctive enough to aid in identification. Because of the severely disoriented structure of the wood, it is difficult to recognize cell structure and features when the wood is sectioned and examined microscopically.

CHAPTER SEVEN

EQUIPMENT AND TECHNIQUE

Each new wood-identification problem is different in some way from previous ones. Equipment and technique vary according to the size and condition of the sample, the species of wood, the identifier's experience and whether the work is being done in the field or in a workshop or laboratory. In the end, however, one of two basic situations develops.

In the first case, you immediately think you recognize the species or can at least place the sample within a certain group of woods. Either by prior experience or by using reference material, you determine which features would confirm this choice, and then you immediately proceed to check for these features.

For example, let's assume you are examining the surface of a structural beam. Based on its use and its gross features — a reddish-brown heartwood color, uneven-grained growth rings, and a characteristic odor — you say to yourself, "Douglas-fir." You remember that the presence of abundant spiral thickenings in conjunction with normal resin canals is a combination of features unique to Douglas-fir. The logical procedure would be to take a tangential section for microscopic examination. This section should show spiral thickenings; it should also include one or more fusiform rays — rays that have horizontal resin canals. If these features are found, your original conclusion is confirmed.

In the second case, you find that you can describe some gross features and physical attributes of the specimen, but they do not immediately suggest a species. This situation calls for a more thorough examination. Typically, you begin with gross features, progress to macroscopic features and, finally, section all

three principal surfaces for microscopic examination. As you proceed, the information gathered may indicate the wood. If not, you can use descriptive literature or an identification key to decide whether you have a possible solution — or whether the sample remains an unknown.

As an example, suppose you have a cigarette-sized stick of wood designated only as an archaeological specimen. It is still relatively intact, but has been stained or weathered to a nondescript grey. All distinctive color and odor have vanished. Because it is water-soaked, you cannot estimate its density or hardness. Its visual characteristics suggest a softwood, but no other gross features are available to take you past this point.

With a knife, you whittle a cross section onto the specimen, cleaning it up further with a razor blade (this procedure is described on pp. 78-79). With a hand lens, you can now see that the wood has resin canals. But they are quite small, sparse and scattered in small, tangential groups of three to eight. This immediately narrows the choice to those conifers with resin canals. The grouping of the resin canals suggests that you can eliminate pines, leaving the spruces and larches, as well as Douglas-fir. You might reason that, as in the first example, abundant spiral thickenings would point to Douglas-fir. If spiral thickenings were absent, you could narrow the possibilities to spruce or larch. You could then separate these by the shape of ray tracheid pitting and the number of pits across the radial walls of the earlywood tracheids. The logical next step, therefore, would be to take a radial section for microscopic examination.

SAMPLE PREPARATION WITH A HANDSAW
A small handsaw used in combination with a bench stop works well in preparing small samples for identification.

SAMPLE PREPARATION WITH A BANDSAW
A bandsaw is especially useful for making the angled cuts necessary to separate a small portion of wood with faces oriented to the principal planes of anatomy.

In both of the above situations, the equipment and the way in which it is used are determined on a case-by-case basis. We now move to a more specific discussion of the equipment and its various applications.

Cutting and Shaping Tools

Tools used in the preparation of samples include handsaws, bandsaws and other tools for rough shaping, as well as knives and razor blades for final surfaces and sectioning.

TOOLS FOR ROUGH SHAPING

A small, fine-toothed saw, such as a miter or dovetail saw, is a key tool in wood identification (see the photo at top). When used in conjunction with a bench stop, the dovetail saw is a very effective means of preparing samples.

Many common shop tools are helpful in reducing pieces of wood to a size that is appropriate for examination. A bandsaw (see the photo above) is especially useful in cutting off portions of a sample and making angled cuts to produce the desired planes.

Other shop tools such as table saws, jointers and planers are also useful for this purpose. It is best to avoid using a sander,

A SIMPLE SPLITTING TOOL
A handy splitting tool for working on small blocks can be made by grinding a tapered edge onto a portion of a steel bar.

A VISE FOR SECTIONING
A modified handscrew clamp becomes a handy vise for holding specimens for sectioning. A layer of rubber is bonded to one face; the other face is notched as shown.

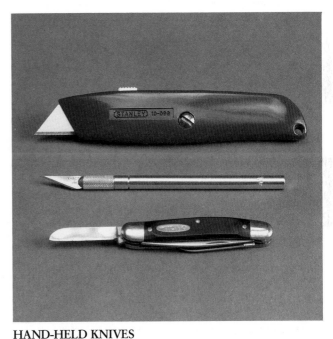

HAND-HELD KNIVES
Hand-held knives can be used for whittling samples to shape or for preparing wood surfaces for examination. From top to bottom: a retractable-blade utility knife, an insert-blade hobby knife and a typical pocket knife.

however, since the grit it leaves might present a problem later when you attempt to cut sections with a razor blade.

It is important to align specimen surfaces with grain direction. In larger pieces this requires the use of various tools for splitting along the grain — froes, hatchets, and chisels. You may find that a mallet helps with chisels and froes. These tools are best used in conjunction with a suitable chopping block or table block.

I have fashioned a small splitting tool by simply sharpening a portion of the edge of a flat steel bar near an end (photo above). It is perfect for tabletop work on small pieces.

Bench planes and block planes are helpful for cleaning up small areas of surfaces to evaluate figure and expose ray fleck. A small vise is helpful for holding specimens while shaping. I use a modified hand clamp to hold smaller pieces for surfacing or sectioning (photo at top right).

KNIVES

A sharp knife is indispensable for whittling pieces to the proper shape or for surfacing pieces for hand-lens examination (photo at bottom right).

A good-quality pocketknife works very well if it is sharpened properly. Knives should be sharpened to a slim bevel, with final honing done on the finest stones, such as hard Arkansas. The objective of sharpening is not only ease of cutting, but also a high-quality surface on the sample. (For a discussion of sharpening technique, see p. 161 of *Understanding Wood,* the companion volume to this book.)

An equally useful alternative to the pocketknife is the insert-blade hobby knife. The replaceable-blade razor knife or utility knife most commonly used for scoring gypsum board and cutting leather is an alternative that eliminates the need for sharpening. For safety, use the retractable-blade type. Even though the blades of most brands are adequate, they are no better than a well-sharpened knife.

RAZOR BLADES

Razor blades are used to surface and section wood for examination . Three common over-the-counter types are useful: single-edged industrial or utility blades, single-edged shaving blades and double-edged shaving blades. These three types vary significantly in shape and honing of the cutting edge.

Industrial blades, which are usually the least expensive, are also the most rugged. They generally have the bluntest edges, although they vary widely in sharpness and grade of steel from brand to brand and even from batch to batch. For hand-lens viewing, the better blades are ideal for routine surfacing of the end grain of all but the densest species. They also do a creditable job of sectioning in most cases. If the sections you cut curl badly, this indicates a blunt blade edge.

Single-edged shaving blades are significantly more expensive than industrial blades, but their edges are much sharper. They are safer to use than double-edged blades. Because the sharpness is achieved by honing the cutting edge more acutely, these blades are also more fragile. Harder woods, if not presoftened, may roll the edge of these blades under, so their use is generally restricted to lower-density woods and longitudinal surfaces.

Double-edged blades, which are the most expensive, usually have the highest-grade steel and the sharpest blades. They are best for cutting sections. But there is a trade-off between sharpness and durability. Although double-edged blades will produce a section with excellent anatomical detail, the slightest misuse of the blade usually wrecks the edge.

I find the cut-rate brands of platinum-chrome blades the best compromise between economy and performance. I generally use them only to whisk off the final ultrathin sections from pieces that have already been softened by boiling (or at least moistened with water), whittled to approximate shape with a pocketknife and surfaced with industrial single-edged blades.

Double-edged blades can be made somewhat safer by taping a temporary cardboard cover over the unused edge. Or they can be broken in half using two pairs of pliers and mounted in a pencil-type handle with a split collet. A heavy paper shim helps hold the blade more tightly.

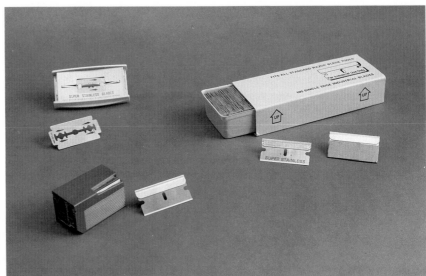

RAZOR BLADES
Razor blades used for surfacing and sectioning include, (clockwise from right), single-edged industrial/utility blades, single-edged shaving blades and double-edged shaving blades.

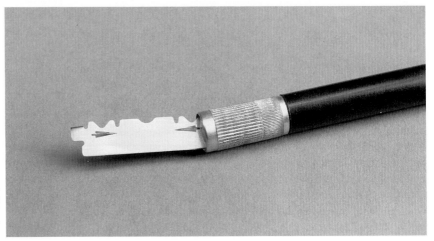

RAZOR-BLADE HOLDER
For greater control and safety, a double-edged razor blade can be broken in half with two pairs of pliers, then mounted in a collet-type hobby-knife handle.

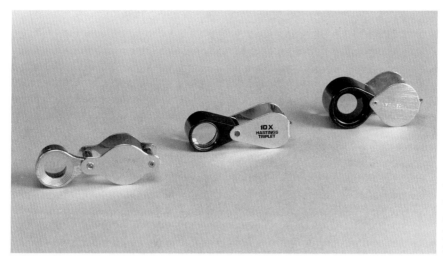

STANDARD HAND LENSES
Three versions of the hand lens used in wood identification. Compared to the economy model on the left, the more expensive lenses provide a larger, clearer field of view with uniform sharpness.

Viewing Equipment

The hand lens and the microscope are the two essential viewing tools in the examination of wood samples.

HAND LENSES

The term **hand lens** is used loosely to include any of a number of low-power, hand-held magnifiers. For wood identification, 10x magnification is the most popular, although 8x and 12x lenses are equally useful. The typical magnifier has its lens mounted in a protective frame, into which the lens is pivoted when not in use (see the photo above); the frame also serves as a handle for the lens.

Because good lighting is important in using the lens, some magnifiers have built-in light sources (see the photo at right). These are particularly handy in museums, antiques shops and other dimly lit settings.

A number of magnifiers, such as those used in the examination of photographs and textiles, have a base that holds the lens at a fixed focusing distance. Some of these magnifiers have the advantage of a calibrated reticle that allows direct measurements on the surface being examined (see the photo at right). However, these magnifiers are inconvenient when the whittled area of a piece of wood is irregular rather than flat and the focusing distance must be adjusted.

MICROSCOPES

Many people avoid using a microscope because they feel it is too complicated. This is unfortunate, since it is untrue—you can learn all the basics of microscope use in a single session.

The standard microscope is little more than a lens magnifier with a frame that holds it in focus on the subject. A fundamental difference between a microscope and a hand lens is that in most instances, light is passed through the subject in a microscope whereas it is reflected off the subject under a hand lens.

This kind of microscope — the kind we will study and use here — is called a transmission **light microscope**. However,

ILLUMINATED HAND LENSES
Hand lenses with built-in illumination are especially useful in dimly lit areas.

MAGNIFIER WITH BASE AND RETICLE
Lenses with calibrated reticles (whose use is described on p. 81) allow direct measurement of features on magnified surfaces.

there are also other types of microscopes, such as **incident-light microscopes,** in which an opaque subject is viewed by light reflected from its surface, and sophisticated, extremely expensive **electron microscopes.**

A magnifier such as a hand lens of fairly high power could be called a microscope, but more accurately, it should be called a simple microscope because it has only a single lens or lens system. The standard hand lens can therefore be thought of as a low-power, hand-held, simple microscope. To achieve higher magnification, a **compound microscope** is used. It comprises two separate lens systems. The one nearer the subject is called the **objective;** the lens system nearer the viewer's eye is called the **ocular** or **eyepiece.**

The objective lens system forms an enlarged image, which is further magnified by the eyepiece lens system. Consequently, the image seen by the viewer has a total virtual magnification equal to the product of the magnification of the two lens systems. For example, a 10x eyepiece used with a 43x objective will produce a 430x magnification. The compound-lens system may also invert the image.

The elementary microscope consists of an objective and an eyepiece held a fixed distance apart on either end of a **body tube.** The simplest of these microscopes are hand-held; they are typically ⅝ in. to ¾ in. in diameter and 4 in. to 5 in. long. The lower end of the body tube extends beyond the objective lens along one side; this extension rests directly on the subject to maintain position and focus. Fine focusing is accomplished by tilting the tube slightly toward or away from the extension tip.

These instruments are commonly 20x to 50x, so they extend surface viewing resolution well beyond that of the simple hand lens. In cross-sectional viewing, for example, they often make it possible to see longitudinal parenchyma in cedars or the smallest resin canals in spruce, both of which would probably go undetected with a hand lens.

My favorite hand-held model is a 50x instrument sold by Edmund Scientific Company that has a reticle with 20-micron divisions. It is excellent for determining texture because tracheid diameters in softwoods or pore diameters in hardwoods can be measured to within about 10 microns. On longitudinal surfaces, fusiform rays can be seen clearly and multiseriate and uniseriate rays are discernible. Even some cell-wall detail is visible. For example, on longitudinal surfaces of yellow-poplar, the scalariform perforations and opposite intervessel pitting are quite clear. Some of these magnifiers have a variable or zoom magnification capability up to 50x; others have built-in light sources (see the photo at right).

The problem with such a high-power hand-held instrument is that it is extremely difficult to keep in steady focus. It is therefore necessary to provide a frame that allows the body tube to be brought into focus over the specimen and kept in alignment with the light passing through the specimen. Thus, what we commonly call a microscope includes not only the body tube and lenses, but also the focus adjustment mechanism, the stage supporting the slide on which the specimen is mounted, the light source or light-reflecting mirror and the overall frame that

ties these parts together. Hereafter, the term "microscope" will denote this composite instrument, even though it would be referred to more accurately as a compound transmitted-light microscope.

The body tube and the stage are lined up and attached to the arm of the frame. A rack-and-pinion gear set moves the tube on some models and the stage on others to bring the subject into focus. Some microscopes have only a single focusing knob. More expensive microscopes have a coarse adjustment knob for approximate focusing and a fine adjustment for more precise focusing. On some models the arm and base form a fixed unit, whereas on others the arm is hinged to the base, enabling the body tube and stage to be tilted to a more comfortable viewing angle.

At the lower end of the body tube most microscopes have a **nosepiece,** which holds two or more objectives of different magnifications. The magnification is changed by rotating the nosepiece to bring a different objective into position. A microscope with an eyepiece of 10x magnification might have objectives of 5x, 10x, 20x and 40x; specimens therefore could be viewed at 50x, 100x, 200x or 400x. These objectives are usually parfocal, meaning that once an object is in focus at one magnification, it will be in approximate focus when a different objective is rotated into position.

**HAND-HELD MICROSCOPE WITH
BUILT-IN LIGHT SOURCE**
This hand-held microscope has a 30x to 50x zoom feature plus a built-in illuminator.

THE MICROSCOPE

A typical compound transmitted-light microscope, showing principal features

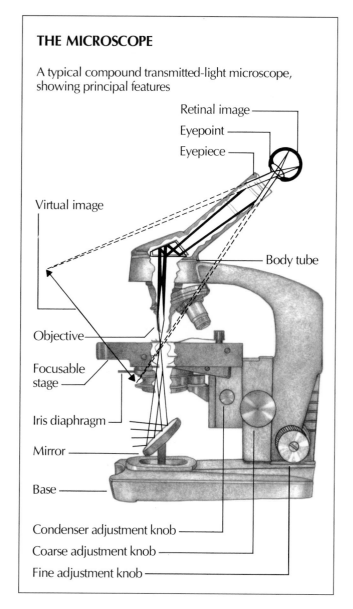

- Retinal image
- Eyepoint
- Eyepiece
- Virtual image
- Body tube
- Objective
- Focusable stage
- Iris diaphragm
- Mirror
- Base
- Condenser adjustment knob
- Coarse adjustment knob
- Fine adjustment knob

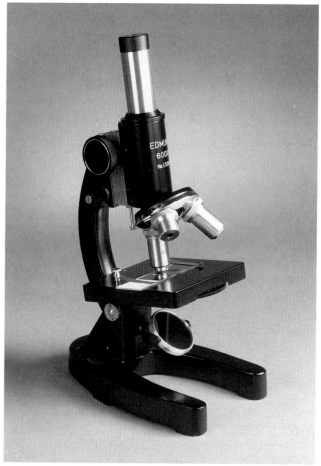

Student models usually cost around $100, and are at the low end of the array of usable microscopes. Although it lacks the convenience, quality and features (such as a built-in illuminator) of more expensive models, this microscope can evaluate all the anatomical features discussed in this book. It is a suitable instrument for getting started, but any serious involvement in microscopic wood identification deserves a higher-quality model.

The microscope **stage** has a hole over which the portion of the glass slide bearing the specimen is placed. Stage clips or slide clips hold down the ends of the slide just firmly enough so that the slide can still be moved laterally to bring the desired portion of the subject into the field of view. More expensive microscopes may have a floating stage; with this feature, the entire stage with the slide on it can be pushed manually in any direction. Other models have a mechanical stage that is positioned mechanically through the use of control knobs.

Below the stage is the light source. In less expensive models, a mirror redirects daylight or lamplight up through the stage opening. A bright lamp will usually suffice as a light source, as will daylight that comes through a window or is reflected from a light-colored surface. Note the difference between ambient daylight and direct sunlight. Never attempt to use direct sunlight: it can injure the retina of your eye. A substage diaphragm helps

regulate the amount of light coming through the stage. In some microscopes, a concave mirror directs light into a substage condenser, another system of lenses that focuses the light onto the specimen. The condenser assembly enhances the contrast within the image and regulates the intensity of illumination.

Your decision as to what kind of microscope to buy is largely a question of cost versus performance. Generally speaking, the greater the cost, the better the instrument. In making compromises, sacrifice convenience rather than viewing quality. For example, some nosepieces carry six objectives rather than two or three. While it is nice to have extra intermediate powers, they are not essential. As another example, you will pay more for a microscope with a mechanical stage. Although this does make lateral adjustment of the slide smoother and more convenient, it in no way affects the ultimate quality of the image — and that is what is most important.

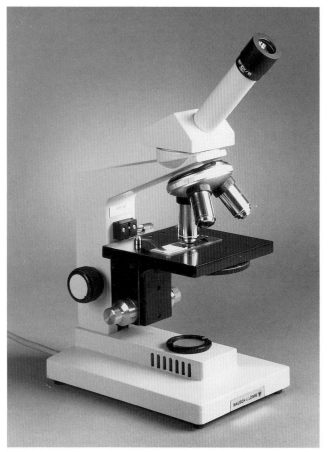

This laboratory model was bought through a discount catalog for under $200 (about 40% of its list price). With magnifications of 40x, 100x and 400x, it does a good job of resolving the routine features studied in wood identification without causing user fatigue. When fitted with a few optional accessories, such as a mechanical stage and a measuring eyepiece, it can be used in a professional capacity.

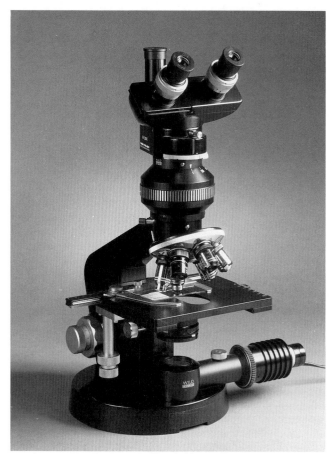

This research-quality microscope is one of a vast array of ultimate instruments whose prices can reach into the tens of thousands of dollars. The features and accessories seem endless —examples include turrets with six objectives or built-in photo and video cameras. With its superb clarity, resolving power and large field of view with even illumination, contrast, and sharpness of focus, a microscope of this quality is a joy to gaze into.

In selecting and using a microscope, resolving power or resolution is more important than magnification. Magnification indicates how many times a subject is enlarged. But some light and image quality is lost during optical magnification, so enlarging an image does not guarantee that proportionately more detail can be seen. **Resolution** refers to the minuteness of detail that can be discerned. This term takes into consideration not only magnification, but also sharpness of image and quality of light transmitted.

For routine wood identification, you do not need magnification of more than 500x, and 300x to 400x is usually adequate if the resolution is good. Inexpensive microscopes often boast magnification of 600x or more but lack satisfactory resolution above 300x. More expensive microscopes tend to have not only greater resolution, but also a larger field of view and more uniform focus over the entire field.

MICROSCOPE ACCESSORIES

Glass slides and cover glasses are used to mount sections for examination (see the top photo on the facing page). Standard **microscope slides** measure 1 in. by 3 in. (25mm by 75mm) and are made of distortion-free glass approximately 1mm thick (metric measurements are almost always used to denote the thickness of slides). Cover glasses are commonly 0.13mm to 0.25mm thick and come in round and rectangular shapes of various sizes. A popular size is 22mm square.

In a pinch, glass slides can be cut from good-quality window glass. When a high-power objective is used, the lower lens surface is very close to the specimen, so cover glasses must be thin. Emergency cover glasses can also be snipped from acetate sheets if the thickness and surface quality are adequate (an irregular surface produces optical distortion). However, neither option is optically equal to regular slides and cover glasses.

You should use only lens tissue to clean optical components,

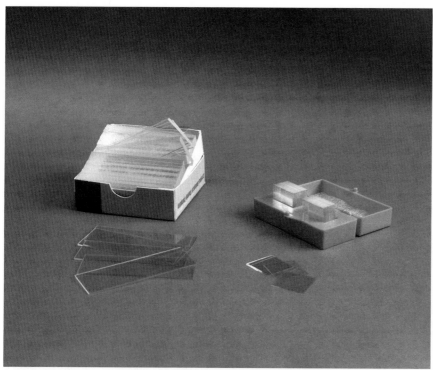

**MICROSCOPE SLIDES AND
COVER GLASSES**
Standard 1-in. by 3-in. glass slides and
22mm-square cover glasses are commonly used to mount sections for microscopic examination.

**LENS TISSUE AND
EYEDROPPER BOTTLE**
A water bottle with an eyedropper cap is
used to prepare temporary slide mounts.
Only lens tissue should be used to clean
lenses.

since impurities and contaminants in ordinary tissue can scratch
or erode the lens surface and the special coatings on the lens.
Camel's-hair lens brushes and cans of compressed, moisture-
free gas are also useful for lens cleaning.

To mount the slides, a small water bottle with an eyedropper cap
is needed. Other useful mounting accessories include a small

camel's-hair or sable artist's brush forceps, toothpicks and ordinary tissue (photo above).

To keep slides from getting mixed up, it is good practice to label
each one as it is prepared. A wax-type glass-marking pencil or
certain felt-tipped pens can be used to write directly on the
glass, or small self-stick labels can be used instead.

Technique

Having looked at the equipment used in wood identification, we can consider how to use it. But first a word about safety.

As we have seen, the preparation of wood for identification routinely involves knives and razor blades. When used as intended and as described here, accidents can be avoided. Carelessness, however, will almost surely result in cuts, possibly severe cuts. Those with previous experience in carving or woodworking know the hazards of sharp cutting tools and have undoubtedly developed a healthy respect for them. However, for those with no previous experience with cutting tools, I cannot overstate the warning that severe injury can result from carelessness.

EXAMINATION OF GROSS FEATURES

Gross features are routinely examined with the unaided eye. Some pieces of wood are ready as found for assessment of color, evenness of grain and figure. However, if the surface is weathered, aged or otherwise discolored, it may be necessary to make a fresh end-grain cut or to whittle down the wood to see the growth-ring orientation (see the photo directly below).

Once the growth-ring orientation has been established, the radial and tangential surfaces can be located. If a bandsaw is available, it is very helpful in cutting these surfaces (see the photo at bottom left). After making the cuts, it may be appropriate to use a hand plane to expose such features as ripple marks on the tangential surface and ray fleck on the radial surface (see the photo at bottom right).

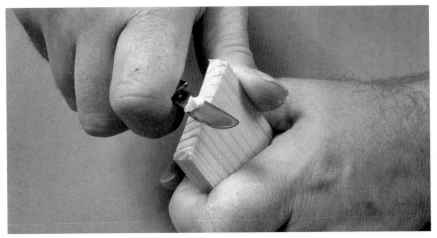

INITIAL SURFACING OF THE END GRAIN
The first step in examining wood is usually to clean up a portion of the end grain with a knife.

BANDSAWING TO CREATE RADIAL SURFACES
After a tangential surface has been established by splitting a piece parallel to the growth rings, perpendicular radial cuts can be made with a bandsaw.

CLEANING UP LONGITUDINAL SURFACES
Split or bandsawn surfaces can be smoothed with a standard hand plane to reveal gross and macroscopic features.

Avoid using abrasive paper or a belt sander to smooth surfaces that will later be sectioned.

This phase should include the simple fingernail indentation test for hardness, a routine check for distinctive odor and a test with a black light, if available, for fluorescence. Once the gross features have been noted, the specimen is ready for the next phase — macroscopic examination.

EXAMINATION OF MACROSCOPIC FEATURES
Macroscopic examination is performed with the 10x hand lens. A typical first step is surfacing an end-grain portion of the sample with a very sharp knife (see p. 78). The specimen should be examined carefully to locate an area of average growth rate.

Such irregularities as juvenile wood, extreme growth, knots, cross grain, reaction wood and decay should be avoided if normal defect-free areas are accessible. A final surface might well be made with a single-edged razor blade (see the photo below).

When using the razor blade, keep your fingers well below the end-grain surface you are cutting. The blade should be moved in a sliding motion to produce a clean cut. Some people prefer to use a pushing stroke, as illustrated; others prefer to pull the blade. Experimenting will reveal which is right for you. Although it is preferable to surface as large an area as possible with a single stroke, it usually takes several overlapping strokes to clean an area up to ¼ in. square. An area of this size is usually adequate for the initial evaluation of macroscopic features.

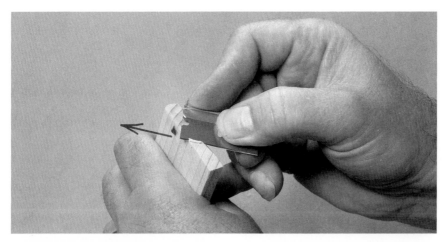

FINAL SURFACING OF THE END GRAIN
To cut an end-grain surface cleanly and safely, hold the blade firmly and move it in the direction of the arrow in a sliding, slicing motion. Make a thin cut on a very small area. Moistening the surface of the wood usually helps.

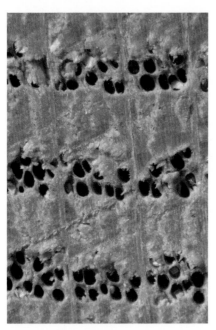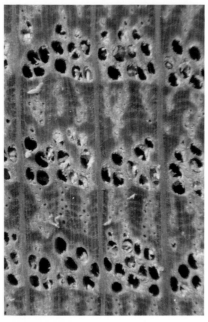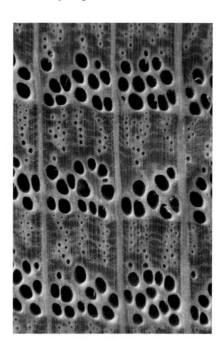

CLEANER CUTS SHOW GREATER DETAIL
These photos show the same transverse surface of a piece of red oak, initially cut with a table saw (left: latewood pores not visible). It was then cleaned up with a pocket knife (center: latewood pores partially visible). And finally it was surfaced with a single-edged razor blade (right: latewood pores individually distinct).

USING THE HAND LENS
To examine the surface of a wood sample, hold the lens close to the eye, then bring the sample toward the lens until the surface comes into focus. Focus is maintained by butting the hands together and bracing them against the cheek.

When using a hand lens to view a cross section, the lens should be held close to the eye, as shown in the photo at left. Bring the sample toward the lens until the surface comes into focus. Maintain focus by butting your hands together and bracing them against your cheek. It is important that the specimen be well-lit, so adjust your position so that the light source fully illuminates the specimen. If necessary, position a lamp so that it illuminates the surface.

The relative size of anatomical features is important in identification work, but judging dimensions from memory can be difficult. A wood that looks medium-textured one day may look coarse-textured a week later. One solution is to make direct comparisions with known specimens (see the photos below). For example, holding a clearly surfaced wood of known texture alongside the unknown can help you judge both texture and cell structure. This is one good reason to maintain a reference collection discussed in Chapter 9.

With experience, routine size estimates — including resin-canal diameter, pore width, maximum ray width and rays per millimeter — can be made with the hand lens. As mentioned

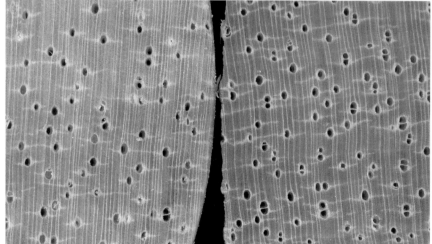

COMPARING END-GRAIN SURFACES
With a hand lens, the difference between the fine texture in eastern redcedar (on the left in the top photo) and the coarse texture in redwood (on the right in the top photo) is evident. At right, an unknown specimen (on the left) is compared to a known sample of ramin (on the right). (10x)

earlier, hand magnifiers used with reticles facilitate measurement and increase precision (see the photos below). Calibrated line widths down to 0.04mm (40 micrometers) are very useful.

A common problem arises when the unevenness or slight concavity of the prepared sample surface prevents the reticle from lying flat. If you plan to use a reticle magnifier, try to whittle a very slightly convex surface.

If you don't have a reticle magnifier, you can estimate dimensions by comparison with common objects, on a larger/smaller basis. I often use a hair from my own head as a "gauge rod."

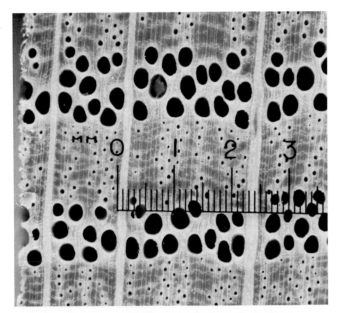

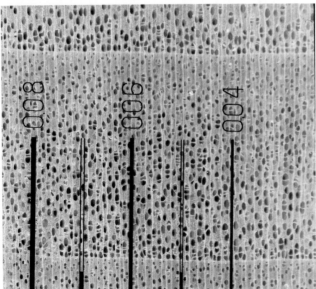

RETICLE SCALES USED TO MEASURE PORE DIAMETER
When a reticle is placed over a transverse surface and the sample is viewed with a hand lens (top), pore diameter can be approximated. (Divisions on the scale are 100 micrometers.) Another type of reticle has bars of labeled thickness (above).

Without knowing the exact diameter of my hair, I have found that pores in maple and gum are generally smaller than the hair diameter, whereas in cottonwood and birch they are larger. Human hair is slightly oval, and its diameter varies. Mine averages 84 micrometers in its wider diameter; most strands fall within 20 micrometers of this value.

Manufactured products tend to be more consistent in diameter. I have a spool of monofilament fishing leader tippet with a rated diameter of 0.004 in.. It actually measures 0.0044 in. (110 micrometers). Alternatively, a strand of copper wire from an electronic cord can be used as a comparative gauge, or as an absolute gauge if it is first measured with a micrometer caliper.

The Forest Products Laboratory at Princes Risborough, Buckinghamshire, in the United Kingdom, has developed a transparent gauge on 35mm acetate film. Based on this, it would seem that a common 35mm photographic slide of a grid might also serve as a useful gauge. It would be necessary to measure the spacing of the grid lines before using it for numerical measurements. Minute features are measured best with a microscope equipped with an eyepiece micrometer, which is discussed under the next heading.

Because hand-lens examination is the most often used procedure in wood identification, it should be practiced until it becomes routine.

EXAMINATION OF MICROSCOPIC FEATURES
Preparation for examination of microscopic features involves a series of preliminary steps, including shaping the specimen, softening the wood, cutting sections and mounting the sections on a slide. How well the slides are prepared will determine how much information they provide.

Shaping the Specimen
For microscopic work, a piece of wood somewhere in size between a sugar cube and a pack of gum is appropriate since it is large enough to be handled easily and small enough to be softened by boiling. Preparation for microscopic examination involves the selection of a portion of the wood that is free of abnormalities.

ACETATE GAUGE AND MILLIMETER SCALE
This clear acetate gauge developed by the Forest Products Research Laboratory contains lines of various widths that can be laid atop or alongside rays and pores to determine width and diameter (shown actual size). The circles are useful for counting pores per unit area.

The piece may already have been prepared as suggested earlier in this chapter. If this has not been done or if there is any question about the orientation of the grain, split planes perpendicular to one another (at this point they need not be parallel to the rays and growth rings) to confirm the assumed grain direction. If these splits reveal any cross grain, the transverse surface should then be resquared to the split surfaces.

The next step is preliminary shaping of the part of each surface that will be sectioned. *It is critical that microscopic sections be oriented accurately along the transverse, radial or tangential planes.*

Begin by carefully whittling the end-grain surface flat by removing very thin flakes of wood tissue, as shown in the photos below. Whittle through any surface discoloration to fresher wood beneath and remove all saw marks or other indications of tissue damage. Next, with a sharp knife, whittle away the area surrounding the chosen portion with slightly beveled cuts, leaving this area as an elevated plane or plateau.

This preparation is obviously less time-consuming on a small piece — say ½ in. square. If you are presented with a larger piece, it might be more practical to split or saw a smaller piece from it. Finish the surface of the transverse cut with a single-edged razor blade.

Next, select radial and tangential locations ⅟₁₆ in. to ⅛ in. wide and flatten them to a smooth surface with a single-edged razor blade. Some reshaping of the piece may be necessary, and small selected areas can be elevated in the manner described above by whittling down the surrounding tissue.

In uneven-grained species, you must decide whether the tangential surface should be developed along a plane of earlywood or latewood. In time, you will learn that in choosing or "splitting out" samples, square pieces with rays running diagonally across the transverse surface have radial and tangential orientation along their corner edges, and that these planes can be created simply by chamfering a ⅛-in. face along those corners, as shown in the photo below.

 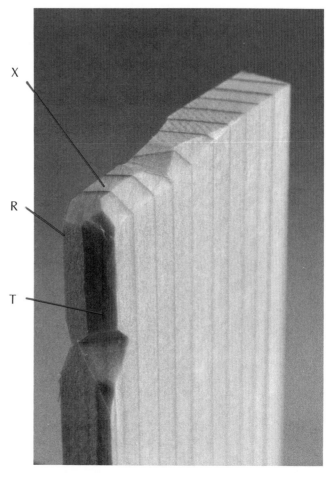

PREPARING A SAMPLE FOR SECTIONING
A sample such as that shown above left need not initially have radial and tangential faces to yield thin sections for microscopic slides, but it should be straight-grained. As shown above right, first clean up the end grain, then bevel the edges to leave a small portion (X) from which transverse sections can be sliced. Next, find edges that can be beveled to produce radial (R) and tangential (T) faces for sectioning. The notches or stop cuts define the areas to be sectioned, making it easier to maintain the proper plane.

Softening

In a few of the softer woods, sections can be cut directly from the sample block. But for most woods, the block should be softened prior to sectioning. With low-density woods such as white pine, spruce, basswood and aspen, flooding the areas to be sectioned with water applied with an eyedropper or camel's-hair brush and allowing the water to soak into the surface for a minute or two are usually sufficient. This technique works for the majority of woods, at least for cutting longitudinal sections. However, cutting will be much more successful with most woods — especially when taking transverse sections — if the wood is first softened by boiling in water (photo below left).

Cover the specimens with water in a Pyrex or enameled container and bring the water to a boil. A general rule is to boil a sample until it sinks, although whether or not this happens depends partially on the size of the piece and on the density and permeability of the wood. The object is to saturate or waterlog the wood for maximum softening. This can take anywhere

from five minutes for softwoods to an hour or more for the densest woods. Additional softening in particularly stubborn samples can be accomplished by placing preboiled samples in a 50/50 solution of alcohol and glycerine for several days. If it is not possible to boil samples, immersing a sample in a container under a running stream of the hottest tap water available is far more effective than simply wetting the sample with cold water.

Sectioning

In samples softened by water at room temperature, sectioning can be done while the surface is still flooded with water. In samples that have been boiled, sectioning should be done as soon as possible after the sample has been withdrawn from the water. A pair of scissor-type wooden kitchen tongs or toast tongs are ideal for removing samples from the water. A small square of spongy rubber, such as that used for carpet padding, can be used as a "potholder" to insulate your fingers if you wish to handhold the sample, but I have found that the improvised handscrew clamp shown in the photo on p. 71 works much better.

When making sections, keep in mind that the section must have flawlessly cut surfaces on both sides. The last preliminary surfacing cut is no less important than the cut that removes the section. Therefore it takes two consecutive flawless cuts to make a high-quality section. The cutting is best done with an unused double-edged razor blade. Cutting strokes should be firm and controlled; the blade edge should move in a smooth sliding action. Be deliberate enough to complete the cut in a single stroke — any hesitation midway will result in ridges or uneven thickness of the section. Avoid any back-and-forth sawing action. Take a couple of preliminary strokes to clean up the surface and to develop a comfortable cutting rhythm and motion. Then make the sectioning cuts (photo below right).

SOFTENING THE SAMPLE BY BOILING
Boiling wood samples in water helps soften them for sectioning. Samples of average density are sufficiently softened when they sink.

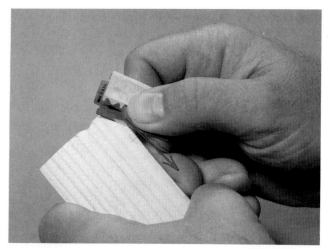

CUTTING WITH A DOUBLE-EDGED RAZOR BLADE
One edge of this double-edged blade has been taped for better control and safety. The sectioning cut is made with a sliding motion. The arrow indicates the direction of blade motion in a pull stroke.

Probably the most common mistake is trying to cut too much. Don't try to imitate the beautiful, large sections you see in books or on commercially prepared slides — they were cut on a special instrument called a **microtome**. Instead, concentrate at first on taking tiny, well-cut flakes ⅛ in. square or smaller. Again, never force the blade — the delicate edge will roll over and that portion of the blade will be ruined.

It is important to realize that handcut sections are never uniformly thin. Consequently, when examining a section, expect to search for regions of appropriate thinness to show the cellular features and desired detail. It may be necessary to take several sections from a surface to increase the odds of exposing the desired detail.

Experience will underscore two basic things about sectioning. First, a clean cutting action exposes features with minimum damage. Second, the areas of a handcut section that show the best detail are usually near the edges of the section, where the thickness feathers off to nothing. This emphasizes the importance of developing a smooth cutting action even as the blade enters and emerges from the cut.

It is impossible to create a perfectly flat surface by hand sectioning with razor blades. Invariably, as multiple overlapping sections are taken from an area, the surface is left with a series of very subtle undulations. Subsequent sections slice the high spots from this undulating surface and almost always have some ideally thin portions along their edges.

It is critical that a section be thin because thick sections are virtually useless. If the surface becomes very irregular because of miscuts, it makes sense to start anew by surfacing another true plane and cutting new sections.

To section a piece of veneer or a small sliver of wood, sandwich it between contrasting species and clamp it between two pieces of even softer wood that have also been boiled. Then cut the sections by slicing across the entire "sandwich."

It is good practice to cut sections over a clean, white sheet of paper to keep track of any flakes that fall. Debris and miscut flakes should be blown away regularly so they do not get mixed up with the "keepers." Sections can be picked up easily by moistening a toothpick or a camel's-hair brush and touching it to them. If the sections are not placed directly on a slide for mounting, they can be kept in water until needed. Using a white saucer will make them easier to see.

Slide Preparation

Before transferring the specimen to a glass microscope slide, place a small self-stick label or tab of invisible tape on one end of each slide. This will enable you to jot down identifying names, numbers and R, T or X designations to radial, tangential or cross sections. An alternative to labeling slides is to use a permanent-type felt-tipped pen or a wax pencil glass marker. The small investment of time in labeling will help avoid mix-ups when working with several pieces of wood or with several views. It is a good habit to pick up slides by the labeled end. This helps prevent annoying fingerprints, which always seem to appear on the cover glass and elsewhere, especially on permanent slides.

A strip of wood 1½ in. to 2 in. wide and ¾ in. thick provides a good rack for the temporary slides (photo below). The rack helps keep the slides in a specific location and elevates them so that the overhanging labeled ends can be grasped easily. A shallow cardboard box or even a book can serve the same purpose.

Simply placing a dry section of wood on a slide is unsatisfactory since it will not stay in place or lie flat; nor is the optical quality as good as when the section is viewed in a liquid medium. The simple but effective temporary water mount described below is effective in all identification work covered in this book.

First, transfer one or several wet or dry wood sections to the slide (see the photos the facing page). Keep them away from the slide's edges and group them closely but not overlapping in a central area. Include no more than can be covered by a single cover glass.

If dry sections have curled, place a small drop of water on them prior to covering; this usually makes them unroll. Sometimes these sections can be teased open by careful manipulation with a pair of sewing needles or toothpicks. Sections need not

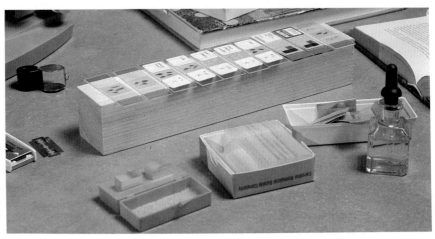

IMPROVISED SLIDE RACK
When making or examining slides, keep them organized on an improvised elevated rack, such as a piece of scrap wood or even a book. The overhanging ends can be grasped easily without touching the specimen area of the slide.

be totally flat, only flat enough so that they flatten further rather than fold over when a cover glass is placed on them. If the sections are still slightly arched under the cover glass, press the cover glass down gently with a needle or toothpick.

Next, using an eyedropper, squeeze a little water onto the slide at the edge of the cover glass. The water will be drawn under the cover glass by capillary action and will automatically spread out. Use enough water to surround all the sections, but not enough to fill the area under the cover glass completely (see the photo at top right).

The object is to create a meniscus under the cover glass whose surface tension draws the cover glass down evenly, creating a minimal but uniform film of water while holding the section flat. If too much water is used, the cover glass will float in a puddle, which makes a mess and gives poor viewing results (as shown in the photo at bottom right).

If you have added too much water, fold a facial tissue to a point and touch it to the puddle near the edge of the cover glass. The tissue will absorb the excess water until the meniscus retreats under the edge.

Air bubbles under the cover glass are inevitable. If they remain in contact with the sections, try tapping or depressing various areas of the cover glass with a toothpick or needle. This may chase out some of the bubbles; you can work around those that remain.

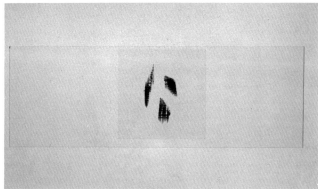

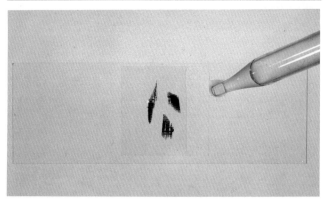

MOUNTING A SLIDE
To make a temporary water-mount slide: Top photo: Arrange one or more sections in the central portion of the slide. Second photo: Place a cover glass over the sections. Third photo: With an eyedropper, gently introduce water at the edge of the cover glass. Use enough water to surround the sections, but not enough to 'float' the cover glass.

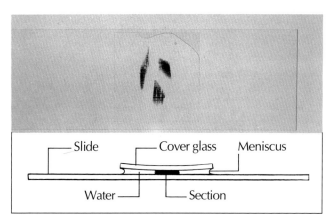

THE RIGHT AMOUNT OF WATER ON A WATER-MOUNT SLIDE
When the correct amount of water is used, it surrounds the section as seen in face view. The water forms a meniscus under the cover glass that holds the glass in place and flattens the section against the slide.

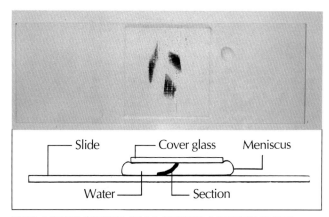

TOO MUCH WATER ON A WATER-MOUNT SLIDE
Too much water leaves messy liquid edges along the cover glass, as shown in the face view. The cover glass is not held in place, and the section is not held flat and level against the slide.

The water will begin to evaporate and may have to be replenished from time to time as the meniscus retreats from the edge of the cover glass and approaches the sections. To prevent evaporation, you can seal the edges of the cover glass with petroleum jelly, but I find this too messy — I prefer to add water as necessary.

If slides must be put aside overnight or even for an hour, the water will evaporate and the dried mount will dislodge at the slightest disturbance. Therefore, tack down the edges of the cover glass with bits of tape or labels to keep the specimen flat and in place (see the photos below). The next time the slide is examined, try viewing it dry. Because the section is flat, it is usually satisfactory for identification. Water can be added at a later time, but this usually results in the entrapment of a myriad of tiny bubbles. One cure for this is to take the slide apart, boil the section briefly and then remount it.

As an alternative to temporary water mounts, a solution of equal parts of glycerine and alcohol (either ethanol or denatured ethanol) or straight glycerine can be used. Warming the slide on a hot plate on the low setting will help expel air bubbles. This makes a mount that lasts for days or even weeks.

USING THE MICROSCOPE

There are many models and styles of microscopes, so any specific instrument should be used in accordance with the manufacturer's instructions. The following discussion provides a general overview of what is involved.

Microscopes with substage illuminators need only to be plugged in and turned on. Other microscopes require a suitable light source, such as a bright lamp. The lamp works best when placed as close to table height as possible, preferably below the level of the microscope stage.

Daylight that comes through an open window or is reflected from a white card placed near a window often works just as well. But remember never to use direct sunlight. Adjust the microscope mirror so that the light is directed upward through the open hole in the stage. When viewed through the eyepiece, the field should be illuminated.

Place a slide on the stage and secure it with the clips. Move the slide to position the section of wood over the center of the hole in the stage. Pivot the lowest-power objective (the shortest one) into position. Using the coarse adjustment knob, move the objective to within 1/8 in. of the cover glass. While looking into the eyepiece, slowly raise the objective (or lower the stage) until the image comes into focus. If you don't see anything, perhaps the specimen is not within the field of view. Adjust the subject so it is directly under the objective and try again.

Once the image is in view, experiment by moving the slide. With a compound microscope, the image is inverted and moves exactly opposite to the way the slide is moved. While this is confusing at first, in time you will grow as accustomed to it as you are to the reverse image you see in the rearview mirror of your car.

Next, find the most distinct features — a good place to look is along the thin edge of the specimen. Try to improve the quality of the image by sharpening the focus, adjusting the substage opening, adjusting the mirror and repositioning the light source. Poor contrast can usually be improved by modifying the light source in some way. Try masking off all but a direct tunnel of light from the source to the mirror. If there is glare, try placing a diffuser such as waxed paper, frosted glass or tissue paper in the path of the light.

A common problem with microscopes requiring external illumination is light that falls directly across the surface of the slide (ideally, all light should be reflected from the mirror). If this happens, try moving the light or shielding the stage by taping a temporary cardboard fence or baffle along the outer edge of the stage. It may also help to turn off all room lights other than the one illuminating the substage mirror.

Find the sharpest detail in the field of view and adjust the slide so the feature is centered. Pivot the next-highest-power objective into place. As mentioned earlier, the subject should still be in approximate focus; only fine adjustment should be necessary to sharpen the image. Recenter the subject and go on to the next-highest-power objective until you are at the highest power.

You may notice two things. First, the illumination is dimmer at a higher power; you may have to adjust the iris diaphragm opening and/or the light source to get the best image. Second, the depth of focus is very shallow — as you adjust the focus up and down, various details jump in and out of focus. Thus, in

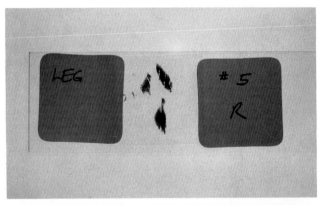

TACKING DOWN THE COVER GLASS
To save a water-mount slide for future use, tack down the cover glass with gummed labels (top) or cellophane tape (above).

using a microscope, along with moving the slide around, you should develop the habit of scanning vertically through the field in order to find the most detail. Otherwise you may completely miss the feature you are looking for, such as a pit pair sectioned exactly through the center.

With the high-power objective in position, note how close the objective is to the cover glass (photo below). It is unwise to examine a slide initially at high power because there is a risk of jamming the objective into the slide while attempting to bring the subject into focus. This usually shatters the cover glass; worse, it may soil or damage the coating on the objective lens. Make a habit of locating and focusing on a subject first at low power, then proceed to successively higher powers. It is also good practice to return the nosepiece to the low-power setting when you finish using the microscope.

Measurements, approximate or precise, can be made during microscopic viewing in various ways. An **eyepiece micrometer** is a scale that is etched into a lens set in the eyepiece; when viewing through the eyepiece, the scale is superimposed on the subject. The value of each division on the scale changes for each objective. Sometimes data on the approximate values of the scale under various power objectives come with the scale. Otherwise, the scale must be calibrated at each magnification. This can be done with a reticle placed on the stage, or with a **stage micrometer**, a special slide with a precise scale etched into it.

To avoid fatigue while using the microscope, find a table and chair height that enables you to feel comfortable when your eye is in front of the eyepiece. If you wear glasses to correct simple nearsightedness or farsightedness, you can use the microscope with or without your glasses. When you remove them, however, you will have to readjust the microscope focus. I am nearsighted, and I feel I can see a little better through the microscope without my glasses, if only because if it eliminates bothersome reflection from them. Still, as often as not, I leave them on so I can make notes. (As a result, I have glasses that are badly scratched from repeated striking against the metal rim of the eyepiece! I have yet to obtain the special rubber protectors that are available to prevent this problem for eyeglass wearers.)

At first, using the microscope continuously for more than 30 minutes can cause discomfort or headaches. This is usually the result of keeping one eye closed while looking through the objective. Our eye muscles are unaccustomed to this, and it causes fatigue. It is a good habit, therefore, to keep both eyes open when looking through either the microscope or hand lens.

In a surprisingly short time, you can train your brain to register only what is seen by the eye in front of the lens. You can learn this most quickly by viewing the magnified image with your "master eye." Almost everyone has a master eye; to determine which eye it is for you, try this simple test. First, select a distant object as a "target." Then, keeping both eyes open, raise either hand with the index finger extended in the manner of pointing a pistol and sight down it until you feel you are "on target." Alternately close one eye and the other. You will discover that you have really lined up the target and your pointing finger with one of your eyes. This is your master eye.

It may help at first to cup your hand over the unused eye or to hold a piece of grey cardboard in front of it. Once you have learned open-eye viewing with one eye, you can readily switch

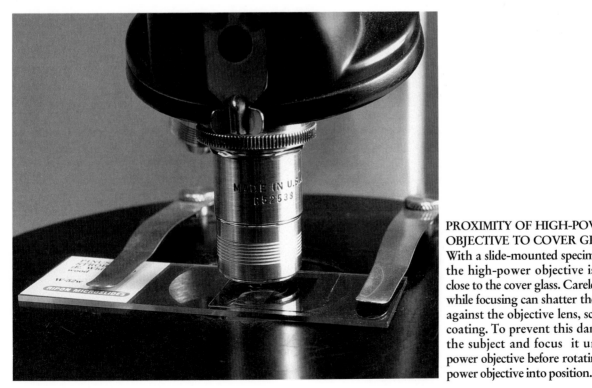

PROXIMITY OF HIGH-POWER OBJECTIVE TO COVER GLASS
With a slide-mounted specimen in focus, the high-power objective is extremely close to the cover glass. Careless adjusting while focusing can shatter the cover glass against the objective lens, scratching its coating. To prevent this damage, locate the subject and focus it under a low-power objective before rotating the high-power objective into position.

to the other — your brain will catch on and cooperate. This way, should one eye tire, you can use the other, allowing continuous viewing with minimal fatigue.

EVALUATING HANDCUT SECTIONS

While examining a handcut section for identification detail, you can also evaluate its quality and consider how you might improve your sectioning technique. Bear in mind that handcut sections will never be as large, uniformly thin and uniformly clear as the illustrations in texts. But certain areas of handcut sections can equal or even surpass the best of these published images. As a general rule, judge a handcut section on its good features and usable details — all handcut sections have some bad areas.

In evaluating a section, first scan the entire section. Start at low power to locate the thin edges that are most likely to show the desired detail; also note where the thick, useless areas are. Never abandon a section as being too thick if it has some feathered edges and it is cleanly cut. Chances are that there will be at least some important detail in these areas. This self-evaluation will improve your technique until you routinely cut sections with clear detail.

There is no escaping the fact that microscopic examination is primarily a game of chance. Rarely can you deliberately section through a particular microscopic structure to locate perfectly an individual detail; rather, you make the cut first, then search for important features. You soon learn, however, that the larger the area with good detail, the greater your chances of finding informative features. Having patience in making good sections and in searching for features is part of the game.

A well-cut section is at least in part thin enough — usually one or two cell layers thick — to show clear detail. In eastern white pine, for example, this would be 20 to 60 micrometers.

The photos at right show not only well-cut sections, but also the common problems encountered in sectioning. It is important to recognize the symptoms of these common problems. Where sections are too thick, entrapped air bubbles show up in cell cavities on transverse sections as black pearls and on longitudinal sections as hotdog-shaped formations. This is less of a problem where a glycerine/alcohol combination rather than water is used for mounting. Ragged cell edges, distorted cells, fractures or ridges through the section all indicate problems with cutting-edge sharpness or cutting technique. When a section is not taken in the true intended plane, the resulting view will be unfamiliar and confusing.

Techniques for more thorough softening of specimens, microtome sectioning, staining of sections and permanent mounting of slides are beyond the scope of this book. Those interested in such advanced techniques will find sources of information on them in the bibliography.

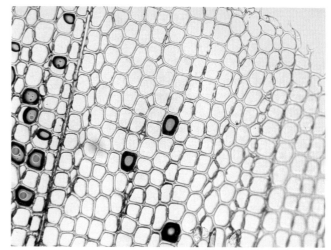

THE BEST PORTION OF A SECTION FOR VIEWING
Above are views of different portions of the same transverse section of eastern white pine. The area along the edge of the section in the top photo provides the best view, since it is thin enough to show cellular detail and is relatively free of air bubbles. The section in the middle photo is also cleanly cut, but this area is thick, and entrapped air bubbles obscure details. The section in the bottom photo is so thick with superimposed layers of cells that individual details are indistinguishable. (100x)

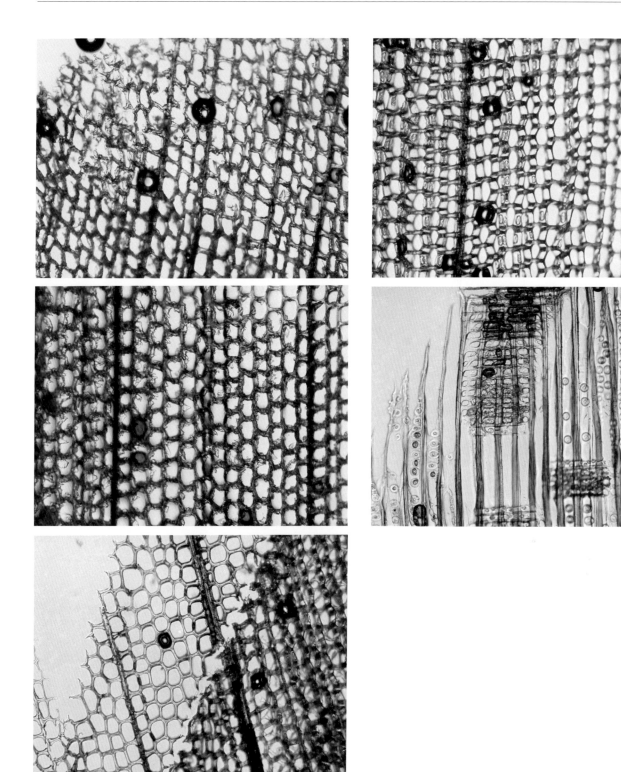

SECTIONING PROBLEMS
The damaged cell structure in the section at top left was caused
by plowing rather than sliding the cutting blade, and the torn,
ragged edges in the photo below it are from a dull blade. The
ridge in the bottom photo indicates hesitation during the cutting stroke. The oblique view of tracheid walls in the top photo
at right indicates that this section is not truly transverse. Similarly, the apparently short rays in the photo below it reveal
that this section is not truly radial. (100x)

CHAPTER EIGHT

TAKING SAMPLES AND KEEPING RECORDS

When I am asked, "How much wood do you need for identification?" two incidents always come to mind. In the first, a home owner phoned to discuss an insect-damage problem that eventually involved wood identification. The caller agreed to send me a small sample of the wood for identification. An envelope arrived a day or two later, but it was empty except for a brief note. I phoned to report the oversight, but the party insisted I recheck the envelope, which I did. "No," I reported, "there's nothing in this envelope but some brown dust in the crease." "Why, yes," the home owner said, "that's it. Can't you tell what kind of wood it is?"

In the second incident a consulting engineer called with an urgent problem: He needed to know whether the heavy structural timbers supplied to a construction job were Douglas-fir, as specified in the contract. I realized that for positive identification, all I really needed to do was check the tracheid walls for spiral thickenings. I suggested he send samples, which he said he would do immediately. Three days later a truck driver appeared at my office with a handcart loaded with 18-in. lengths of 12x12s. This was an expensive shipment, considering that enough material to check a few tracheids could have been sent in an ordinary envelope. Ever since these two incidents I have made a point of discussing just what size and manner of specimen is needed.

Sample Size

There is no single answer to the question of how large a sample must be taken for identification. Obviously it is best when the person identifying the wood has a chance to see the entire object in question. Where this is possible, the whole object becomes the sample and can be assessed for gross features. Characteristic figure, color or odor may immediately suggest a species or at least narrow the possibilities so that only a tiny sliver is needed to confirm a choice or to separate two or more woods.

Short of actually seeing the wood, a good-quality photograph of gross detail is sometimes quite valuable. For example, if the wood panel in a photograph is clearly a ring-porous hardwood with conspicuous rays up to 4 in. high, it can probably be assumed to be white oak. In my experience, however, photographs taken by someone who is not familiar with identification features usually lack useful detail. It is always better to have the entire object, or at least a sample of it, however small, for examination. Still, there are instances in which a photograph of a wood object is the only evidence available, particularly where the original object has been lost, stolen or destroyed by fire.

It is tempting to ask for as large a sample as possible when arranging to have a sample taken and sent for identification. But, in most cases, this would result in something wastefully large. Instead, I usually begin by asking for a sample the size of a pack of chewing gum. Even though a sample this size may not be large enough to display both heartwood and sapwood or to reveal the overall figure of the original piece, it still has many advantages. It is small enough to be mailed easily, yet large enough to be handled conveniently during shaping and sectioning. A portion can be used to determine specific gravity or even to provide a future reference sample if it turns out to be a new addition to your collection.

A gum-pack-sized piece can be taken comfortably in some situations, as when sampling structural timbers, checking a doubtful lumber shipment or identifying shipwreck remains. But with routine identification work on objects such as furniture, musical instruments or sculpture, we must settle for a far smaller sample. A piece the size of the eraser on a pencil is usually more than enough to provide adequate radial, tangential and transverse sections. But it is more difficult to handle and cut. In the end, where damage to the object must be minimized, a ⅜-in. long piece the diameter of a kitchen match will usually do the job, and even a toothpick-diameter piece is worth a try.

It is important to realize that in an amazing number of instances a tiny sliver will yield successful results. In fact, sometimes you can take a definitive section from the object with virtually no damage at all. Thus, in working with valuable objects, it is best to start with the smallest sample that seems likely to have a chance of providing the necessary detail, rather than to take large samples just to be sure that there will always be plenty of tissue available.

In summary, appropriate sampling is dictated by the individual situation. Whenever possible, it is best to examine the gross features of the entire object, then to select the best sample location and size for the follow-up macroscopic and microscopic investigations. However, as often as not, restricted access or the extreme value of an object limit sampling to tiny fragments or direct sectioning. In these instances, microscopic evaluation alone is the basis for identification.

Guidelines for Taking Samples

Take only good samples. Don't waste time collecting substandard material. Scrapings, shavings and sawdust are virtually useless; diagonal, angled and shattered fragments run a close second. A well-chosen, cleanly severed sample is worth a pocketful of shavings.

To take a small sample with minimal damage, as is often necessary with furniture, study the larger piece closely and figure out the placement of the growth rings and the grain direction before deciding where to cut the sample. After choosing an appropriate location, make one or more stop cuts to define the ends of the sample. If possible, plan the cuts along the radial and tangential planes. Stop cuts can usually be made with a pocketknife or chisel, or with a fine saw when working at the edge of a piece. Next, score longitudinally to define the rest of the sample and to guide the split as you pry the sample free with a knife or chisel. Caution: If you are working with a valuable piece, be sure to read the section on the next page on historic objects first (and see the drawings on this page and the next).

TAKING A SAMPLE FROM A LONGITUDINAL EDGE OF AN OBJECT

To remove a small sample from the edge of a board, first use a knife to notch in a pair of stop cuts (A). With the knife tip, score the edges of the sample area to guide the split (B). Finally, engage the knife edge in the bottom of one of the stop cuts and gently pry the sample free with a slight twisting motion of the knife (C).

TAKING A SAMPLE FROM AN END-GRAIN SURFACE

To remove a sample from an end-grain surface, first use a narrow chisel to make a pair of wedge-shaped parallel holes (A). Use a thin knife blade to connect the walls of the two holes and define the sample (B). Finally, use the narrow chisel to undercut and pry out the sample (C).

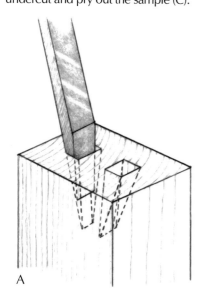
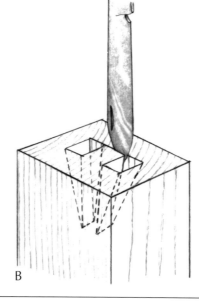
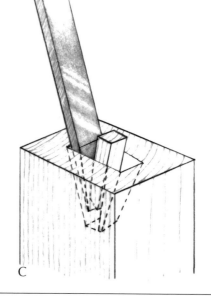

A   B   C

Avoid reaction wood, cross grain, knots and juvenile wood. Also avoid decay, crushed or dented areas, areas that are contaminated with oil or dirt and areas that are saturated with glue or finish. If possible, stay away from the very bottoms of the feet of chairs, tables, stools, etc. — these areas are typically damaged, decayed or fouled by dirt and oil.

When you take microsections directly from an object but do not plan to mount them for microscopic examination until later, treat them with special care. They should be placed in glassine packets, coin envelopes, paper packets or small vials, and labeled (see the top drawing on the facing page).

Taking Samples from Historic Objects

We all share the responsibility for protecting historic objects and artifacts. The question of how much wood can be removed from a historic object or how much the object can be defaced in the name of wood identification should be left to curatorial and conservation professionals. On the other hand, you should be able to give them some idea of how large a piece you need. The decision whether or not to proceed is then in their hands.

Perhaps the best time to take samples from antiques and other historic objects is during restoration when these objects are disassembled, since you can usually find an area that will be hidden when the object is put back together. Taking a sample from the underside of a table is illustrated in the bottom drawing on the facing page.

Guidelines for Keeping Records

If you work with wood products, antiques, forensic evidence or archaeological specimens, or in other fields where you are regularly engaged in wood identification, it is very important to develop a systematic, efficient routine not only for the sampling process itself, but also for keeping records. Notes taken during identification can provide critical information at some later time; moreover, your records enable you to justify your results, and they will prove invaluable to others who want to examine your work.

Begin a notebook immediately. Don't be overly concerned with neatness as long as your writing is legible. Date every entry and write down everything pertinent — later you will be thankful that you did. When taking a sample from a large object, examine the entire piece and record all significant features. This information may prove invaluable later if it is recorded in meaningful language. Comments like "dimpling present on 95% of the tangential surfaces," or "pith approximately 1/8 in. in diameter, dark brown and chambered" are extremely useful. But comments like "wild grain on tabletop," or "nail holes in one end" have little meaning in the identification of the wood.

Gather all available information about the source or geographical origin of the piece, even if you feel this information may not be entirely accurate. If the origin is alleged but not proven, simply qualify the entry: "said to be imported from France," or, "attributed to Connecticut Valley craftsmen based on style and hardware." Always describe as precisely as possible

WRAPPING AND LABELING A MICROSECTION

Thin sections for microscopic examination can be wrapped as shown for safe storage. A self-stick label seals the packet and describes its contents.

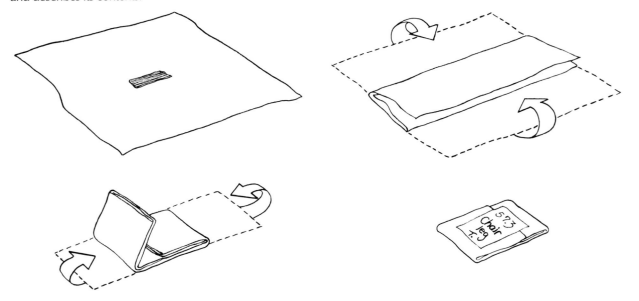

TAKING MICROSECTIONS FROM A TABLE

The growth rings in the leg on the underside of this table make it possible to find the tangential face, which can then be cleaned with a razor blade for direct sectioning (A). Growth-ring placement visible in the end grain of glue block B suggests that a direct radial section can be taken along the edge of the block. Growth-ring orientation in glue block C, however, indicates the block would yield only a tangential section directly.

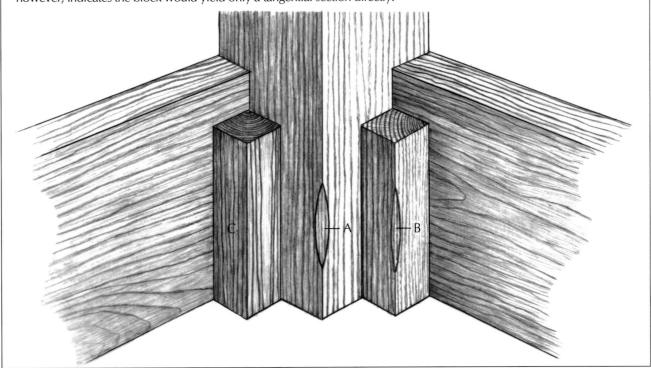

the location or furniture part from which you took the sample. If the object is part of a collection, record any accession numbers that appear on it.

Above all, remember that the best identification record documents not only the species of the wood, but also the basis for that determination. For example, a record that says simply that a piece of wood is *Pinus resinosa*, with no description of how that conclusion was reached, does not indicate the basis for the answer. Was this conclusion reached by visual judgment of gross features alone, or were more reliable microscopic features ex-

amined? On the other hand, if your records indicate that the sample was identified as *Pinus resinosa* based on dentate ray tracheids and windowlike ray parenchyma and crossfield pitting, as well as an assumed southeastern New Hampshire origin, the basis for the determination is clear. The anatomical data clearly separate the specimen from *P. strobus* and *P. rigida,* and the assumed origin, if accurate, separates it from *P. sylvestris.*

Use a routine form or log sheet for recording source information, field descriptions and macro/micro detail gathered during examination (as shown in the example below).

Wood Science and Technology
Holdsworth Natural Resource Center
University of Massachusetts
Amherst, MA

WOOD IDENTIFICATION RECORD

Voucher No. _____

Project _____

Specimen label/tag # _____

Source information _____

_____

_____

Date_____ Identified as _____

Identified by _____ _____

_____ _____

Specimen size, description, condition _____

_____

_____

Gross and macroscopic features _____

_____

_____

_____

Microscopic features _____

_____

_____

_____

Remarks _____

_____

_____

_____

CHAPTER NINE

WOOD COLLECTIONS AND OTHER VISUAL REFERENCE MATERIALS

Visual reference materials such as wood samples, permanent slides and photographs are invaluable for learning about wood features; they also aid in the actual identification process. A personal collection of known wood samples provides direct, tangible information about wood characteristics. If you have a known sample of a species, you can study its various surfaces, whittle it, section it and evaluate its features in a side-by-side comparison with an unknown sample. In lieu of actual wood samples or microscope slides of wood tissue, photographs such as those seen in this book also help in exploring the gross, macroscopic and microscopic features of wood.

Personal Collections

Probably the most important use of a collection of known samples is in providing a reference against which an unknown wood can be compared. This comparison may involve a quick check of visual features, a more careful study of the end grain with a hand lens or sectioning and slide-making to examine microscopic anatomy.

A primary objective in establishing and building a personal collection is to custom-tailor it to include those woods that are common in your region or that reflect your interests, particularly if those woods cannot be found in standard reference books. For example, an archaeologist's personal collection for a particular region might include not only timber trees, but also shrubs and vines. It is surprisingly difficult to find published information on regional woody shrubs and vines, like mountain laurel, blueberry, grape and wisteria. A craftsperson might include such woods as mesquite, avocado, desert ironwood and sumac. Antiques dealers would likely want to establish reference collections that contain the woods used in the period or regional pieces they handle.

There is no universal sample size, although the International Wood Collectors' Society (IWCS) promotes a standard of ½ in. by 3 in. by 6 in. Your own collections, however, will eventually contain some odd sizes. Shrubs like barberry may never attain a 3-in. diameter. And the feather crotch figure on a 2-ft. long slab of black walnut would be destroyed by cutting it down to size. The routine samples kept for anatomical study and comparison need not be any larger than the IWCS standard size; somewhat smaller samples will often suffice.

Although a single sample might be typical of a given species, one sample does little to indicate the normal range of variability within a species. Therefore, you should try to include as many samples for each species as necessary to represent the common range of variability. Duplicate samples showing extremes of growth rate are important, especially for ring-porous hardwoods.

As the collection grows, keeping samples organized and accessible can become a challenge, especially if some samples are irregular in size or shape. A handy way of storing and organizing groups of smaller samples is by drilling a small hole near the edge of each sample and stringing it on a stiff wire ring (see photo at right). By fashioning the ring so that it opens like a giant safety pin, samples can be added or removed easily.

It pays to decide on a numbering system for samples before a collection is started. One convenient approach is to use a checklist of trees that is alphabetized by genus — for example, E. L. Little's *Checklist of United States Trees,* or the British Standards Institution's *Nomenclature of Commercial Timbers* (both are listed in the Bibliography on pp. 212-215). Simply number each entry in the checklist in sequence and then apply these numbers to your corresponding samples. One method for numbering is to use whole numbers for genera followed by decimal numbers for particular species within the genus. The examples below illustrate numbering a checklist.

1. *Abies* spp. (fir)
   1.1 *A. amabilis* (Pacific silver fir)
   1.2 *A. balsamea* (balsam fir), etc.
2. *Acacia* spp. (acacia)
   2.1 *A. berlandieri* (guajillo)
   2.2 *A. choriophylla* (cinnecord), etc.
3. *Acer* spp. (maple)
   3.1 *A. barbatum* (Florida maple)
   3.2 *A. circinatum* (vine maple), etc.

Under this system, samples will always be in a systematic and logical numerical order, and the numbered checklist becomes the key to the woods. For a quick common-name key to the woods in the collection, jot down the sample number next to its common name in the index of the book you are using each time you add a wood to the collection. No matter which checklist is chosen, it will probably be necessary to insert new numbers later to cover the occasional species — shrubs, for example — that are not included.

Commercially Available Wood Samples

Purchasing a set of labeled wood samples is a good way to get a collection started (see the top photo on the facing page). A list of suppliers and sources is included in Appendix V. Some sources, such as those listed in *World of Wood,* the bulletin of the International Wood Collectors' Society, offer individual samples as well as sets.

Some dealers in cabinet woods offer labeled promotional samples of the species they sell, either individually or in sets, sometimes free of charge. Be cautious about commercial yard lumber as a source of identification material; stray species of lumber often get intermixed or substituted unintentionally.

Labeled woods may seem expensive; they typically cost up to $35 for a set of 20 common woods, and up to $10 for single samples of exotic woods such as snakewood. Sets of veneers are more widely available and are temptingly less expensive. They are helpful in showing certain gross features like color and figure, but because they are commonly only 1/28 in. to 1/40 in. thick, they do not provide meaningful cross-sectional views; useful longitudinal sections can usually be taken with careful planning and cutting.

WIRE SAMPLE HOLDER
A large clip ring fashioned from a wire coat hanger makes a handy organizer for irregular wood samples, such as this assortment from the red oak group.

WOOD SAMPLE SETS
Purchasing a labeled set of the more common woods is a good way to begin a collection of reference samples.

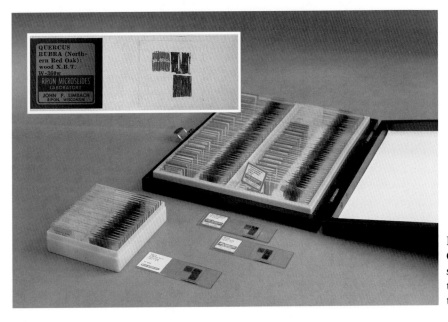

REFERENCE SLIDES
Commercially available prepared slides, sold individually or in sets, include transverse, radial and tangential sections that are stained and permanently mounted.

Prepared Slides

An excellent aid in learning microscopic detail is prepared slides (see the photo above). Prepared slides also provide a good reference base for comparison of microscopic features in the final stages of an identification problem. Commercially available slides have permanently mounted sections that were cut on a microtome and then stained. Each slide usually has three principal sections — transverse, radial and tangential — mounted under a single cover glass and labeled with the wood's scientific name.

Studying slides is an excellent way for the beginner to get acquainted with wood microstructure in general and become familiar with the microscopic identification features of the more commonly encountered woods. A slide collection of about a dozen species makes a good starter set; this collection can be added to over time.

Because it eventually becomes easy to make temporary slides of handcut sections taken from reference samples, it may be worth saving temporary slides when they show an unusual feature clearly or when they have required tedious preparation, such as long boil-softening.

Institutional Wood Collections

A formal collection of carefully documented wood samples is called a **xylarium** (plural **xylaria**). Note that the important feature that distinguishes a xylarium from a group of wood samples is that xylarium samples include herbarium materials. Such collections are found in universities, botanical gardens, government laboratories and other institutions. The xylarium at the U.S. Forest Products Laboratory at Madison, Wisconsin, has nearly 100,000 samples. Collections with approximately 40,000 specimens are housed at the Smithsonian Institution in Washington, D.C., and at the State University of New York College of Environmental Science and Forestry at Syracuse. Major collections are also located at Harvard University in Cambridge, Massachusetts, and at North Carolina State University in Raleigh.

In England, principal xylaria are found at Oxford University, at the Royal Botanical Gardens at Kew in Richmond, Surrey, and at the Princes Risborough Laboratory in Princes Risborough, Aylesbury, Buckinghamshire.

William Stern's *Index Xylariorum, Institutional Wood Collections of the World, 3* (1988), lists over 125 xylaria (see the Bibliography on pp. 212-215). Xylaria are maintained primarily for research purposes; however, they may accommodate any serious user with a demonstrated need for reference material. Xylaria sometimes include files of prepared slides of sections of selected species that are available for examination by visitors or can be taken out on loan.

As previously discussed, some woods of very closely related species cannot be separated on the basis of wood characteristics alone. To make this separation, more must be known about the leaves, twigs, flowers, fruit and bark of the trees from which the wood came. Herbarium samples contain specimens of these materials in carefully pressed and dried form. The term **herbarium** refers to a large collection of plant samples or to the place where such a collection is housed. Each sample in the herbarium has a voucher or file that includes data about the size, age, geographical origin, altitude and site characteristics of the tree, as well as information as to whether the tree was forest-grown or plantation-grown and what part of the tree — base, stem, crown or limb — was collected.

Photographs

Although there is no substitute for actual samples of wood or prepared slides of tissue, photographs of wood at various magnifications are of great value in wood identification. Wood figure, for example, can never be explained as well verbally as it can be shown in a photograph. A few books, such as B. J. Rendle's *World Timbers* and William A. Lincoln's *World Woods in Color*, have excellent color plates. Macroscopic photographs are very useful in showing the appearance of cross sections. Published photographs usually show typical, representative samples of a species.

High-resolution micrographs of wood tissue are also abundant in the literature; they are extremely helpful for identification by comparison. Yet to the practitioner, they can involve certain unfortunate trade-offs. For example, if taken at low power, micrographs may show a large area of tissue but in so doing may fail to show the most minute critical features. If taken at high power to show detail, obviously only a very limited area can be included.

When studying photographs — as opposed to using a microscope to examine tissue sections — the viewer cannot move to out-of-view adjacent areas or change the magnification. It is unfortunate that many publications apparently strive for uniformity by presenting micrographs mostly taken at the same magnification, even though, in practice, identification of different species requires that specific features be observed at different magnifications. Nevertheless, published macrographs and micrographs probably offer the most important and most accessible reference base for wood identification. In addition to the many photographs presented in this book, the reader will find a wealth of additional photographs among the references listed in the Bibliography.

CHAPTER TEN

HARDWOOD IDENTIFICATION

This chapter is devoted to the identification of the more common North American hardwoods — with an occasional brief mention of similar or related European species. It begins with an initial screening process that allows an unidentified sample to be placed in one of three major categories. Within each category a combination of photographs and verbal descriptions completes the separation process.

The woods in this chapter and the next (softwoods) are arranged in groups primarily on the basis of similar appearance. This has two advantages. First, it helps to locate, by visual comparison between an unidentified sample and the photographs provided, all those woods that look reasonably like the unknown. Second, it brings together and compares woods that are commonly confused. Once an unknown is placed within a particular group of woods, the separation schemes provided will distinguish among the individual woods in that group. Each wood has a complete descriptive summary; therefore, a tentative identification can be verified by checking all the major characteristics listed for that wood against the unknown.

Once an unknown sample is determined to be either a hardwood or a softwood, there is a choice as to how to proceed. The initial schemes provided can be used to place the wood in one of the major groups. Or you can try to find that group by browsing through the photographs to find the closest set of look-alikes. In either case, once you find the right group, follow the prescribed directions for separating the species within that group.

I strongly recommend reading this entire chapter and the next at least once before working with your first unknown sample. This will give you a clearer sense of the relationships among groups of woods and of how the separation and identification procedures are ordered. Many of the anatomical terms defined in previous chapters are used for the first time here. Unfamiliar terms can also be checked in the Glossary on pp. 200-211.

Initial Screening

All the hardwoods discussed in this chapter are initially placed into broad categories according to their pore size and distribution. Thus, the first step in identifying hardwoods is an examination of a cleanly cut cross-sectional surface with a hand lens to assess the size and distribution of pores. The traditional classifications are ring-porous, semi-ring-porous (semi-diffuse-porous) and diffuse-porous, as illustrated in the photos on p. 100.

An important comment before proceeding, however: If the presence of very narrow growth rings indicates that the tree from which the sample was taken grew slowly, particularly in ring-porous woods, the normal width of the earlywood and important latewood characteristics may be missing. It is therefore best to make observations in the area containing the widest growth rings.

Once you have determined the pore classification, turn to the appropriate section in this chapter.

Although most woods fall clearly into one pore class or another, an occasional specimen may be borderline. In this book, I have tried to resolve at least some of these cases by applying the following additional qualifications:

In order for a wood to be considered ring-porous, the earlywood pores must be relatively large and form a distinct row. For this reason, I consider cherry diffuse-porous, not ring-porous. Although its earlywood has a distinct row of pores, they are quite small. For a wood to be semi-ring-porous, the largest earlywood pores must be distinctly larger than the latewood pores, as in black walnut or butternut. Some woods show a definite pore-size gradation from earlywood to latewood, but even their earlywood pores are small; in other words, the pores grade from small to very small, as in willow and cottonwood. I have chosen to group these with the diffuse-porous woods, with the notation that they appear semi-diffuse-porous.

Once the initial pore classification of an unknown wood has been decided, turn to the appropriate section as indicated in the photos below.

GROUP I: RING-POROUS HARDWOODS

Most ring-porous woods are easy to recognize. However, not every piece of wood offers a perfect fit every time. Some species that are usually strongly ring-porous, such as catalpa, produce occasional wood that appears semi-ring-porous due to an earlywood/latewood transition that is more gradual than normal.

If an unknown wood is ring-porous, begin by trying to place it in one of the four ring-porous groups (RP I-1, I-2, I-3, I-4). Classify on the basis of latewood features, namely, pore arrangement and/or parenchyma and ray size.

Ring-Porous Subgroup I-1: Chestnut and Oak

Woods in this group have radially arranged latewood pores. They can be further separated into subgroups based on ray size. The oaks have huge multiseriate rays—the largest rays of all temperate woods—that can be seen readily with the unaided eye. Oaks also have fine, mostly uniseriate rays that are barely visible with a hand lens. Other woods that have radial pore arrangement but lack the large rays of oak fall into the chestnut group.

THE CHESTNUT GROUP
This group includes two North American woods: American chestnut *(Castanea dentata)* and giant chinkapin *(Castanopsis chrysophylla)*, also called golden chinkapin (see the photos on p. 102). In both, the rays are always uniseriate, although a few may be biseriate (a ray is considered biseriate if any portion of it is biseriate).

HARDWOOD CLASSIFICATION
Hardwoods are classified as ring-porous, as seen in white ash (left), semi-ring-porous, as seen in black walnut (center), or diffuse-porous, as seen in yellow birch (right).

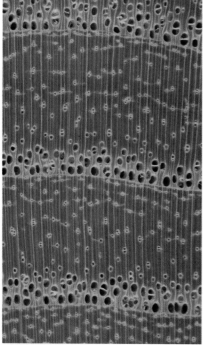

See Group I: ring-porous hardwoods above.

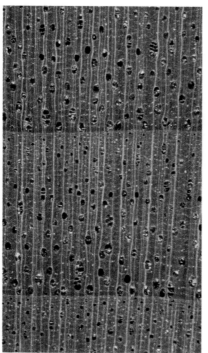

See Group II: semi-ring-porous or semi-diffuse-porous hardwoods, p. 114.

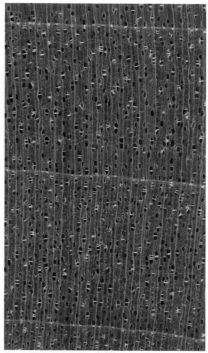

See Group III: diffuse-porous hardwoods, p. 117.

INITIAL BREAKDOWN OF
RING-POROUS HARDWOODS
Check an unknown ring-porous hardwood by determining the latewood pore
arrangement.

Ring-Porous Subgroup I-1:
Chestnut and Oak
Latewood pores in more or less radial arrangement, in obvious radial rows, or in
loose dendritic or flamelike patches; rays
either huge or barely visible even with
hand lens.
(See p. 100.)

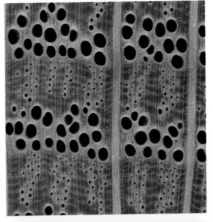
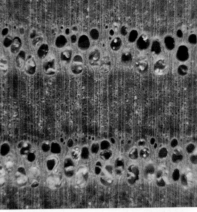

Ring-Porous Subgroup I-2:
Elm and Hackberry
Latewood pores in regular wavy tangen-
tial patterns, appearing uniformly
throughout latewood.
(See p. 104.)

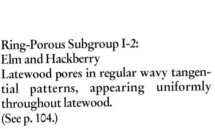
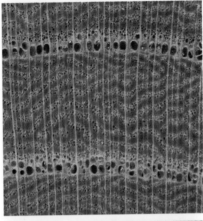
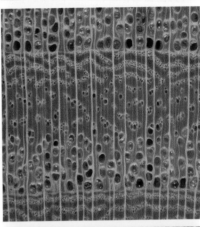

Ring-Porous Subgroup I-3:
Confusing Ring-Porous Woods
Latewood pores appear neither in strict
radial nor in tangential arrangement, but
rather as solitary pores, pore multiples, or
pore/multiples with aliform parenchyma,
or joined in outer latewood by confluent
parenchyma.
(See p. 106.)

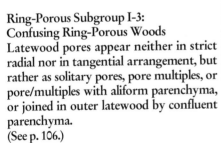
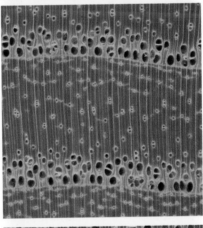
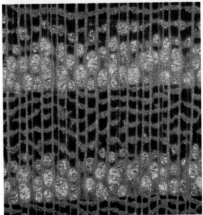

Ring-Porous Subgroup I-4:
Hickory and Pecan
Latewood pores mostly solitary or in
compact radial multiples. Latewood fiber
mass lined with regular banded parenchyma. (Row of earlywood pores discontinuous.)
(See p. 112.)

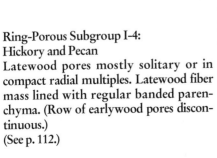
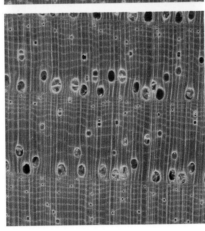
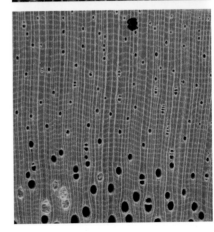

These are the only ring-porous woods with almost exclusively uniseriate rays; this separates them from other species, like sassafras and black ash, that initially resemble them. A microscopic check of a tangential section will quickly show the uniseriate rays (photo below right).

Chinkapin is usually light brown, often with a pinkish tinge, whereas American chestnut is a greyish brown or medium to dark brown. Neither has a characteristic odor. The two are separated best by pore distribution. In American chestnut, the earlywood zone consists of a continuous row of large pores up to six wide that are surrounded by lighter-colored tissue. The latewood pores are small and numerous, arranged in erratic flamelike or dendritic patches stretching radially across the latewood.

In chinkapin, the large earlywood pores are in a single, intermittent row along the growth-ring boundary. They are quite round, in contrast to the typically oval pores of chestnut. Between the earlywood pores is darker fiber mass that is continuous with the fiber mass of the latewood. The latewood pores of chinkapin are in distinct radial fanlike or dendritic patches that appear to erupt from each earlywood pore. Both chestnut and chinkapin have tyloses in medium abundance. European sweet chestnut (*Castanea sativa*) is similar to American chestnut.

THE OAK GROUP
The oaks are widely distributed throughout the world. The wood of oaks is typically ring-porous, with the exception of live oak (*Quercus virginiana*) and tanoak (*Lithocarpus densiflorus*). But the distinguishing feature of oak is its very large, 20+ multiseriate rays. No other North American wood has rays this big — this is a feature that can be immediately recognized on a cleanly cut surface. The oaks also have many narrow uniseriate rays.

When we examine a cross-sectional surface with a hand lens, we see that both the earlywood pores and the radial arrangements of latewood pores are surrounded by lighter-colored tissue. This tissue is a mixture of parenchyma and

CHESTNUT AND CHINKAPIN
Both chestnut and chinkapin have a flamelike, more-or-less-radial arrangement of latewood pores and imperceptible uniseriate rays. The interrupted row of round earlywood pores in chinkapin readily separates it from chestnut, which has a continuous row of multiple oval pores.

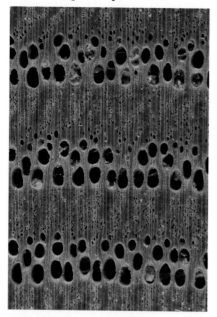

AMERICAN CHESTNUT
Castanea dentata

*Average Specific Gravity: 0.43
Heartwood Color: Greyish brown, or medium to dark brown
Pore Distribution: Ring-porous
Earlywood: Several pores wide; pores oval, surrounded by lighter tissue
Latewood: Pores very numerous, in radial, dendritic or wandering patches
Tyloses: Present
Rays: Barely visible with lens

**Rays: Amost exclusively uniseriate

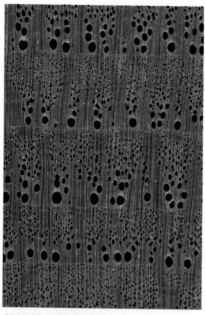

GIANT CHINKAPIN
Castanopsis chysophylla

Average Specific Gravity: 0.46
Heartwood Color: Light brown, often pinkish
Pore Distribution: Ring-porous
Earlywood: Pores large, round, in single row, intermittent with dark fiber
Latewood: Pores in distinct fanlike or dendritic patches, appearing to erupt from each earlywood pore
Tyloses: Present
Rays: Fine, barely visible with lens

Rays: Almost exclusively uniseriate

UNISERIATE RAYS IN CHESTNUT
Rays in chestnut are uniseriate, or occasionally partially biseriate. (100x)

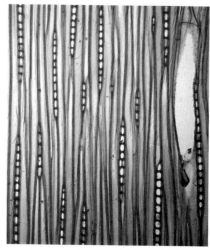

* Throughout the chapter, macroscopic features are listed in the text block immediately under the species name.
** Microscopic features are below the macroscopic and separated from them by a line.

vasicentric tracheids. In the latewood between pore groupings we see darker masses of the thick-walled fibers that give oaks their characteristic high density and strength. Faint tangential lines of banded parenchyma as well as uniseriate rays are visible across the fiber mass. These features are common to all oaks.

Separating the many oak species is another matter — the best we can do is assign the wood to either the red oak or the white oak group (Table 10.1, at right).

The red oak and white oak groups can usually be separated by considering ray height, the presence or absence of tyloses, and latewood pores (see photos at right). Ray height is apparent in the initial examination. Rays among red oaks are commonly 3/8 in. to 5/8 in. high, with a maximum height of 1 in.; rays taller than 1 in. are not often seen in red oaks. Among white oaks, rays taller than 1¼ in. are usually found, and heights up to several inches sometimes occur. I came across one piece of white oak in a 17th-century Hadley chest with a single ray 9 in. high.

But ray height is probably no more than 90% to 95% reliable. Among the exceptions to the ray-height rule, the occasional red oak with rays taller than 1 in. is more common than a white oak whose rays never exceed 1¼ in. Obviously ray height cannot be used to separate oak pieces shorter than about 1½ in.

White oak heartwood usually has abundant tyloses; rarely are they completely absent. In red oaks, by contrast, tyloses are usually absent or scattered, although the occasional piece may have areas of fairly abundant tyloses.

Latewood pores are the most reliable basis for separating red and white oaks. As shown in the photo at near right, latewood pores in red oak are large, individually distinct and few enough so that they can be readily counted. By contrast, latewood pores in white oaks, shown in the photo at far right, grade downward in size to very small at the outer edge of the growth ring and are usually so numerous and indistinct that they are impossible to see and count individually.

Further investigation of microscopic cross sections shows that the larger,

TABLE 10.1: SOME MAJOR SPECIES OF THE RED OAK AND WHITE OAK GROUPS (QUERCUS)

Red Oak Group		White Oak Group	
Q. coccinea	Scarlet oak	Q. alba	White oak
Q. falcata	Southern red oak	Q. bicolor	Swamp white oak
Q. kelloggii	California black oak	Q. garryana	Oregon white oak
Q. palustris	Pin oak	Q. lyrata	Overcup oak
Q. rubra	Northern red oak	Q. macrocarpa	Bur oak
Q. velutina	Black oak	Q. petraea	Sessile oak
		Q. prinus	Chestnut oak
		Q. robur	European oak
		Q. stellata	Post oak

RED OAK AND WHITE OAK

The oaks are recognized by their ring-porous arrangement and extremely large rays. When viewed with a hand lens, a transverse surface of red oak has fewer distinct latewood pores, as compared to the numerous, indistinct latewood pores in white oaks. Tyloses are usually abundant in white oak heartwood, sparse to absent in red oaks.

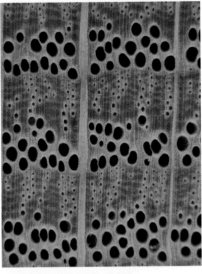

NORTHERN RED OAK
Quercus rubra

Average Specific Gravity: 0.63
Heartwood Color: Light brown, usually with flesh-colored cast
Pore Distribution: Ring-porous
Earlywood: Up to four rows of large, solitary pores
Latewood: Pores solitary in radial lines, few and distinct ("countable")
Tyloses: Absent or sparse in earlywood
Rays: Largest rays conspicuous; tallest less than 1 in.

Rays: Narrow rays uniseriate or in part biseriate
Latewood vessels: Thick-walled

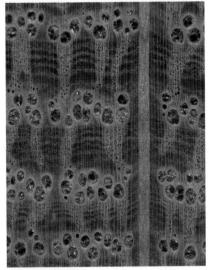

WHITE OAK
Quercus alba

Average Specific Gravity: 0.68
Heartwood Color: Light to dark brown to greyish brown
Pore Distribution: Ring-porous
Earlywood: Up to four rows of large pores
Latewood: Pores small, solitary or in multiples, in spreading radial arrangement, numerous and indistinct ("uncountable"), grading to invisibly small with lens
Tyloses: Abundant
Rays: Largest rays conspicuous; tallest greater than 1¼ in.

Rays: Narrow rays uniseriate or in part biseriate
Latewood Vessels: Thin-walled

more distinct red oak latewood pores are thick-walled (thicker than the earlywood pores), whereas the indistinct white oak pores are thin-walled (photos below).

This contrasting nature of latewood pores is highly reliable and should be used as the basis for separation whenever possible. It clearly has a great advantage over ray height in the identification of small specimens. And it is more reliable than the presence or absence of tyloses when it is not certain whether the specimen is sapwood or heartwood (see Table 10.2, below).

Ring-Porous Subgroup I-2: Elm and Hackberry

The important and clearly distinguishing feature of the elm/hackberry group is the arrangement of latewood pores in wavy bands. As mentioned in Chapter 4, the term ulmiform, which is used to describe this pore arrangement, is derived from the word *Ulmus* (elm) since elms display the pattern so characteristically. Typically, these wavy bands are more than one pore wide,

LATEWOOD PORES IN OAKS
In red oaks (left), latewood pores are solitary, with thick walls. In white oaks (right), the smaller, more numerous latewood pores have thin walls and may occur in multiples. (125x)

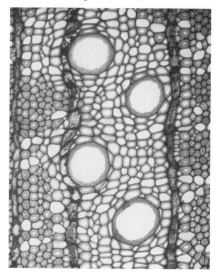 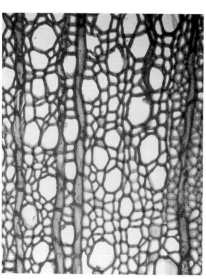

HACKBERRY
The multiple row of earlywood pores and wavy bands of latewood pores in hackberry imitate the appearance of slippery elm (see photos on the facing page). The cream to yellowish-grey heartwood of hackberry contrasts with the reddish brown of slippery elm. A microscopic check of the greater ray seriation (7+) in hackberry provides a more positive separation from elms, whose ray seriation is never greater than 7.

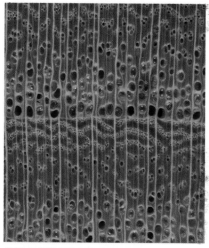

COMMON HACKBERRY
Celtis occidentalis

Average Specific Gravity: 0.53
Heartwood Color: Cream, light brown or light greyish brown, with a yellowish cast
Pore Distribution: Ring-porous
Earlywood: More than one pore wide
Latewood: Pores in wavy bands
Rays: Distinct

Rays: Heterocellular, widest up to 10-13 seriate

TABLE 10.2: SUMMARY OF IDENTIFICATION FEATURES FOR RED OAK AND WHITE OAK

Feature	In Red Oak	In White Oak
Rays (a)	Tallest less than 1 in. high, mostly ⅜ in. to ⅝ in. high	Tallest more than 1¼ in. high; can be up to several inches high
Tyloses (a)	Absent or sparse and sporadic	Abundant in heartwood
Latewood pores (b)	Distinct with lens; few enough to be counted	Grading in size to indistinctly small; too numerous and indistinct with lens to be counted
	Vessel walls thick (thicker than earlywood vessels)	Vessel walls thin (not thicker than earlywood vessels)

(a) Less reliable features.
(b) Most reliable features.

continuous and well established across the entire latewood.

Woods in this group can be separated by considering the placement and number of earlywood pores and the character of the rays. The earlywood band of pores of slippery elm *(Ulmus rubra)* is more than one pore wide; by contrast, it is only a single pore wide in American elm and the hard elms. American elm *(U. americana)* has a fairly continuous row of large, uniformly sized, evenly spaced earlywood pores. Woods of En-

glish elm *(U. procera)* and wych elm *(U. glabra)* are similar. Earlywood pores in the hard elms are typically small, indistinct and intermittent along the growth-ring boundary. Among the hard elms, woods of rock elm *(U. thomasii),* winged elm *(U. alata)* and cedar elm *(U. crassifolia)* are not readily separated.

Although both slippery elm and common hackberry *(Celtis occidentalis)* have earlywood bands of multiple pores, heartwood color is one basis for separating the two. In hackberry the heartwood

is light brown or light greyish brown, commonly with a yellowish cast. The heartwood of slippery elm is dark brown to reddish brown.

Microscopic ray characteristics are another reliable means of separating these two woods. In slippery elm the rays are homocellular and up to 7 seriate (but commonly 4-5 seriate), whereas in hackberry they are heterocellular and up to 13 seriate (but mostly 5-8 seriate). The wood of sugarberry *(Celtis laevigata)* is very similar to hackberry.

ELM
Latewood pores in elms are arranged in wavy bands (ulmiform arrangement) that are the trademark of this wood. Slippery elm has a multiple band of earlywood pores; American elm has a continuous row of earlywood pores that are easily visible to the eye. Winged elm, a hard elm, has earlywood pores in an interrupted row that are invisible without a hand lens.

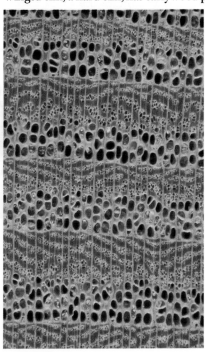

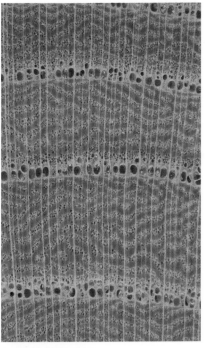

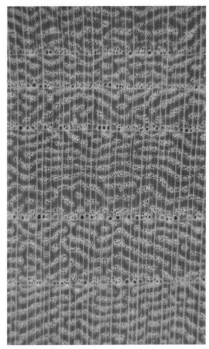

SLIPPERY ELM
Ulmus rubra

Average Specific Gravity: 0.53
Heartwood Color: Red to dark brown or reddish brown
Fluorescence: Heartwood fluoresces dim yellow-green
Pore Distribution: Ring-porous
Earlywood: 2-6 pores wide
Latewood: Pores in wavy bands
Tyloses: Sparse in earlywood
Rays: Not distinct without lens

Rays: Homocellular, 1-7 (mostly 4-5) seriate

AMERICAN ELM
Ulmus americana

Average Specific Gravity: 0.50
Heartwood Color: Light brown to brown or reddish brown
Pore Distribution: Ring-porous
Earlywood: Pores large, in continuous row
Latewood: Pores in wavy bands
Tyloses: Present in earlywood, but usually sparse
Rays: Not distinct without lens

Rays: Homogeneous, 1-7 (mostly 4-6) seriate

WINGED ELM
Ulmus alata

Average Specific Gravity: 0.66
Heartwood Color: Light brown to brown or reddish brown
Pore Distribution: Ring-porous
Earlywood: Pores small and indistinct in an intermittent row
Latewood: Pores in wavy bands
Tyloses: Sparse in earlywood
Rays: Not distinct without lens

Rays: Homogeneous, 1-7 seriate

Knowing the origin of the specimen can also be very helpful in pinning down the woods of this group. American elm and slippery elm are widely distributed throughout the eastern and central United States and the southern edge of the central and eastern Canadian provinces. Rock elm is concentrated in the north-central states, winged elm in the south-central and southeastern states and at the southern tip of Ontario and hackberry in the north-central and northeastern states. Sugarberry is restricted to the southeastern part of the United States.

The chart below summarizes the separation scheme for the elm/hackberry group.

Ring-Porous Subgroup I-3: The Confusing Ring-Porous Woods

Of the many ring-porous woods, those in this group can be the most difficult to identify because they share some major characteristics. One such characteristic is the row of multiple earlywood pores. Another is the latewood pores, which are neither radially grouped as in oaks nor in strong tangential or wavy bands as in elm; rather, they range from solitary or small radial multiples to clustered. Sometimes the clusters merge into short-to-long tangential groups that are confusingly similar to ulmiform groupings.

Woods found at the extremes of this ring-porous subgroup are unlikely to be mistaken for each other. For example, it is hard to confuse white ash with Osage-orange. Yet within the group, nearly every wood is confusingly similar to one or more others. For example, white ash is confused with black ash, black ash is confused with sassafras and, toward the other end of the list, honeylocust is similar to Kentucky coffeetree. This is illustrated in the cross-sectional views in the photos on p. 108 and pp. 110-111.

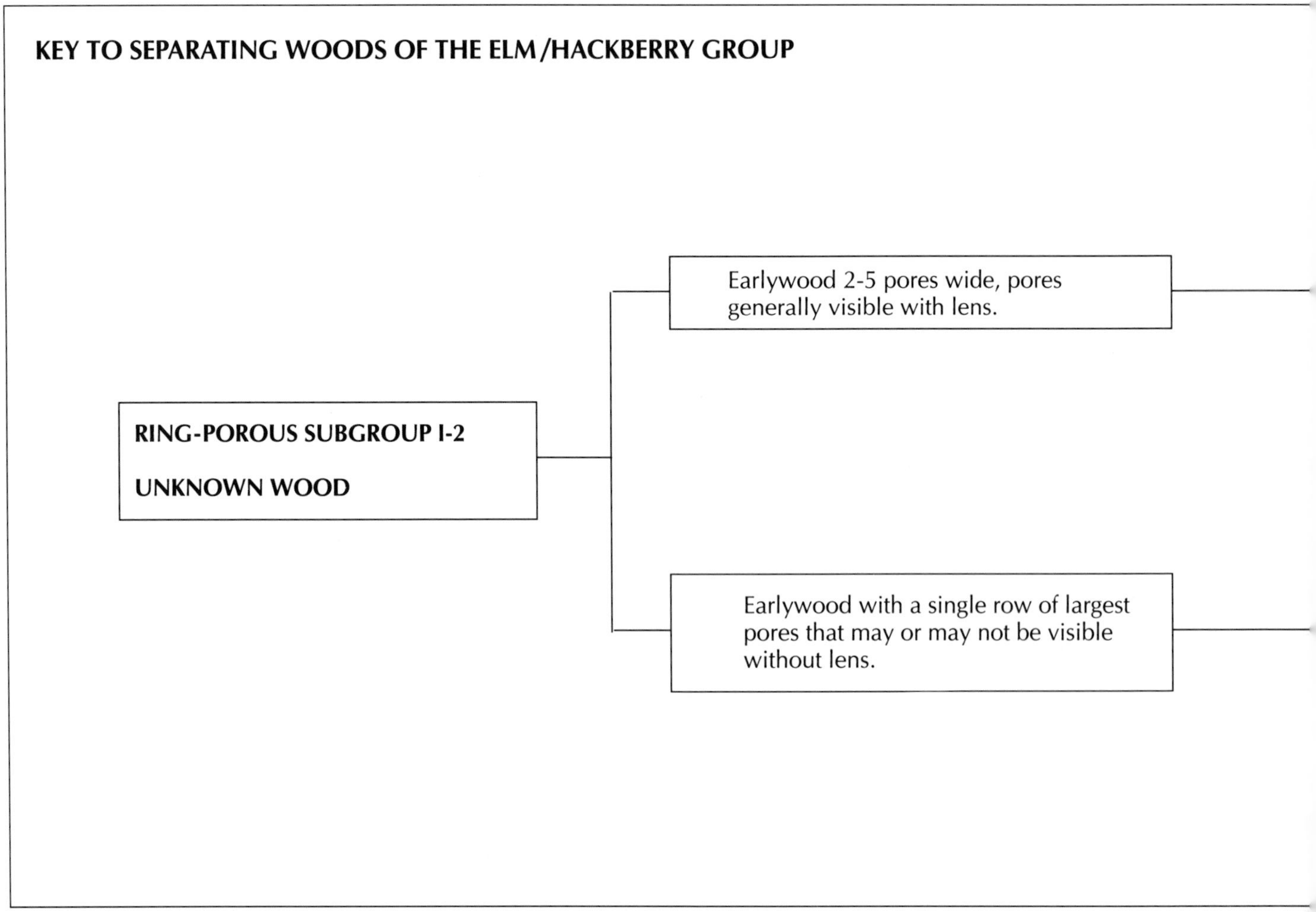

KEY TO SEPARATING WOODS OF THE ELM/HACKBERRY GROUP

RING-POROUS SUBGROUP I-2

UNKNOWN WOOD

Earlywood 2-5 pores wide, pores generally visible with lens.

Earlywood with a single row of largest pores that may or may not be visible without lens.

As a start in narrowing the possibilities within this difficult subgroup, consider the following:

1. If the wood is bright and fresh:
   A) Check for characteristic odor. Only sassafras and catalpa have noteworthy fragrances.
   B) Note the heartwood color. For example, the heartwood is orange-to-red in coffeetree, grey-brown in sassafras and green in sumac.
   C) Where possible, note the width of the sapwood. For example, black locust and catalpa never have sapwood more than three rings wide.

2. Evaluate density and hardness. Most of these woods are hard to extremely hard, but catalpa is soft.

3. Note the presence or absence of tyloses and whether they sparkle in iridescent colors. For example, tyloses are absent from honeylocust but extremely abundant in black locust.

4. If a black light is available, check for fluorescence. Only sumac, honeylocust, coffeetree and black locust fluoresce.

5. If you have a microscope, check for :
   A) Ray seriation. For example, sumac rays are generally 1 or 2 seriate, while honeylocust rays are up to 14 seriate.
   B) Latewood pores. Note whether they are solitary or occur in multiples or clusters, and whether they are thick- or thin-walled. Ashes, for example, have thick-walled latewood pores, catalpa's are thin-walled.
   C) Spiral thickenings in the latewood vessels as viewed in longitudinal sections. For example, spiral thickenings are absent from sassafras but present in catalpa and sumac.

The summary of features in Table 10.5 on pp. 136-137 should also help suggest possible species.

Heartwood cream to light brown or yellowish grey. Largest rays up to 10-12-seriate, always some rays wider than 7-seriate.	HACKBERRY (*Celtis occidentalis*) SUGARBERRY (*Celtis laevigata*)	HACKBERRY
Heartwood brown to reddish brown. Largest rays generally 4-5-seriate, never wider than 7-seriate.	SLIPPERY ELM (*Ulmus rubra*)	SOFT ELM
Earlywood pores visible to the eye (maximum pore diameter more than 200 um), mostly in a continuous row.	AMERICAN ELM (*Ulmus americana*)	
Earlywood pores not visible without lens (maximum pore diameter less than 200 µm), mostly in a discontinuous row.	WINGED ELM (*Ulmus alata*) CEDAR ELM (*Ulmus crassifolia*) ROCK ELM (*Ulmus thomasii*)	HARD ELM

WHITE ASH

Of the woods in Subgroup I-3, ash is probably the most common. Wood known as white ash is most often *Fraxinus americana*, but it is indistinguishable from and often confused with green ash (*F. pennsylvanica*). Both species are widely distributed throughout the central and eastern United States and in southeastern Canada. In addition, the woods of Oregon ash (*F. latifolia*) and European ash (*F. excelsior*) are essentially similar.

White ash is characterized by a band of multiple earlywood pores surrounded by lighter-colored tissue. In the first-formed latewood, the pores are solitary or occur in short radial multiples, typically surrounded by a sheath of lighter-colored vasicentric parenchyma. Although the vasicentric parenchyma cells are in

ASH AND SASSAFRAS

Compared to white ash, black ash has narrow sapwood and brown heartwood color. Its growth rings are usually narrow and its latewood pores are seldom connected, as they are in white ash. Like white ash, sassafras normally has latewood pores with aliform and confluent parenchyma, but it is distinct in its grey heartwood color and spicy aroma. A good microscopic check for sassafras is its large, yellow oil cells (see the photo on the facing page).

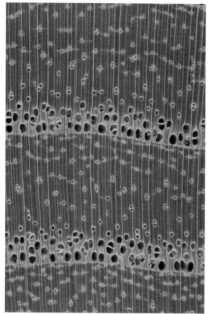

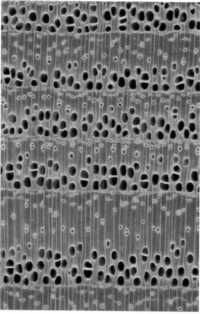

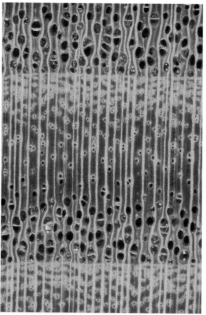

WHITE ASH
Fraxinus americana

Average Specific Gravity: 0.60
Heartwood Color: Light brown or greyish brown
Sapwood Color: Creamy white (may be very wide)
Pore Distribution: Ring-porous
Earlywood: 2-4 pores wide; pores moderately large, surrounded by lighter tissue
Latewood: Pores solitary and in radial multiples of 2-3, surrounded by vasicentric parenchyma or connected by confluent parenchyma in outer latewood
Tyloses: Fairly abundant (some vessels open)
Rays: Not distinct to eye, but clearly visible with hand lens

Rays: 1-3 seriate
Latewood Pores: Thick-walled

BLACK ASH
Fraxinus nigra

Average Specific Gravity: 0.49
Heartwood Color: Greyish brown to medium or dark brown
Sapwood Color: Creamy white (fairly narrow)
Pore Distribution: Ring-porous
Earlywood: 2-4 pores wide; pores large, surrounded by lighter tissue
Latewood: Pores solitary and in radial multiples of 2-3, not numerous, rarely with aliform or confluent parenchyma in outer latewood
Tyloses: Fairly abundant (some vessels open)
Rays: Not distinct to eye, but clearly visible with hand lens

Rays: 1-3 seriate
Latewood Pores: Thick-walled

SASSAFRAS
Sassafras albidum

Average Specific Gravity: 0.45
Heartwood Color: Grey or greyish brown
Heartwood Odor: Spicy odor and taste
Pore Distribution: Ring-porous
Earlywood: 3-8 pores wide; pores medium to large
Latewood: Pores solitary and in radial multiples of 2-3, grading to very small with aliform parenchyma in outer latewood
Tyloses: Fairly abundant, with sparkle
Rays: Barely visible to the eye

Rays: 1-4 seriate
Oil Cells: Rays and longitudinal parenchyma contain scattered oil cells that appear very large with bright yellow contents
Latewood Pores: Thick-walled

thin layers, they are visible against the darker fiber mass. Toward the outer latewood, there is an increasing tendency for aliform parenchyma to develop and intergrade with confluent parenchyma, which connects the pores laterally. The result is variable-length interrupted tangential or wavy bands, especially in the outermost latewood. Note that these solitary pores and radial multiples connected by parenchyma are different from the continuous clusters or bands of pores seen in elm and black locust.

Radial surfaces of white ash and hackberry are easily confused at first glance, because both share a creamy white or yellowish color and are ring-porous with a multiple earlywood pore zone. However, they can be separated by careful examination of latewood pores and rays. The pores are ulmiform in hackberry and discontinuous in ash, even though they are connected by confluent parenchyma. The rays are indistinct and 1-3 seriate in ash, distinct and up to 12 seriate in hackberry. Only occasional rays in white ash exceed 2 seriate, whereas rays of European ash are commonly 3 (and occasionally 4) seriate.

BLACK ASH

Black ash (*Fraxinus nigra;* photo at center on the facing page) is sometimes confused with white ash. In white ash, the sapwood can be very wide, in some cases occupying the entire mature stem, whereas in black ash, it is narrow, rarely in excess of 1 in. Moreover, the light-colored sapwood of white ash contrasts with the medium-brown or greyish-brown heartwood of black ash. In black ash, the latewood pores are sparser, and only occasionally do confluent parenchyma connect pores in the outer latewood. The narrow rings caused by slow growth rate are common in black ash. These narrow rings have little latewood, making latewood pore distribution difficult to evaluate. White ash, a denser wood, has wider growth rings and well-represented latewood. In recent years, however, an affliction called ash dieback has caused extremely slow growth in white ash in some regions. The resulting wood can be difficult to identify.

SASSAFRAS

Black ash is more apt to be confused with sassafras (*Sassafras albidum;* photo at right on the facing page). Both woods have similar greyish-brown heartwood and a similar appearance on longitudinal surfaces. In the latewood of sassafras, however, tangential extensions of aliform parenchyma from latewood pores are the rule rather than the exception. The tyloses of sassafras are fairly abundant and tend to sparkle. Sassafras has a pleasant root-beer-like fragrance and taste. If the sample is aged and you suspect that it has lost its taste and smell, examine a radial section microscopically. The presence of oversized yellow oil cells among the rays in a radial section confirms that it is sassafras (see the photo below).

CATALPA

Of the woods in this group, catalpa (photo at left on p. 110) is the least dense and is noticeably soft. Its heartwood is medium brown to medium dark brown. It may also be greyish, and it sometimes has a faint lavender cast. Catalpa never retains more than two

OIL CELLS IN SASSAFRAS

Oil cells (A) are regularly scattered along the ray margins in sassafras. They are readily recognized by their extreme size and yellow oil contents.

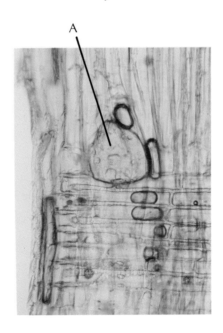

growth rings of sapwood; if an unknown contains sapwood with more than two rings, it is not catalpa. The tyloses of catalpa sparkle, and the wood has a faint but characteristic odor that is often described as spicy or musty. Latewood pores tend to be larger toward the earlywood; as they approach the outer latewood, they grade to very small and occur in clusters or nests. These clusters commonly merge to form either interrupted or reasonably continuous tangential bands in the outer latewood.

The earlywood is usually well defined, but occasionally it makes a gradual transition to latewood, suggesting a semi-ring-porous classification. Southern catalpa (*Catalpa bignonioides),* though essentially indistinguishable from northern catalpa (*C. speciosa),* tends to be semi-ring-porous.

RED MULBERRY

Red mulberry (*Morus rubra;* see the middle photo on page p. 110) is, at first glance, very similar to catalpa, but the heartwood is more orange or russet brown and turns deep russet brown on exposure to the atmosphere. Larger latewood pores are solitary and in multiples in first-formed latewood; they grade to small and clustered, often merging laterally in the outer latewood. Tyloses are moderately abundant and sparkle, and vessels sometimes contain white deposits. Whereas catalpa has no noticeable ray fleck, this feature is quite conspicuous in the heartwood of red mulberry, where it appears as a light fleck against darker background on radial surfaces. In cross sections, rays are not visible in catalpa but are distinct to the eye in red mulberry.

BLACK LOCUST
AND OSAGE-ORANGE

The visibility of rays in cross sections also distinguishes red mulberry from black locust (*Robinia pseudoacacia)* and Osage-orange (*Maclura pomifera).* In the latter two species the rays can barely be seen with the eye, although they show up very distinctly as crisp bright lines with a hand lens. In red mulberry there is only a slight color contrast between earlywood and latewood; in black locust and Osage-orange the earlywood pores of the heartwood are completely packed with tyloses. Thus, individual pores are indistinct and the earlywood zone takes on a frothy yellow or creamy appearance. This contrasts with the dense, very dark brown fiber mass of the latewood, giving the wood an overall two-tone effect.

In both red mulberry and black locust the lighter-colored latewood pores and rays show up clearly against the dark fiber mass. The latewood pores of both species are small and clustered, and merge into tangential, intermittent bands and wavy lines. In both species, the high density and extreme hardness of the wood become obvious when you try to cut a cross section with a razor blade.

The similarity of principal features in

CATALPA, RED MULBERRY AND OSAGE-ORANGE

Catalpa is the least dense species in Subgroup I-3. Its softness, brown heartwood color and unique odor are characteristic. The rays of red mulberry are plainly visible on a transverse surface, and they produce conspicuous ray fleck on radial surfaces. Osage-orange, the densest of the group, has inconspicuous rays, and its pores are packed with tyloses. Osage-orange does not fluoresce under ultraviolet light, but it does contain water-soluble heartwood extractives.

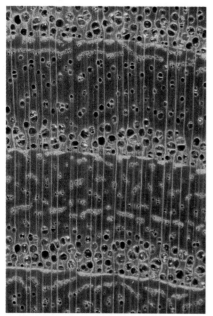

NORTHERN CATALPA
Catalpa speciosa

Average Specific Gravity: 0.41
Heartwood Color: Greyish brown, sometimes with faint lavender cast
Sapwood: Never more than two growth rings wide
Odor: Faint but unique pungent, spicy odor
Pore Distribution: Ring-porous, or grading to semi-ring-porous
Earlywood: 3-8 pores wide; pores large
Latewood: Pores variable, large near earlywood, grading to very small in outer latewood, numerous in nests, which commonly merge into continuous or interrupted bands
Tyloses: Variably abundant, with sparkle
Rays: Indistinct to eye, but clearly visible with lens

Rays: 1-6 (mostly 2-3) seriate

RED MULBERRY
Morus rubra

Average Specific Gravity: 0.66
Heartwood Color: Orange-yellow to golden or russet brown, turning deep russet on exposure
Pore Distribution: Ring-porous
Earlywood: 2-8 pores wide; pores moderately large, may contain white deposits
Latewood: Pores in nestlike groups, merging laterally to form wavy or interrupted bands, especially in outer latewood
Tyloses: Variably present, with sparkle
Rays: Plainly visible to eye on transverse surface; radial surface shows conspicuous ray fleck

Rays: 1-8 (mostly 5-7) seriate

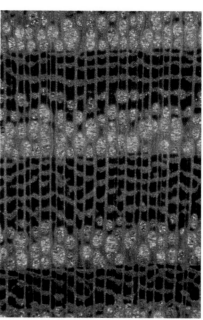

OSAGE-ORANGE
Maclura pomifera

Average Specific Gravity: 0.81
Heartwood Color: Golden yellow to orange or orange-brown, darkening on exposure; contains water-soluble dye, so shavings turn warm water yellow when soaked
Pore Distribution: Ring-porous
Earlywood: 2-3 pores wide; pores medium-large
Latewood: Pores small, in clusters, often merging or connected by confluent parenchyma into wavy or concentric bands in outer latewood
Tyloses: Extremely abundant, packing vessels
Rays: Barely visible to eye

Rays: 1-6 (mostly 2-4) seriate
Crystals: Occur occasionally in longitudinal parenchyma cells

black locust and Osage-orange — high density, abundant tyloses, two-color effect, latewood pore arrangement and ray size — makes it difficult to tell these two woods apart. With a microscope, however, black locust can be confirmed by the presence of vestured intervessel pits, which do not occur in Osage-orange. Also, crystals are present in the longitudinal parenchyma of Osage-orange, whereas they are absent from black locust.

If a microscope is not available, either of two "gimmicks" can be used to separate these species. The first is a test that detects the water-soluble extractives in Osage-orange. To make the test, place a small handful of shavings in a white cup or dish, then cover with warm tap water and allow them to soak for two hours. Osage-orange will tint the water yellow

BLACK LOCUST, HONEYLOCUST AND KENTUCKY COFFEETREE

Abundant, packed tyloses give the pores of black locust a light color that contrasts with the dense, dark background fiber mass. The rays are equally light and conspicuous. Black locust has no soluble heartwood extractives, but all three woods show brilliant yellow fluorescence under ultraviolet light. Kentucky coffeetree and honeylocust have similar reddish-brown to reddish-orange-brown heartwood, but honeylocust has latewood pores that connect tangentially and more distinct rays than Kentucky coffeetree.

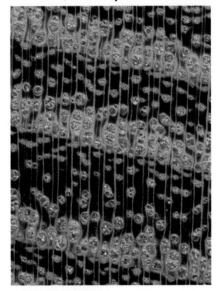 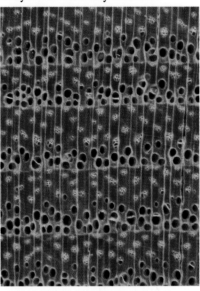

BLACK LOCUST
Robinia pseudoacacia

Average Specific Gravity: 0.69
Heartwood Color: Olive or yellow-brown to dark yellow-brown; dark russet brown with exposure
Fluorescence: Bright yellow
Sapwood: Never more than 3 growth rings wide
Pore Distribution: Ring-porous
Earlywood: 2-3 pores wide; pores large
Latewood: Pores in nestlike groups, which merge into interrupted or somewhat continuous bands in outer latewood; latewood fiber mass appears dense and dark in contrast to yellowish, tyloses-filled pores and rays
Tyloses: Extremely abundant, with yellow cast and sparkle, solidly packing vessels and making adjacent pores indistinct
Rays: Barely visible to eye, but distinct with lens

Rays: 1-7 (mostly 3-5) seriate

HONEYLOCUST
Gleditsia triacanthos

Average Specific Gravity: 0.66
Heartwood Color: Light red to reddish- or orange-brown
Fluorescence: Bright yellow
Pore Distribution: Ring-porous
Earlywood: 3-5+ pores wide; pores large
Latewood: Pores solitary, in radial multiples, and, especially in outer latewood, in nestlike groups and connected into bands by confluent parenchyma
Tyloses: Absent
Gum: Pores regularly contain reddish gum, especially evident on longitudinal surfaces
Rays: Variable in size, the largest conspicuous to the eye

Rays: 1-14 (mostly 6-9) seriate

KENTUCKY COFFEETREE
Gymnocladus dioicus

Average Specific Gravity: 0.60
Heartwood Color: Light red to reddish or orange brown
Fluorescence: Bright, deep yellow
Pore Distribution: Ring-porous
Earlywood: 3-6 pores wide; pores large
Latewood: Pores nested and sometimes merging into short irregular tangential bands in outer latewood
Tyloses: Absent
Gum: Reddish gum only rarely present
Rays: Uniformly fine, barely visible to eye

Rays: 1-7 (mostly 3-5) seriate

(it was used to dye military uniforms a dull khaki in World War I). Black locust shavings leave little or no color.

The second test is for fluorescent response to ultraviolet light. Osage-orange does not fluoresce under a black light, but the heartwood of black locust shows a brilliant yellow fluorescence (see Table 5.2 on p. 53).

HONEYLOCUST AND KENTUCKY COFFEETREE

Honeylocust (*Gleditsia triacanthos*) and Kentucky coffeetree (*Gymnocladus dioicus;* photos on p. 111), although detectably different from the other woods in this group, are easily confused with each other. The heartwood of both species commonly has an orange or reddish-orange hue; it also ranges from yellowish to brown. Neither wood has tyloses, as can be confirmed with a hand lens.

However, the pores of honeylocust do routinely contain red gum deposits, which are seen as deep red specks on longitudinal surfaces. These deposits are absent or very sparse in coffeetree. The two woods are more reliably separated by the size and arrangement of latewood pores, as well as by their rays. In both woods the first-formed latewood pores are large and distinct as seen with a hand lens. They are either solitary or in multiples. But they grade to very small and occur in clusters toward the outer latewood.

In coffeetree, the small latewood pores are seen with a hand lens to occur in separate nestlike clusters, with only a slight tendency to merge into short wavy bands along the outer latewood. In honeylocust, the pores in the outer latewood are fewer — either solitary or in small multiples — and are usually indistinct with a hand lens. But they are surrounded by lighter parenchyma. The parenchyma forms distinct interrupted or continuous wavy bands in the outer latewood.

In coffeetree, the rays are practically invisible to the eye, although they are distinct and appear uniformly thin with a hand lens. With a microscope, however, they are found to be 1-7 (mostly 3-5) seriate. In honeylocust, the largest rays are distinctly visible to the eye and can be seen with a hand lens to vary in size. Microscopic examination shows them to be mostly 6-9 seriate, with the widest up to 14 seriate. Both woods show yellow fluorescence under ultraviolet light. Honeylocust is the brighter of the two; in coffeetree, the earlywood shows a diminished response, making its growth rings more conspicuous.

SUMAC

The earlywood of staghorn sumac is a multiple band of very small pores. The unique heartwood shades of green, yellow-greens and russet show brilliant but variable fluorescence under ultraviolet light.

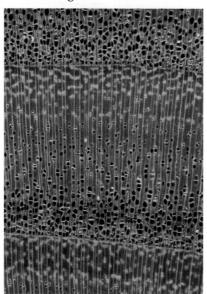

STAGHORN SUMAC
Rhus typhina

Average Specific Gravity: 0.47
Heartwood Color: Olive green to greenish yellow, often with orange or russet brown earlywood
Fluorescence: Variable yellow to bluish
Pore Distribution: Ring-porous
Earlywood: 2-6 pores wide; pores fairly small
Latewood: Pores solitary or in radial multiples, but merging into continuous bands in outer latewood
Tyloses: Variably abundant with iridescent sparkle
Rays: Not conspicuous, even with lens

Rays: 1-3 (mostly 1-2) seriate

STAGHORN SUMAC

The final wood in this group is staghorn sumac (*Rhus typhina;* photo below left). Fresh specimens are recognized quickly on the basis of color. The narrow sapwood is milky white, but the heartwood is a dusty yellowish green — or greenish yellow — reflecting the dominant latewood hue. Careful examination of the earlywood bands reveals a brownish or russet tinge. On a cross section, each growth ring appears darker along the outer latewood.

Earlywood pores are imperceptibly small to the unaided eye. They are clearly seen with a lens, but they are small compared to those of most ring-porous woods. We see the wood as ring-porous more because this 2 to 6-pore wide band is so concentrated than because the pores are large. Rays are visible but inconspicuous with a hand lens. They are the narrowest of this group; a microscopic check reveals that they are only 1-2 seriate.

Under ultraviolet light, staghorn sumac heartwood shows striking fluorescence, each ring glowing a bright greenish yellow along the outer latewood, grading through dull yellow to a very deep blue in the earlywood. The sapwood has a dim medium-blue fluorescence.

Ring-Porous Subgroup I-4: Hickory and Pecan

Woods in Subgroup I-4 belong to the genus *Carya*. All have the distinguishing feature of banded parenchyma in fairly long, straight, more or less continuous tangential lines. With a hand lens, these appear as bright as the rays. The combination of parenchyma and rays creates a netlike or weblike pattern on the dense fiber-mass background of the latewood. The earlywood pores are large and form an intermittent row along the growth-ring boundary. Between earlywood pores are masses of dense fiber, continuous with the fiber mass of the latewood, unlike the earlywood of species such as white ash, where lighter-colored tissue mass fills any earlywood areas not already occupied by pores. The latewood pores, which are thick-walled, are relatively few, occurring as solitary pores or in radial multiples of two or more. They are individually distinct when viewed with a hand lens.

The above features make it possible to differentiate these woods of the *Carya* group from others. Within the genus, however, separation of the individual species is another matter. The *Carya* species at best can be divided into two subgroups, the true hickories and the pecan hickories (Table 10.3 below).

As seen in the table below, true hickories are denser — and thus stronger — than pecan hickories; the woods are used in practical applications accordingly. But these density differences alone are not great enough to be reliable for identification. A more positive separation can be made on the basis of parenchyma distribution and pore classification. In true hickories banded parenchyma does not occur in the earlywood zone (see the photo at near right), whereas in the pecan hickories the lines of banded parenchyma occur well into the earlywood pore zone.

The true hickories are typically ring-porous, the largest pores being restricted mainly to a single interrupted row along the growth-ring boundary; thus the earlywood is really only one pore wide. The transition to latewood is defined by a fairly abrupt reduction in pore size. The pecan hickories, by contrast, have less well defined earlywood and a gradual gradation in pore size across the growth ring. In this way they are more typically semi-ring-porous, which is why I have placed the *Carya* group at the end of the ring-porous woods. Pecan hickory could just as well be included in the next section on semi-ring-porous woods.

TRUE HICKORY AND PECAN HICKORY

The fine tangential lines of banded parenchyma, together with the radial lines of the rays, form a meshlike pattern against the background latewood fiber mass, which is a trademark of *Carya*. The true hickories (such as shagbark hickory) are more typically ring-porous; the pecan hickories appear more semi-ring-porous. In true hickory, the banded parenchyma is absent in the zone of earlywood pores; in pecan hickories, banded parenchyma occurs even in the earlywood.

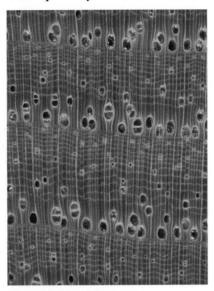

SHAGBARK HICKORY
Carya ovata

Average Specific Gravity: 0.72
Heartwood Color: Light to medium brown or reddish brown
Pore Distribution: Ring-porous
Earlywood: Mostly an intermittent single row of thick-walled pores with fiber mass where interrupted
Latewood: Pores not numerous, solitary and in radial multiples of 2-5
Tyloses: Moderately abundant
Parenchyma: Banded parenchyma and rays form a reticulate pattern distinctly visible against the background fiber mass with a hand lens (but banded parenchyma absent from earlywood zone)

Rays: 1-4 seriate
Latewood Pores: Thick-walled

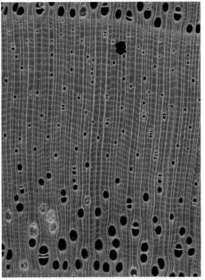

PECAN
Carya illinoensis

Average Specific Gravity: 0.66
Heartwood Color: Light to medium reddish brown
Pore Distribution: Semi-ring-porous (or ring-porous in narrow rings)
Pores: Large, interrupted along growth ring boundary by areas of fiber mass; gradual transition to smaller latewood pores, which are solitary or in radial multiples of 2-3
Tyloses: Moderately abundant
Parenchyma: Banded, distinct with hand lens, forming reticulate pattern with rays (banded parenchyma usually also in earlywood zone)

Rays: 1-4 seriate
Latewood Pores: Thick-walled

TABLE 10.3: WOODS IN THE HICKORY AND PECAN GROUP *(CARYA)*

	True Hickories	Specific Gravity		Pecan Hickories	Specific Gravity
C. glabra	Pignut hickory	0.75	*C. aquatica*	Water hickory	0.62
C. laciniosa	Shellbark hickory	0.69	*C. cordiformis*	Bitternut hickory	0.66
C. ovata	Shagbark hickory	0.72	*C. illinoensis*	Pecan	0.66
C. tomentosa	Mockernut hickory	0.72	*C. myristiciformis*	Nutmeg hickory	0.60

GROUP II: SEMI-RING-POROUS OR SEMI-DIFFUSE-POROUS HARDWOODS

As the title of this group suggests, these woods have a pore distribution that is neither ring-porous nor diffuse-porous but somewhere in between. Here we find pores that are noticeably larger in the earlywood than in the latewood but that typically show a gradual size transition from earlywood to latewood. The earlywood zone is not clearly delineated from the latewood zone.

Only a few woods consistently show the pore distribution that typifies this category. These include live oak, tanoak, persimmon and the walnuts, including butternut. It is important to emphasize that in many species, samples of the wood can intergrade from the categories of ring-porous or diffuse-porous into the "semi-" category. If they are usually ring-porous, I think of them as semi-ring-porous. If they are usually diffuse-porous, I call them semi-diffuse-porous. However, there is a wide range of opinion regarding these names.

As previously mentioned, the true hickories in the *Carya* group are typically ring-porous, whereas the pecans are semi-ring-porous. In some species, like northern catalpa, which is usually ring-porous, individual samples may appear semi-ring-porous. Among certain woods that are typically diffuse-porous, larger earlywood pore size in some samples strongly suggests semi-ring-porosity. Examples are willow and cottonwood.

Semi-Ring-Porous Subgroup II-1: Live Oak and Tanoak

Among the semi-ring-porous woods, live oak (*Quercus virginiana*) and tanoak (*Lithocarpus densiflorus*) are distinct in that they have rays that are large enough to be clearly visible to the eye on transverse surfaces, as shown in the photos at right. With a hand lens, it is apparent that

LIVE OAK AND TANOAK

Both live oak and tanoak are semi-ring-porous (the largest pores are scarcely visible to the eye) and have occasional large conspicuous rays. In tanoak, latewood pores occur in radially arranged streamerlike patches of lighter-colored tissue; between the patches, irregular lines of banded parenchyma cross the fiber mass. In live oak, latewood pores are in a more irregular arrangement, and tangential bands of parenchyma are absent in the latewood.

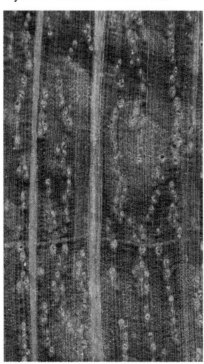

LIVE OAK
Quercus virginiana

Average Specific Gravity: 0.88
Heartwood Color: Brown to grey brown
Pore Distribution: Semi-ring-porous; growth rings not always distinct
Pores: Largest pores not distinct without lens; gradual transition to smaller pores in latewood, generally in radial arrangement
Tyloses: Sparse
Parenchyma: Faintly visible in broken, irregular lines in latewood
Rays: Largest multiseriate rays conspicuous to eye

Rays: Narrow rays uniseriate

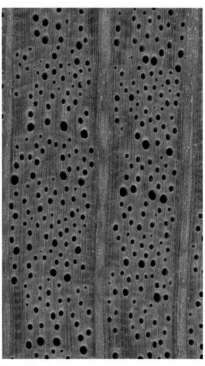

TANOAK
Lithocarpus densiflorus

Average Specific Gravity: 0.65
Heartwood Color: Light reddish brown, aging upon exposure to dark reddish brown
Pore Distribution: Semi-ring-porous; growth rings not distinct
Pores: Largest pores barely visible, distinct only with lens, arranged in streamerlike patches of lighter-colored tissue that may extend across several growth rings
Tyloses: Present
Parenchyma: Tangential lines of parenchyma visible with lens across areas of fiber mass between pore zones
Rays: Large rays present, but not always conspicuous

Rays: Narrow rays uniseriate

rays in these two woods are of two distinctly different sizes — the large, conspicuous multiseriate rays are separated by many smaller, mainly uniseriate rays.

Unlike most oaks, which are typically ring-porous woods with large, obvious earlywood pores, the largest pores in live oak are barely visible to the unaided eye. Its rings are usually indistinct, delineated more by denser marginal latewood than by pore-size variation. Although the pores are larger in the earlywood zone of the growth ring, the transition to the smallest latewood pores is gradual. Tyloses are sparse, and the pores are arranged in groups that are somewhat irregular, although they have a general radial arrangement. Broken lines of banded parenchyma are only faintly visible in the latewood.

The large rays are up to ½ in. high and show on tangential surfaces as conspicuous dark lines against the brown to grey-brown heartwood. Live oak is unusually heavy and dense, with an average specific gravity of 0.88. It is a native of coastal areas of the southern states along the Atlantic Ocean and the Gulf of Mexico. Tanoak is found in southwest Oregon and in California. Although not a true oak, it doubtless derives its common name in part from its large, oaklike rays. (Its common-name prefix alludes to the historical use of its bark as a source of tannin for the hide-tanning process.) In contrast to live oak, its wood is moderately heavy, with an average specific gravity of 0.65, and it has a light, reddish-brown heartwood that darkens with age.

In cross section, the pores of tanoak, which are barely visible to the eye, occur in streamerlike radial patches of lighter-colored tissue. These appear to be continuous across several rings, sometimes with so little pore-size variation as to suggest classification as a diffuse-porous wood. Growth rings are only faintly evident because of the denser fiber mass along the outer latewood. With a hand lens we can see that there are irregular but distinct tangential lines of banded parenchyma between the radial patches of pores; this helps distinguish tanoak from live oak. The large, multiseriate rays

occur in varying numbers. The largest are up to 1 in. or more high, but, in contrast to live oak, they are often inconspicuous because of the lack of color contrast with background tissue.

Semi-Ring-Porous Subgroup II-2: Persimmon

The very wide sapwood of common persimmon (*Diospyros virginiana;* photo at right) is creamy white when freshly cut, but it typically darkens to yellow or light grey upon aging. The small core of heartwood may be dark to nearly black — persimmon is in the same genus, *Diospyros,* as ebony. Growth rings are not conspicuous to the eye, and longitudinal surfaces may be lacking in growth-ring figure. Although vessels are barely visible to the eye, vessel lines are quite evident. At first glance, they give this wood an appearance that can be confused with coarse-textured birch, cottonwood or ramin.

In cross section, the pores of persimmon are distinct with a hand lens. Vessels are very thick-walled, occurring as radially oval solitary pores or in radial multiples of two to three. The gradual gradation in pore size from larger in the earlywood to fairly small in the latewood is typical of the semi-ring-porous arrangement. Overall, the pores appear to be few in number against the uniform fiber mass. This fiber mass forms a background against which the fine 1-3 seriate rays and fine lines of banded parenchyma can be seen with a hand lens.

Persimmon is most often confused with the pecan hickories due to its similar pore distribution and ray/parenchyma network. But it can be separated by either of two important features. Persimmon has ripple marks, which are lacking in hickory. These ripple marks, which are due to storied structure, are best seen with a hand lens on tangential surfaces. Under a microscope, a tangential section shows very fine intervessel pitting — 2 to 4 micrometers in diameter — with confluent apertures connecting up to several pits. This contrasts with the larger — 6 to 8 micrometers in diameter — alternate pits of hickory.

PERSIMMON

Persimmon exemplifies the semi-ring-porous subgroup. On a transverse surface, it displays a fine-lined network of banded parenchyma and rays that is confusingly similar to that of pecan hickories. However, positive separation is provided by the ripple marks on tangential surfaces of persimmon, or microscopically by the much smaller-diameter intervessel pits of persimmon.

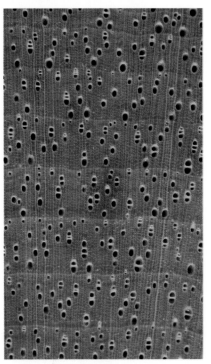

COMMON PERSIMMON
Diospyros virginiana

Average Specific Gravity: 0.74
Heartwood Color: Core dark, nearly black
Sapwood Color: Creamy white, darkening to yellow or light grey with age (very wide)
Pore Distribution: Semi-ring-porous; growth rings not distinct
Pores: Largest pores medium to large, thick-walled; appear to be relatively few; solitary or in radial multiples
Tyloses: Occasionally present
Parenchyma: Fine tangential lines visible with lens
Rays: Fine, visible with lens
Ripple Marks: On tangential surfaces due to storied structure

Rays: Storied, 1-3 seriate
Latewood Pores: Thick-walled
Intervessel Pitting: Alternate, minute (2-4 µm) with confluent apertures

Semi-Ring-Porous Subgroup II-3: Walnut and Butternut

The genus *Juglans* includes several species that we know as walnut, as well as butternut (photos at right). In the United States, black walnut (*Juglans nigra*) is probably the most valuable and certainly one of the most popular domestic hardwoods. Several other species of walnut with similar wood are also locally important. These include northern California walnut (*J. hindsii*) and southern California walnut (*J. californica*). Another important timber species of the genus is butternut (*J. cinerea*).

With the hand lens, the cellular features on transverse surfaces read with confusing similarity in both black walnut and butternut. Since the first earlywood pores are fairly large, vessel lines show quite prominently on longitudinal surfaces at a considerable distance. Although this feature is common to black walnut and butternut, it helps separate them from a few confusing look-alikes such as willow. The pores decrease very gradually from large in the earlywood to fairly small in the outer latewood, which is typical of the semi-ring-porous classification. This distribution is often quite striking on tangential surfaces (see the photo on the facing page).

It is especially difficult to distinguish walnut from butternut in slowly grown pieces, where the narrow rings repeat the larger earlywood pores in closely spaced succession. In both woods, the pores are sometimes solitary and sometimes in radial multiples of two to several. Other similar features, as seen with a hand lens, are the fairly abundant tyloses, fine rays (indistinct without a hand lens) and short tangential lines of banded or diffuse-in-aggregates parenchyma. These latter are visible with a hand lens, but are more distinct in some samples than in others.

Due to the similarities among features seen with the hand lens, it helps to consider several gross features when separating black walnut and butternut. The first is color. Typical black walnut heartwood is a rich, dark chocolate brown; sometimes there is a purple tinge when the wood is freshly cut. By contrast, butter-

WALNUT AND BUTTERNUT

Walnut and butternut are similar in that they are both semi-ring-porous, with faint, short bands of diffuse-in-aggregates parenchyma visible with a hand lens. The chocolate-brown heartwood and higher density of black walnut distinguishes it from butternut, whose heartwood is medium brown.

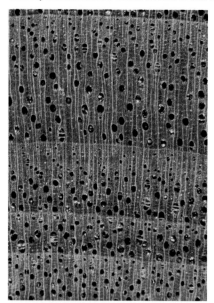

BLACK WALNUT
Juglans nigra

Average Specific Gravity: 0.55
Heartwood Color: Medium brown to deep chocolate brown
Pore Distribution: Semi-ring-porous
Pores: Earlywood pores fairly large, decreasing gradually to quite small in outer latewood; pores solitary or in radial multiples of 2 to several
Tyloses: Moderately abundant
Parenchyma: Short tangential lines of banded or diffuse-in-aggregates parenchyma visible with lens
Rays: Fine, visible but not conspicuous with lens

Rays: 1-5 seriate; ray cells appear round in tangential view
Crystals: Occur sporadically in longitudinal parenchyma cells
Thickenings: Reticulate thickenings ("gashlike pits") occasionally found in latewood vessels

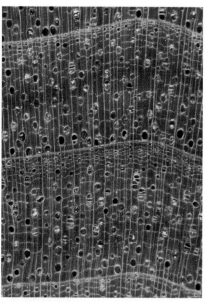

BUTTERNUT
Juglans cinerea

Average Specific Gravity: 0.38
Heartwood Color: Medium or cinnamon brown, often with uneven streaks of color; growth rings usually fluted
Pore Distribution: Semi-ring-porous
Pores: Earlywood pores fairly large, decreasing gradually in size to quite small in outer latewood; pores solitary or in radial multiples of 2 to several
Tyloses: Moderately abundant
Parenchyma: Short tangential lines of banded or diffuse-in-aggregates parenchyma visible with lens
Rays: Fine, visible but not conspicuous with lens

Rays: 1-4 seriate; ray cells appear oval in tangential view
Crystals: Never present in longitudinal parenchyma
Thickenings: Reticulate thickenings never present in vessels

nut is typically lighter, in shades of chestnut or cinnamon brown. It often shows considerable variegation (unevenness of pigment). When a cleanly cut end-grain surface of typical black walnut is moistened, it deepens in color to rich sepia — the color of baker's chocolate — whereas wet butternut end grain usually maintains its brown color with a distinct orange cast. Still, these color ranges overlap occasionally — darker pieces of butternut can be confused with lighter pieces of black walnut.

Density and hardness are also important distinguishing features. The softer, lighter-weight butternut is more easily dented with the thumbnail, in contrast to black walnut, which is harder. Bear in mind, however, that their specific-gravity ranges overlap. Butternut usually has fluted rings, so look for an irregular or jagged ring figure on tangential surfaces. This feature is far from conclusive, however, since black walnut occasionally has fluted rings too (see the photos on p. 64.

Microscopic features are especially valuable in identifying these woods. In black walnut, rays are up to 5 seriate; viewed tangentially, the cells are typically rounded. In butternut, the rays are up to 4 seriate, and the cells are vertically oval or flattened. In black walnut, reticulate thickenings occur sporadically on the walls of latewood vessel elements. These so-called gashlike pits (see the photo on p. 37), which do not occur in butternut, are best found by scanning radial sections of latewood.

In addition, some samples of black walnut may contain large, diamond-shaped crystals within longitudinal parenchyma (see the photo on p. 40). When looking for crystals, any convenient longitudinal surface can be used as a section.

Another very important species of walnut, European walnut (*Juglans regia*) — variously known as English walnut, French walnut and Circassian walnut, according to its source — is confusingly similar to black walnut (*J. nigra*). When these two woods are placed side by side, the European walnut usually appears lighter and less uniform in color than the more evenly deep-brown black walnut. Moreover, the European walnut seems more diffuse-porous, since it lacks the pore-size contrast of black walnut. Isolated samples, however, are best separated microscopically. Since neither "gashlike pits" nor crystals in longitudinal parenchyma occur in European walnut, the presence of either confirms black walnut. It requires careful judgment to identify a sample as European walnut because neither feature is found.

GROUP III: DIFFUSE-POROUS HARDWOODS

As with the ring-porous hardwoods, a few diffuse-porous species have strikingly large rays that clearly distinguish them from the other species in the category. We shall begin with these since their identification is relatively easy. Among the remaining subgroups, however, there are some confusing similarities, or perhaps more accurately, a lack of differences. In some cases it is best to examine microscopic features to make a reliable separation within a group or between a pair of woods.

Diffuse-Porous Subgroup III-1: Beech and Sycamore

Among diffuse-porous woods, the two that can be identified by eye most of the time are American beech (*Fagus grandifolia*) and sycamore (*Platanus occidentalis*). Both have large and fairly numerous rays that are easily visible on any surface (see the photos on p. 118). On smoothly cut cross sections, the rays appear as conspicuous bright lines; with a hand lens they are clearly noded. Ray fleck is striking on radial surfaces (see the photos at the top of p. 119).

Even though beech and sycamore can usually be identified by gross features, the two woods can be confused. For example, a cross section of either species will show pores that are numerous and either solitary or in multiples. The multiples are both radial and tangential, merging in some cases into small clusters. The pore size and distribution is fairly uniform through most of the ring. Toward the latewood edge, however, pores decrease abruptly in size and number, resulting in a distinct narrow-to-wide band of latewood. This band appears to be denser than the surrounding wood.

Beyond these general similarities, however, there are some important differences that enable us to separate the species readily. In a beech cross section, the latewood band has a distinctly darker shade. In sycamore, by contrast, the late-

PORE SIZE DISTRIBUTION IN BLACK WALNUT
The semi-ring-porous structure of black walnut is evident in the gradation of vessel lines from earlywood to latewood within each growth ring on this tangential surface.

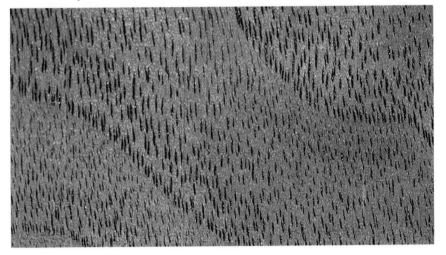

wood appears lighter in color, which is an anomaly among most woods.

Closer study of the rays gives us further basis for separating these woods. In sycamore, the rays are much more uniform in size and more closely spaced when viewed tangentially (see the photos at the bottom of the facing page). In beech the largest rays are generally larger and are more widely and irregularly distributed. In cross section, the rays of beech look much more uneven in size.

Either wood can be checked microscopically. The rays of sycamore range in width to a maximum of about 14 seriate; beech rays are divided into two size classes: narrow 1-5 seriate, and large 15-25 seriate. Intervessel pits in sycamore are typically oval and more widely spaced. The pitting in beech is often crowded and angular and sometimes transversely elongated.

The wood of the plane tree or London plane (*Platanus acerifolia*), which grows in Europe, is very similar to that of American sycamore (*Platanus occidentalis*). Nor can European beech (*Fagus sylvatica*) be readily separated from American beech (*Fagus grandifolia*), although American beech sometimes has crystals in its longitudinal and ray parenchyma cells which are not seen in European beech.

BEECH AND SYCAMORE SURFACES
Both beech and sycamore are diffuse-porous woods with a zone of denser latewood in the outer ring; their large, conspicuous rays are noded at the growth-ring boundary. Ray size and spacing are fairly uniform in sycamore, but much more variable in beech.

AMERICAN BEECH
Fagus grandifolia

Average Specific Gravity: 0.64
Heartwood Color: Creamy white with reddish tinge to medium reddish brown
Pore Distribution: Diffuse-porous; growth rings distinct
Pores: Small, solitary and in irregular multiples and clusters, numerous and evenly distributed throughout most of the ring; narrow but distinct latewood in each ring due to fewer, smaller pores
Rays: Largest rays conspicuous on all surfaces; darker ray fleck against lighter background on radial surfaces

Rays: Largest rays 15-25 seriate; uniseriate rays common
Crystals: Occasional in longitudinal parenchyma and ray cells

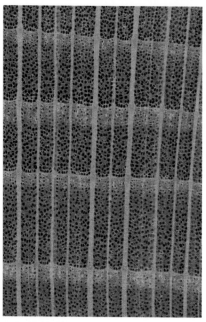

SYCAMORE
Platanus occidentalis

Average Specific Gravity: 0.49
Heartwood Color: Light to dark brown, usually with a reddish cast
Pore Distribution: Diffuse-porous; growth rings distinct due to unusual lighter color of latewood
Pores: Small, solitary and in irregular multiples and clusters, numerous and evenly distributed throughout most of the growth ring; latewood zone evident by fewer, smaller pores and lighter color
Rays: Easily visible without lens on all surfaces, appearing uniform in size and evenly spaced on transverse and tangential surfaces, producing conspicuous dark ray fleck on radial surfaces

Rays: Largest up to 14 seriate; uniseriate rays not common

Diffuse-Porous Subgroup III-2: Hornbeam and Alder

American hornbeam (*Carpinus caroliniana*), commonly called blue beech, and red alder (*Alnus rubra*) are the two woods in this book that have aggregate rays. Each wood has a normal distribution of invisibly narrow rays, but the aggregate rays are very conspicuous (see the photos on p. 120). This is as much a result of their erratic placement as of their large size. The placement is different from the reasonably uniform spacing of large rays in the oaks, beech and sycamore. The aggregate rays appear in cross sections as bold radial lines, less conspicuous in some samples than in others. On longitudinal surfaces they appear as darker lines or streaks against the background of fairly uniform cell structure.

Both species are diffuse-porous and fine-textured; their pores are indistinct without a hand lens. The narrow rays are also invisibly small. The two woods can usually be separated on the basis of gross and macroscopic features and confirmed through microscopic examination.

Red alder, the principal hardwood species of the coastal Pacific from central California to Alaska, is pale tan when freshly cut. It darkens gradually with age to a light reddish brown or flesh color. It is moderately soft, with an average specific gravity of 0.41. The small pores are seen with a hand lens as solitary or in multiples of a few to several — in this re-

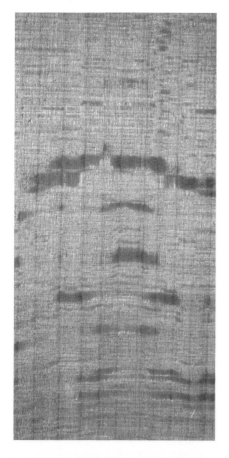
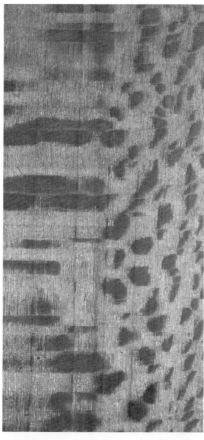

**BEECH AND SYCAMORE
RADIAL SURFACES**
Although both beech (far left) and sycamore (left) display striking ray fleck, the
more even size and spacing of rays in sycamore is evident.

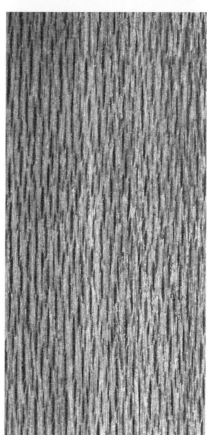

**BEECH AND SYCAMORE
TANGENTIAL SURFACES**
In beech (far left), rays appear variable in
size and irregular in distribution; in sycamore (left), the rays are more uniform in
size and distribution, and appear to be
more closely spaced. Microscopic examination reveals that the largest rays in
beech are 15-25+ seriate; maximum ray
width in sycamore is about 14 seriate.

AMERICAN HORNBEAM AND RED ALDER

American hornbeam and red alder are fine-textured, diffuse-porous woods that have scattered conspicuous aggregate rays. The heartwood of red alder ages to a light reddish brown or flesh color; American hornbeam heartwood is pale yellowish or brownish white, commonly with a grey cast. A microscopic check reveals that American hornbeam has vessels with spiral thickenings and perforation plates that are simple or scalariform, with very few bars. Red alder has scalariform perforation plates with 20+ bars and its vessels lack spiral thickenings.

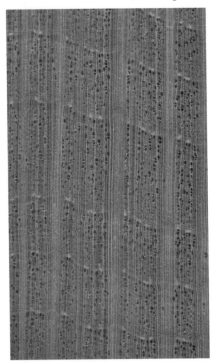

AMERICAN HORNBEAM
Carpinus caroliniana

Average Specific Gravity: 0.70
Heartwood Color: Light with brown, gray, or yellow cast
Pore Distribution: Diffuse-porous; growth rings with irregular contours or flutes, delineated by narrow, lighter tissue in outer latewood
Pores: Small, sometimes more numerous in initial earlywood, mostly arranged in radial multiples and forming loose radial strings
Rays: Large aggregate rays few to fairly numerous and visible without lens

Rays: Narrow rays 1-4 (mostly 1-2) seriate
Spiral Thickenings: Present
Perforation Plates: Mostly simple, or scalariform with up to 5 thick bars
Intervessel Pitting: Alternate

RED ALDER
Alnus rubra

Average Specific Gravity: 0.41
Heartwood Color: Pale tan when freshly cut (not distinct from sapwood) darkening with age to light reddish brown or flesh color
Pore Distribution: Diffuse-porous
Pores: Small, solitary and in mostly radial multiples
Rays: Large aggregate rays widely scattered but easily seen without lens

Rays: Narrow rays uniseriate, occasionally biseriate
Spiral Thickenings: Absent
Perforation Plates: Scalariform with up to 20+ thin bars
Intervessel Pitting: Alternate to opposite

gard, they are confusingly similar to aspen or fine-textured birch.

Microscopically, vessels in red alder are seen to have scalariform perforation plates with up to 20 or more thin bars. The intervessel pitting is alternate to opposite, and the pits are small, oval and spaced apart from each other. The narrow rays are homocellular and uniseriate.

Grey alder *(A. incana)*, a native of Europe, and common alder *(A. glutinosa)* are anatomically similar to red alder.

American hornbeam is pale yellowish or brownish white, commonly with a greyish cast. The wood is heavy and hard, with an average specific gravity of 0.70. These relatively small trees have fluted stems, so growth rings are commonly irregular. They are distinct to the eye, with a narrow band of whitish tissue along the outer latewood. In some pieces, as seen with a hand lens, the pores along the first-formed earlywood are more numerous and form a loosely organized row; the remaining pores occur mostly in multiples of two or more and merge into longer radial multiples or irregular radial chains.

Microscopic examination of American hornbeam shows vessels with spiral thickenings and mostly simple perforations, with occasional scalariform perforations with up to five thick bars. Intervessel pitting is alternate, the large pits rounded or angular and crowded together. The narrow rays are 1-4 (mostly 1-2) seriate.

Diffuse-Porous Subgroup III-3: The Confusing Diffuse-Porous Woods

These hardwoods are the most difficult to identify because they generally lack striking figure, are fine-textured and have inconspicuous or invisible rays. Although any one wood in this group may be quite clearly different from another—for example, cherry from holly—there is also one or more species to which it is confusingly similar. There is no obvious systematic way to order these woods. I have chosen to group those that tend to be confused with each other close together, so that distinguishing differences — where they exist — can be highlighted. I have also attempted to separate the overall subgroup into groupings based on the visibility of rays on transverse surfaces.

But there are two important caveats. First, the normal range of anatomical variation includes a few samples that will depart drastically from the average description. For example, although the rays in a typical piece of soft maple are moderately distinct with a hand lens, on occasion they may scarcely show; in another piece they may be unusually conspicuous. Second, the identification of diffuse-porous hardwoods is the least reliable when the wood is examined with the naked eye or hand lens. As with conifers, an identification is best confirmed with a microscope. On the plus side, very small, toothpick-sized samples of diffuse-porous woods often contain the key anatomical features for positive identification.

Table 10.6 on pp. 138-141 summarizes some of the features that are helpful in the separation process and that can be seen with the naked eye, hand lens or microscope. The table can be used for quick reference to find definitive differences between tentative possibilities. For example, it tells us that dogwood is significantly heavier than aspen, and that birch has scalariform perforation plates, whereas those in maple are simple. It can also be used as a starting point for identification within the diffuse-porous group. For example, if you examine an unknown wood and find spiral thickenings, you can immediately eliminate those woods in which spirals are not found. If that same wood is obviously soft and of low density, basswood and buckeye are likely candidates. Comparing these woods further with the chart would indicate that they differ in ray seriation. Therefore, the logical next step would be to check to see if the rays are exclusively uniseriate, as in buckeye, or multiseriate, as in basswood.

THE MAPLES

North American maples (see the photos on p. 123) are divided into hard maples and soft maples. The hard-maple group, whose specific gravity ranges from 0.57 to 0.63, consists principally of sugar maple *(Acer saccharum)*, but also includes black maple *(A. nigrum)*. In the soft-maple group, whose specific gravity range is 0.46 to 0.54, the most common species is red maple *(A. rubrum)*. Other soft maples are silver maple *(A. saccharinum)* and boxelder *(A. negundo)*. An unknown maple can usually be placed in one of these subgroups; within a group, however, species cannot be identified with certainty.

Wood of the maples ranges in color from creamy white to very light brown. Soft maples, red maple in particular, may have an overall greyish cast or streaks.

Growth rings are distinct due to the narrow zone of darker brown cells at the outer margin of the growth ring. On tangential surfaces the rays are visible as tiny dark lines, reasonably uniform in size and spacing (see the photos below). They vary from quite distinct in hard maples to barely visible in some soft maples. Usually, however, they can be seen readily with a hand lens.

On radial surfaces, the rays produce a conspicuous ray fleck that appears dark against a lighter background (see the photo at bottom left on p. 122).

As seen on a cross-sectional surface with a hand lens, the pores are small and solitary or in radial multiples of two to several. They are fairly uniform in size and quite uniform in distribution. The rays are very distinct, with the largest about as wide as the diameter of the

RAY SIZE DISTRIBUTION IN MAPLE AND BEECH

Although rays in maple (left) are usually visible to the unaided eye on tangential surfaces, they appear much smaller, more numerous and more uniformly distributed than they are in beech (right).

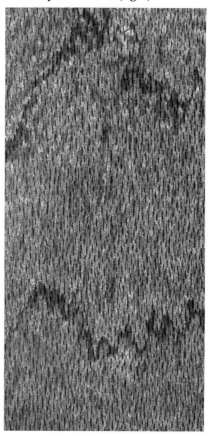
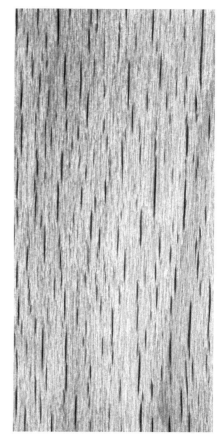

largest pores (they are a shade wider in some sugar maples, the same width or slightly narrower in soft maples).

Under the microscope, vessels show simple perforations and fine, very evenly spaced spiral thickenings (see the photo below, far right). Vessels may have yellow to orange contents. Intervessel pitting is alternate. A tangential microscopic view is the best means of separating hard and soft maples because it allows ray shape and seriation to be observed. In hard maples, the rays appear to be of two distinct sizes. The smaller rays are uniseriate, and the larger are 5-8 seriate, with blunt bullet-shaped ends (see the photos on p. 45). In soft maples, the rays are variable in width up to 5 seriate, and they end in slender tapered points.

Bigleaf maple (A. macrophyllum) is similar to the soft maples in its ray seria-tion, but its heartwood usually has a pinkish cast and slightly larger pores. Field maple (A. campestre) and Norway maple (A. platanoides), both from Europe, are characteristic of the soft maples. A. pseudoplatanus, called "sycamore" in England, closely resembles hard maples. In England, maple wood that is artificially dyed grey is called "harewood;" it is often used in marquetry.

DOGWOOD

Flowering dogwood (Cornus florida) (see the photo on the facing page) is easily confused with maple. As with maple, the wood is fairly hard, with an average specific gravity of 0.73, and it has small, evenly distributed pores. They are mostly solitary but also occur in multiples of two or more. The broadest rays are as wide or wider than the pores.

However, they are usually not as distinct to the eye or with a hand lens as they are in maple, nor are the growth rings as conspicuous. The sapwood in dogwood ranges from a creamy color with a pinkish tinge to flesh-colored. Heartwood is medium to dark brown.

The microscopic features of dogwood positively distinguish it from other woods. The vessels have scalariform perforation plates with up to 20+ bars. Opposite-oval intervessel pitting sometimes intergrades with scalariform pitting. The heterocellular rays are up to 8 seriate. The multiseriate rays have small, rounded, procumbent cells throughout the body of the ray, but uniseriate tails of up to several upright cells. A similar species, Pacific dogwood (Cornus nuttallii), has rays that are much shorter and up to 5 seriate.

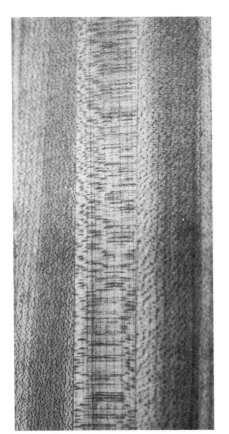

RAY FLECK IN MAPLE
Maple has a distinctive dark-against-light ray fleck, as shown in this photographically reproduced surface of high-pressure laminate.

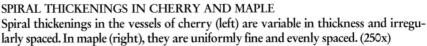

SPIRAL THICKENINGS IN CHERRY AND MAPLE
Spiral thickenings in the vessels of cherry (left) are variable in thickness and irregularly spaced. In maple (right), they are uniformly fine and evenly spaced. (250x)

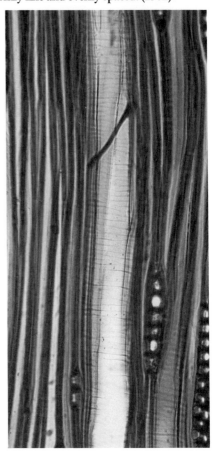

HARD MAPLE, SOFT MAPLE AND DOGWOOD

As seen with a hand lens, the width of rays in maple is approximately equal to the largest pore diameters. Although the greyish cast of soft maple will often distinguish it from the creamy white to light reddish brown of hard maple, a microscopic check of ray seriation — up to 5 in soft maple, up to 8 in hard maple — gives a more reliable separation. Pores in maples are commonly in multiples. In dogwood, pores are mostly solitary. Perforation plates are simple in maples, scalariform with 20+ bars in dogwood.

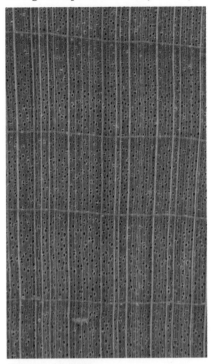

SUGAR MAPLE
Acer saccharum

Average Specific Gravity: 0.63
Heartwood Color: Creamy white to
   light reddish brown
Pore Distribution: Diffuse-porous;
   growth rings distinct due to darker
   brown narrow latewood line
Pores: Small, with largest
   approximately equal to maximum
   ray width in cross section; solitary or
   in radial multiples; very evenly
   distributed
Rays: Visible to eye on tangential
   surface as very fine, even-sized,
   evenly distributed lines; on radial
   surfaces, ray fleck usually
   conspicuous

Rays: Largest 7-8 seriate; uniseriate
   rays numerous
Spiral Thickenings: Fine, evenly spaced
Perforation Plates: Simple
Intervessel Pitting: Alternate, large

RED MAPLE
Acer rubrum

Average Specific Gravity: 0.54
Heartwood Color: Creamy white to
   light reddish brown, commonly
   with greyish cast or streaks
Pore Distribution: Diffuse-porous
Pores: Small, solitary and in radial
   multiples, very evenly distributed;
   largest as large or slightly larger
   than widest rays on cross section
Rays: May be visible on tangential
   surface as very fine, even-sized and
   evenly spaced lines; on radial
   surfaces, ray fleck usually
   conspicuous

Rays: 1-5 seriate
Spiral Thickenings: Fine, evenly spaced
Perforation Plates: Simple
Intervessel Pitting: Alternate, large

FLOWERING DOGWOOD
Cornus florida

Average Specific Gravity: 0.73
Heartwood Color: Dark brown
Sapwood Color: Creamy with flesh or
   pinkish cast (fairly wide sapwood)
Pore Distribution: Diffuse-porous
Pores: Very small, mostly solitary with
   some radial multiples
Rays: Approximately as wide or wider
   than largest pores

Rays: 2-8 seriate; in tangential view,
   multiseriate rays have uniseriate
   tails of up to several upright cells
Spiral Thickenings: Absent
Perforation Plates: Scalariform with up
   to 20+ fine bars
Intervessel Pitting: Alternate to
   opposite, sometimes intergrading
   with scalariform

HOLLY, CHERRY AND YELLOW-POPLAR

In holly, cherry and yellow-poplar, rays are visible on a transverse surface without a hand lens. The wood of holly is near white, or with a grey or bluish cast or streaks. With a hand lens, its transverse surface shows very small pores in chains that may cross the indistinct growth-ring boundaries. Cherry has heartwood color in shades of light to dark cinnamon or reddish brown; its growth rings are evident by a line of slightly larger pores along the earlywood side of the boundary. In yellow-poplar, the greenish shades of heartwood color give background contrast to the cream-colored rays, noded as they cross the bright line of terminal parenchyma at the growth-ring boundary.

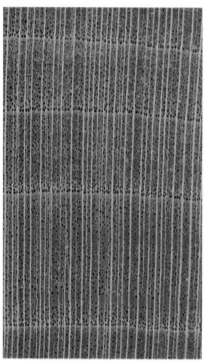

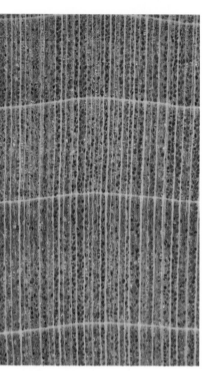

AMERICAN HOLLY
Ilex opaca

Average Specific Gravity: 0.56
Heartwood Color: Near white with grey or blue-grey cast or streaks
Pore Distribution: Diffuse-porous
Pores: Extremely small, mostly arranged in chains that commonly extend across growth-ring boundaries

Rays: Widest visible without lens: with lens, widest rays exceed pore diameters
Rays: 1-5 (mostly 5) seriate, heterocellular; uniseriate rays homocellular up right
Spiral Thickenings: Present in vessels, also in fibers
Perforation Plates: Scalariform with up to 15+ bars
Intervessel Pitting: Opposite, in spiral rows following spiral thickenings

BLACK CHERRY
Prunus serotina

Average Specific Gravity: 0.50
Heartwood Color: Light to dark cinnamon or reddish brown
Pore Distribution: Diffuse-porous; growth rings sometimes distinct because of narrow zone or row of numerous slightly larger pores along initial earlywood
Pores: Pores through growth ring solitary and in radial or irregular multiples and small clusters
Gum Defects: Common
Rays: Not visible on tangential surface; conspicuous light ray fleck on radial surfaces; distinct bright lines across transverse surface, conspicuous with lens

Rays: 1-6 (mostly 3-4) seriate
Spiral Thickenings: Variable from thick to very thin, irregularly spaced
Perforation Plates: Simple
Intervessel Pitting: Alternate
Contents: Vessels with orange to ruby red contents

YELLOW-POPLAR
Liriodendron tulipifera

Average Specific Gravity: 0.42
Heartwood Color: Green, or yellow to tan with greenish cast
Sapwood Color: Creamy white (often wide)
Pore Distribution: Diffuse-porous; growth rings delineated by distinct light cream or yellowish line of marginal parenchyma
Pores: Small, solitary, but mostly in radial or irregular multiples and small clusters
Rays: Distinct on cross section with lens; produce conspicuous fine light ray fleck on radial surface

Rays: 1-5 (mostly 2-3) seriate, heterocellular
Spiral Thickenings: Absent
Perforation Plates: Scalariform with few to several thick bars
Intervessel Pitting: Opposite, oval to rectangular

HOLLY

American holly (*Ilex opaca*) is another fairly dense wood—its average specific gravity is 0.56—that is occasionally confused with maple because its rays are visible to the naked eye. However, when viewed with a hand lens, the rays are generally not as distinct as in maple. The growth rings of holly are indistinct due to a lack of contrast; although the outer latewood has a narrow band of denser fiber, it lacks darker pigment as in maple. The wood—both sapwood and heartwood—is well known for its whiteness, described variously as ivory white or greyish white; it is valued for inlays. However, the wood often becomes discolored by blue-grey streaks of stain.

The most important feature that can be seen with a hand lens is the pores, which are uniformly quite small. Some are solitary, but most occur in multiples that are arranged in long linear series or pore chains that sometimes extend across growth-ring boundaries. Holly is commonly the textbook example used to show pore chains.

Microscopically, holly has scalariform perforation plates, with 15+ bars, along withspiral thickenings in the vessels. The fairly small intervessel pits are in opposite arrangement, the rows following the angle of the spiral thickening. Fibers also have spiral thickenings. As well as uniseriate homocellular rays, holly has heterocellular rays that are 3-5 (mostly 5) seriate.

English holly (*Ilex aquifolium*) has characteristics that are similar to American holly, and the two are difficult if not impossible to tell apart.

CHERRY

Black cherry (*Prunus serotina*) is a moderately dense hardwood whose average specific gravity is 0.50. Its distinctive heartwood ranges in color from light to dark cinnamon to reddish brown. Its rays are quite distinct on transverse surfaces, especially against the darker heartwood background. On pieces that are entirely sapwood, which is whitish to pale tan, the distinct rays are easily confused with those of maple.

With a hand lens, however, the differences are clear. In cherry, the pores form an apparent row along the earlywood, which prompts some to call the wood ring-porous. But the pores are not distinctly larger than those in the rest of the growth ring. This concentration of pores, although slight, is a major contributor to the growth-ring figure of cherry as seen on tangential surfaces. Throughout the body of the growth ring, the pores are both solitary and in variable multiples or small clusters. This contrasts to the predominantly radial multiples in maple. Insect-caused gum spots are common in cherry and result in a familiar feature on longitudinal surfaces (see the photo on p. 62).

On tangential surfaces the rays are not visible, as they are in maple. But on radial surfaces, a striking ray fleck is revealed (see the photo below). This lighter-colored fleck against a darker heartwood background is a most familiar figure in cherry, but it is sometimes confused with yellow-poplar.

With a microscope, we find that cherry has rays that are up to 6 (mostly 3-4) seriate. Vessels have simple perforations and alternate intervessel pitting, features that are confusingly similar to maples. One distinguishing feature is the spiral thickenings of the vessels, which are variable in thickness, unevenly spaced and at inconsistent angles. This contrasts with the very uniformly thin, closely and evenly spaced spiral thickenings in maple (see the photos on p. 122).

European cherry (*Prunus avium*) is similar to black cherry, as are several other woods in the genus *Prunus,* including plums, peaches and apricots. Also in the *Rosaceae* family are the genera *Malus* (apple) and *Pyrus* (pear), whose woods have narrower rays—1-3 (mostly 2) seriate and sparse spirals. The term "fruitwood" is used as a convenient collective term for the indistinguishable species in this group.

YELLOW-POPLAR AND MAGNOLIA

Yellow-poplar (*Liriodendron tulipifera*) is probably the most common of the traditional secondary hardwoods. It is also known as tulip tree and tulip poplar because of its tuliplike yellow flowers. Yellow-poplar grows tall and straight, and old-growth trees have yielded clear (knot-free) boards up to 3 ft. wide. The sapwood, which is a flat creamy white, is fairly wide. As a result, boards consisting entirely of sapwood are frequent, and have earned for this wood the alternate common name of whitewood. By contrast, heartwood color is distinct but varies from tree to tree. The most common color is green,

RAY FLECK IN CHERRY
A drawer front displaying striking light-against-dark ray fleck in black cherry.

ranging from moss green to yellow-green to tan with a greenish cast. An occasional piece has a purple cast, or is streaked with dark brown or black. The wood is moderately soft, with an average specific gravity of 0.42.

In cross-sectional view, yellow-poplar growth rings are delineated by light-colored whitish or yellowish marginal parenchyma (see the photo below). This line is usually conspicuous to the eye, but occasionally it is so thin as to be distinct only with a hand lens.

The light-colored rays are also visible to the eye and are distinct with a hand lens. This is especially true of the darker heartwood, where they appear to be slightly wavy as they pass radially among the pores (this contrasts sharply with the arrow-straight rays of maple and cherry). They are conspicuously noded as they cross through the band of marginal parenchyma.

The pores of yellow-poplar are fairly small, although they can vary somewhat in average size from one sample

to the next. They seem to occupy the majority of the exposed surface. Some are solitary but most are in irregular multiples and clusters.

On tangential surfaces, figure is subtle at best, sometimes displaying lines of marginal parenchyma. But the rays are not visible. On radial surfaces, however, a light-colored ray fleck may show strikingly, especially against the darker heartwood (photos below). This light-on-dark ray-fleck figure can be deceivingly suggestive of black cherry, especially where pieces have darkened or faded with age or have been stained artificially to imitate cherry's color. The best means of telling these two woods apart by eye or with a hand lens is by checking for the light-colored line of marginal parenchyma that is present in yellow-poplar but absent from cherry.

Microscopically, yellow-poplar is quick and easy to check. Its perforation plates are scalariform, commonly with only two to four bars, although some plates have up to a dozen or more.

Intervessel pitting is opposite, with fairly large, oval to rectangular pits that on rare occasion intergrade to scalariform. The pits have distinct borders and apertures (see the photos at the bottom of the facing page). Because the pores in yellow-poplar are so irregularly arranged in multiples and clusters, virtually any longitudinal section shows perforation plates and intervessel pitting in face view when examined microscopically. Because it is not necessary to get a perfect tangential or radial cut to check perforations or intervessel pitting, you can section directly from furniture parts along any exposed longitudinal surface to confirm yellow-poplar.

A proper radial cut (top photo on the facing page) shows the rays to be heterocellular. These horizontally elongated, thin-walled, procumbent cells are in obvious contrast to the squarish or rectangular upright cells at the margins of the rays that have extremely thick walls and large rounded

GROWTH RINGS IN YELLOW-POPLAR
The growth-ring pattern on this yellow-poplar veneer results from the very thin layers of lighter-colored terminal parenchyma cells.

RAY FLECK IN YELLOW-POPLAR
The light-against-dark ray fleck evident on this yellow-poplar board (left) could be mistaken for that of cherry (right) if the yellow-poplar were stained a cherry-brown color.

PERFORATION PLATES AND RAY CELLS IN YELLOW-POPLAR
Scalariform perforation plates with only a few or several bars (A) are important identification features of yellow-poplar. The portion of the ray visible clearly shows a row of thick-walled upright cells (B) and thinner-walled procumbent cells (C). (175x)

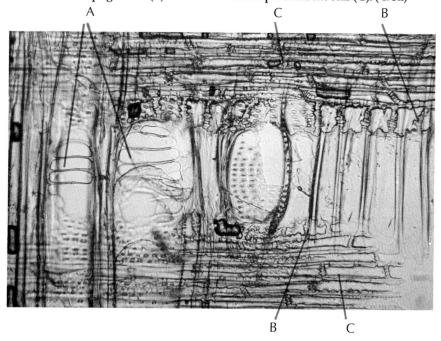

pits. A tangential section shows rays to be in the 1-5 (mostly 2-3) seriate range.

The woods of the genus *Magnolia* are strikingly similar to yellow-poplar and cannot be reliably separated from yellow-poplar by eye. In the lumber trade, boards of cucumbertree *(M. acuminata),* which is also called simply "cucumber," are usually combined with and sold as "yellow poplar." Southern magnolia *(M. grandiflora)* is likewise confusingly similar in appearance to yellow-poplar. The only reliable separation is by microscopic examination, since magnolias have linear (scalariform) intervessel pitting, in comparison to the opposite pitting characteristic of yellow-poplar (compare the photos at left and right below).

Separation of the two *Magnolia* species is less clear-cut. Perforation plates are mainly simple in cucumbertree, but in southern magnolia they are both scalariform and simple. Southern magnolia has spiral thickenings in its vessels; it also has wider rays, 1-5 seriate, in comparison to 1-3 (mostly 2) seriate in cucumbertree.

BASSWOOD
American basswood *(Tilia americana;* center photo on p. 128) is a fairly soft, low-density hardwood with an average specific gravity of 0.37. It ranges in color from creamy white or pale tan (sapwood) to pale brown (heartwood). Basswood has a faint but characteristic musty odor.

In cross-sectional view, growth rings are delineated by marginal parenchyma that is conspicuous in some samples but quite faint in others. In some samples there are small, blurry whitish spots along the growth-ring boundary. Rays are usually faintly visible to the eye; with a hand lens they are distinct and noded where they cross through the growth-ring boundary.

Pores are small and fairly evenly distributed. Sometimes, however, they are more numerous at the beginning of the growth ring, then sparse and smaller toward the outer margin of the ring. Some pores are solitary, but most occur in irregular multiples and clusters. The combination of distinct marginal paren-

INTERVESSEL PITTING IN YELLOW-POPLAR
The oval to rectangular opposite intervessel pitting is a prominent feature of yellow-poplar. Note that rays are commonly 2-3 seriate. (250x)

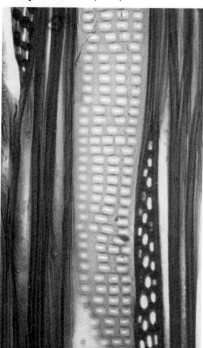

INTERVESSEL PITTING IN MAGNOLIA
The scalariform intervessel pitting of magnolia provides positive separation from yellow-poplar. (250x)

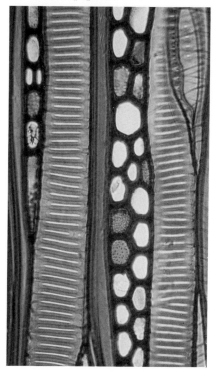

chyma and distinct noded rays may appear confusingly similar to the sapwood of yellow-poplar.

Vessels in basswood have simple perforation plates and alternate intervessel pits that usually appear crowded. Vessels also have thick spiral thickenings.

Ray cells are homocellular and 1-6 seriate. In tangential view, the ray cells appear to be compressed into vertically elongated ellipses or hexagons, which I call "coffin-shaped" cells (see the bottom left photo on the facing page). These are quite distinctive in comparision to the

rounded cells of soft maple or the slightly oval ray cells of yellow-poplar.

Microscopically, basswood can be distinguished from yellow-poplar sapwood by its simple perforations, alternate intervessel pitting and spiral thickenings. (Yellow-poplar sapwood has scalariform

MAGNOLIA, BASSWOOD AND BIRCH

Magnolias resemble yellow-poplar in their green or greenish yellow heartwood and the bright lines of terminal parenchyma. But with a microscope, magnolia is positively separated by its linear (scalariform) intervessel pitting (which is opposite in yellow-poplar). Although basswood is soft and has a characteristic musty odor, it lacks other definitive macroscopic features. It is best confirmed microscopically by its vessels with thick spirals and laterally compressed ray cells. Birches are moderately dense, and have pores (solitary and in radial multiples) that are clearly wider tangentially than ray widths. Some appear to have whitish contents. It is best to confirm birch microscopically by its scalariform perforation plates and extremely small intervessel pits.

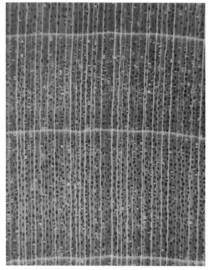

SOUTHERN MAGNOLIA
Magnolia grandiflora

Average Specific Gravity: 0.50
Heartwood Color: Yellow, medium to dark greenish yellow, or dark green
Pore Distribution: Diffuse-porous; growth rings delineated by fine to distinct line of marginal parenchyma
Pores: Small, solitary or in radial multiples or clusters of 2 to several
Rays: Distinct with lens on transverse surface, producing light, fine ray fleck on radial surfaces

Rays: 1-5 seriate, heterocellular
Vessels: In *M. grandiflora*, vessels with spiral thickenings and scalariform perforation plates; in *M. acuminata*, vessels lack spiral thickenings, having mostly simple perforations
Intervessel Pitting: Scalariform

AMERICAN BASSWOOD
Tilia americana

Average Specific Gravity: 0.37
Heartwood Color: Creamy white to pale brown
Odor: Faint but characteristic musty odor
Pore Distribution: Diffuse-porous; growth rings indistinct or faintly delineated by marginal parenchyma, sometimes with blurry whitish spots along the growth-ring boundary
Pores: Small, mostly in irregular multiples and clusters
Rays: Distinct but not conspicuous on transverse surface with lens

Rays: 1-6 seriate; ray cells appear laterally compressed in tangential view; rays have bright yellow cast
Spiral Thickenings: Thick, evenly spaced
Perforation Plates: Simple
Intervessel Pitting: Alternate

YELLOW BIRCH
Betula alleghaniensis

Average Specific Gravity: 0.62
Heartwood Color: Light to dark brown or reddish brown
Pore Distribution: Diffuse-porous
Pores: Small to medium, solitary or in radial multiples of 2 to several; with lens, pore diameters clearly greater than ray width; some pores appear to be filled with whitish substance

Rays: 1-5 seriate
Spiral Thickenings: Absent
Perforation Plates: Scalariform with many fine bars
Intervessel Pitting: Alternate, very small with confluent apertures

perforations and opposite intervessel pitting, and it lacks spirals.) The multiseriate rays of basswood make it easy to separate from aspen, cottonwood and buckeye, which have only uniseriate rays.

White basswood *(Tilia heterophylla)* is indistinguishable from American basswood. European lime *(T. vulgaris)* is anatomically similar but somewhat denser (its average specific gravity is 0.48) than American basswood.

BIRCH
The birches include several species. Some of the more common are yellow birch *(Betula alleghaniensis),* with an average specific gravity of 0.62; sweet birch *(B. lenta),* with an average specific gravity of 0.65; and paper birch *(B. papyrifera),* with an average specific gravity of 0.55. Birches usually have whitish or light yellow sapwood and heartwood ranging from light brown to dark brown or reddish brown. The growth rings may not be very distinct and are terminated by a narrow zone of slightly denser and darker fiber mass.

The rays are barely visible to the unaided eye, but they are distinct with a hand lens on cross-sectional surfaces. The wood is uniformly diffuse-porous, with pores either solitary or in multiples that are mainly radial. Average pore diameter may vary from sample to sample among or within species. The pores are usually distinct and well separated; some may appear to have whitish contents. The pore diameters are clearly larger than the width of the rays, which provides a quick and reliable means of separating this wood from maple, whose pore diameter and ray width are about the same (compare the photos on p. 128 of yellow birch with the photos on p. 123 of sugar maple and red maple).

Birch can always be confirmed microscopically. The intervessel pitting is alternate, but the pits are extremely small and numerous, giving the intervessel wall the appearance of a pebbly surface. These pits may be connected by confluent apertures; that is, there may be fine lines running from one pit aperture to the next. Once this very small alternate

intervessel pitting is seen in one sample, it will be easily recognized in others. Intervessel pitting in birch is most easily found in tangential sections because most multiples are radial. Rays in birch are 1-5 seriate.

The second important microscopic feature is the scalariform perforation plates with up to 20 or more bars. These are best seen in a radial section, which also shows ray-vessel pitting. These pits, another easily recognized feature of birch, are very small, numerous and similar in appearance to the intervessel pits.

Although birches cannot be separated from one another with certainty on the basis of cell structure, it should be noted that pith flecks are usually numerous in paper birch, but sparse or absent from yellow birch (see the photo on p. 62).

It is also important to note that European birch (principally *B. pubescens* and *B. pendula*) cannot be distinguished from North American species.

EASTERN HOPHORNBEAM
Eastern hophornbeam *(Ostrya virginiana)* is sometimes confused with birch, and the lumber of these two trees is sometimes mixed together. Eastern hophornbeam is fairly dense, with an average specific gravity of 0.70, and has light sapwood. Its heartwood ranges from very light brown to reddish brown. On cross-sectional surfaces, the growth rings are inconspicuous and rays are indistinct. With a hand lens, rays are visible and tiny whitish spots are seen in the outer latewood along the growth boundary. Pores are both solitary and in multiples of two to several, commonly aggregated in short radial or diagonal chains or in flamelike groups. Banded parenchyma are faintly visible with a hand lens.

In the vessels, perforation plates, although usually simple, are occasionally scalariform with three to six bars. Intervessel pitting is alternate with medium-sized rounded to angular pits. Spiral thickenings are also present in the vessels. These spiral thickenings and larger intervessel pits provide a positive separation from birch. The rays of eastern hophornbeam are 1-3 seriate, compared to 1-5 in birch.

RAY CELLS IN BASSWOOD
Ray cells in basswood appear to be laterally compressed.

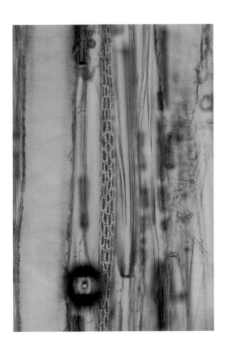

INTERVESSEL PITTING IN BIRCH
This portion of a vessel element in birch shows its extremely small intervessel pitting. Most pits are connected by confluent apertures.

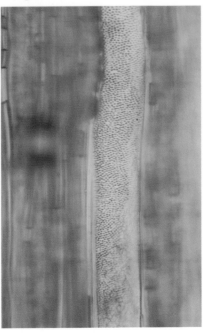

CALIFORNIA-LAUREL

California-laurel, also called Oregon-myrtle (*Umbellularia californica*), is a moderately dense diffuse-porous hardwood with an average specific gravity of 0.55. Its sapwood is near-white to light brown, and its heartwood is light brown to greyish brown, sometimes with darker streaks. The wood has a characteristic spicy odor that is quite pronounced when the wood is freshly cut but fades with time and exposure.

Growth rings are fairly distinct because of a narrow band of darker fiber at the growth-ring boundary. Rays are not easily seen with the unaided eye, but are fairly distinct with a hand lens. The small pores are solitary or in radial multiples of two to several. They are spaced well apart and are evenly distributed throughout the growth ring; some pore multiples extend

---

BIRCH, HOPHORNBEAM AND CALIFORNIA-LAUREL

Paper birch resembles other birches, but commonly has abundant pith flecks. Hophornbeam resembles birch at a glance, but examination with a hand lens reveals whitish specks along the outer latewood and pores that form into short radial or diagonal chains or flamelike aggregates. Its vessels have spiral thickenings and medium-large alternate intervessel pits. California-laurel is light brown to greyish brown, and is typically streaked with varying darker shades. It has a characteristic spicy odor. With a lens, the pores and pore multiples are seen to be encircled by a whitish sheath of parenchyma.

PAPER BIRCH
Betula papyrifera

Average Specific Gravity: 0.55
Heartwood Color: Light brown to dark or reddish brown (pith flecks usually conspicuous and numerous)
Pore Distribution: Diffuse-porous
Pores: Small to medium, solitary or in radial multiples of 2 to several; with lens, pore diameters clearly greater than ray width; some pores may appear to be filled with whitish substance
Rays: Distinct with lens on transverse surface

Rays: 1-5 seriate
Spiral Thickenings: Absent
Perforation Plates: Scalariform with many fine bars
Intervessel Pitting: Alternate, very small with confluent apertures

EASTERN HOPHORNBEAM
Ostrya virginiana

Average Specific Gravity: 0.70
Heartwood Color: Light brown to brown tinged with red
Pore Distribution: Diffuse-porous; growth rings usually distinct due to whitish spots along outer latewood
Pores: Small, solitary and in multiples commonly aggregated into radial, diagonal or flamelike groups; banded parenchyma faintly visible with hand lens
Rays: Distinct with lens on transverse surface

Rays: 1-3 seriate
Spiral Thickenings: Present
Perforation Plates: Mostly simple; occasionally scalariform, 3-6 bars
Intervessel Pitting: Alternate, medium to large

CALIFORNIA-LAUREL
Umbellularia californica

Average Specific Gravity: 0.55
Heartwood Color: Light brown to greyish brown with darker streaks
Odor: Characteristically spicy
Pore Distribution: Diffuse-porous
Pores: Small to medium, evenly distributed, solitary or in multiples of 2 to several, each pore or multiple surrounded by a lighter sheath of vasicentric parenchyma
Rays: Fine, clearly visible with hand lens

Rays: 1-4 (mostly 2) seriate
Oil Cells: Scattered oil cells common
Spiral Thickenings: Sporadically present
Perforation Plates: Simple
Intervessel Pitting: Alternate, medium to large

across the growth-ring boundary. As seen with a hand lens, each pore or pore multiple is surrounded by a lighter-colored layer of vasicentric parenchyma.

California laurel is confirmed microscopically by its simple perforation plates and alternate intervessel pitting with medium-sized, rounded to angular pits. Spiral thickenings are sporadic. Rays are heterocellular and 1-4 (mostly 2) seriate; they contain scattered oil cells.

Among the remaining woods in this section, none have rays that are visible to the naked eye; viewed with a hand lens they may just barely be visible, although not distinct.

WILLOW

Black willow (*Salix nigra*) is a moderately soft hardwood with an average specific gravity of 0.39. Its sapwood is very light in color, and its heartwood ranges from light brown to medium grey-brown, frequently showing streaks of variable color.

Willow commonly has interlocked grain. Although some samples appear diffuse-porous, the gradation in pore size across the growth ring in most samples suggests classification as a semi-diffuse-porous wood. The first-formed early-wood pores are barely visible to the eye, but grade to invisibly small toward the

---

WILLOW, COTTONWOOD AND ASPEN

Willows and poplars are moderately soft, diffuse-porous woods without distinct characteristic features. Rays are invisibly fine. In willows and cottonwoods, large pores in the earlywood may give the wood a semi-diffuse-porous appearance. Aspens have the smallest pores; their wood is typically creamy or whitish, compared to the greyish brown of cottonwood and the reddish brown of willow. Microscopically, all three woods have simple perforations, large alternate intervessel pitting and uniseriate rays that are homocellular in poplars, heterocellular in willow.

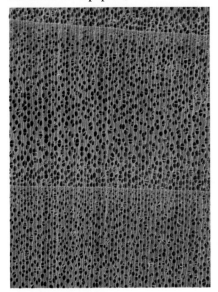

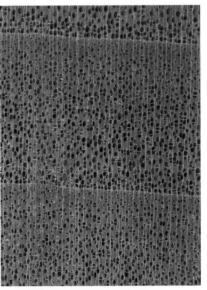

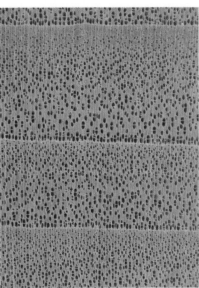

BLACK WILLOW
Salix nigra

Average Specific Gravity: 0.39
Heartwood Color: Light brown to pale reddish or greyish brown
Pore Distribution: Diffuse-porous to semi-diffuse-porous
Pores: Medium to small, usually with apparent size gradation from earlywood to latewood, solitary and in multiples of 2 to several
Rays: Very fine, barely visible with hand lens

Rays: Uniseriate, heterocellular
Spiral Thickenings: Absent
Perforation Plates: Simple
Intervessel Pitting: Alternate, medium to large

EASTERN COTTONWOOD
Populus deltoides

Average Specific Gravity: 0.40
Heartwood Color: Greyish to light greyish brown, sometimes with olive cast
Odor: Moist wood with foul odor
Pore Distribution: Diffuse-porous or semi-diffuse-porous
Pores: Medium to small, usually with apparent size gradation from earlywood to latewood, solitary and in radial multiples of 2 to several
Rays: Very fine, not easily seen with lens

Rays: Uniseriate, homocellular
Spiral Thickenings: Absent
Perforation Plates: Simple
Intervessel Pitting: Alternate, large

QUAKING ASPEN
Populus tremuloides

Average Specific Gravity: 0.38
Heartwood Color: Creamy white to light greyish brown
Pore Distribution: Diffuse-porous
Pores: Small to very small, with gradation in pore size from earlywood to latewood, solitary and in multiples of 2 to several
Rays: Very fine, not easily seen with hand lens

Rays: Uniseriate, homocellular
Spiral Thickenings: Absent
Perforation Plates: Simple
Intervessel Pitting: Alternate, large

outer latewood. The pores are solitary or in multiples that are mainly radial.

The rays of willow are very fine, invisible to the eye and just barely visible with a lens. A tangential section examined microscopically reveals that the rays are exclusively uniseriate. A radial view shows that the rays are heterocellular (see the photo below); the upright cells, which occur at the margins of the rays or within the body of the ray, have large, rounded, crowded pits connecting them to the vessel elements.

Willow also has simple perforations in

HETEROCELLULAR RAYS IN WILLOW

In this radial view of willow, the upright cells with large pits identify a heterocellular ray, which is important in distinguishing willow from poplars. (250x)

the vessels and alternate intervessel pitting with distinct medium-sized, rounded to angular pits.

In summary, *Salix* is a huge genus with dozens of species in North America and elsewhere, including white willow (*S. alba*), which is common in the British Isles. Wood specimens of the various willows cannot be reliably separated.

POPLARS

The genus *Populus* includes several common species (see the photos on p. 131), many of which we know as cottonwoods, aspens or, simply, poplars (see Table 10.4 below). The woods of all species of *Populus* are very similar. They can usually be recognized as belonging to either the cottonwood group or the aspen group, principally on the basis of the slightly larger size of the pores in cottonwoods, but also on such physical properties as color and odor. Even so, this separation cannot always be made with certainty.

The poplars are moderately soft to soft, with average specific gravities ranging from 0.33 to 0.40. The light-colored sapwood zone is indistinct. It merges gradually into heartwood that is typically light greyish brown in aspen. Cottonwood heartwood is greyish to light greyish brown, sometimes with an olive cast; it is darkest in swamp cottonwood. The wood of aspen may be lustrous, while cottonwood's is usually dull. When moist, cottonwood has a foul odor, which fortunately disappears as the wood dries.

On cross-sectional surfaces, growth rings are distinct due to narrow margins of slightly darker, denser fibers, although they are not conspicuous. Within a ring, the pores are larger, more numerous and crowded in the earlywood, grading to small in the outer latewood. This suggests a semi-diffuse-porous arrangement. A few pores are solitary, but most occur in multiples. The multiples are typically radial, but frequently also form tangential multiples or merge into small clusters.

The average size of the largest pores best serves to separate cottonwood from aspen. In aspen, all pores are invisibly small. In cottonwood, the largest pores are usually barely visible to the unaided eye, which makes them confusingly similar to willow's. It is worthwhile making a direct comparison to known samples of aspen and cottonwood to assess pore-size ranges. The rays of *Populus* are extremely fine; they are invisible to the unaided eye and barely visible even with a hand lens.

Microscopic anatomy is virtually the same among all the species, except for vessel (pore) diameter, which is larger among cottonwoods. Vessels have simple perforations and no spiral thickenings. Intervessel pitting is alternate, with large, clearly defined pits that are usually rounded or slightly angular due to crowding (see the top photo on the facing page). In tangential sections, rays are seen as uniseriate, and the cells appear to be laterally compressed and flattened.

In radial sections, the rays are found to

TABLE 10.4: SELECTED SPECIES OF *POPULUS* COMMON IN NORTH AMERICA AND EUROPE

	North America:		Europe:	
Cottonwoods/Poplars	*P. balsamifera*	Balsam poplar	*P. alba*	White poplar
	P. deltoides	Eastern cottonwood	*P. canescens*	Grey poplar
	P. heterophylla	Swamp cottonwood	*P. nigra*	Black poplar
	P. trichocarpa	Black cottonwood		
Aspens	*P. grandidentata*	Bigtooth aspen	*P. tremula*	European aspen
	P. tremuloides	Quaking aspen		

be homocellular with procumbent cells. Cells along the ray margin where they come into contact with vessels have large, crowded pits that look like bunches of grapes, virtually filling the cell. Because these pits are so distinctive, they are an easily recognized feature for identifying *Populus*. The homocellular composition of the rays, as seen in radial sections (photo below right), distinguishes *Populus* from *Salix* (willow), in which the rays are heterocellular (compare the photo at bottom right on this page with the photo on p. 132).

Aspen is sometimes confused with basswood, but the two can be distinguished quickly by comparing ray width, which is uniseriate in aspen and multiseriate in basswood, or spiral thickenings, which are absent in aspen and present in basswood.

BUCKEYE AND HORSECHESTNUT

Of the several species of buckeye found in the United States, two of the more common are yellow buckeye (*Aesculus octandra;* photo at left on p. 135) and Ohio buckeye (*A. glabra).* The wood is relatively soft, with an average specific gravity of 0.36, and has near-white sapwood and creamy to yellowish white heartwood. The growth rings are imperceptible in most samples, although they are faintly visible in some due to a thin light line of marginal parenchyma.

Rays in buckeye are invisibly fine to the naked eye and are barely visible even with a hand lens. The very small pores are not visible except with a hand lens. They are numerous and fairly evenly distributed, and are either solitary or in multiples of two to several. The multiples occur radially, tangentially and in small clusters. In short, there are few striking features in buckeye that can be seen with either the eye or the hand lens.

Microscopic examination reveals vessel elements that have simple perforations and alternate intervessel pitting with medium-large pits. Spiral thickenings are usually absent, but occasionally they appear in latewood vessels. The rays are either homocellular or heterocellular, but they are entirely uniseriate.

Woods of the two buckeye species are quite similar. However, yellow buckeye usually has visible ripple marks due to its storied rays and other longitudinal cells. This feature is seldom present in Ohio

buckeye. The wood of horsechestnut (*A. hippocastanum),* a native of Europe but widely planted in the United States, closely resembles buckeye. But spiral thickenings in horsechestnut are regularly present in the vessel elements, and the wood is denser, with an average specific gravity of 0.45. Buckeye is probably most often confused with aspen. When present, ripple marks or spiral thickenings indicate *Aesculus.* Also, the ray-vessel pitting, as seen microscopically in a radial section, is smaller in buckeye than in aspen.

SWEETGUM

Sweetgum *(Liquidambar styraciflua;* photo at center on p. 135) is a moderately dense diffuse-porous hardwood with an average specific gravity of 0.52. Its alternate common name, redgum, derives from the heartwood, which is often reddish brown, frequently with streaks of varying shades. Wood with these streaks is marketed as "figured redgum" (see the photo on p. 63). The heartwood is also found in varying shades of grey and greyish brown. The sapwood is near-white, sometimes with a pinkish cast, and is marketed separately

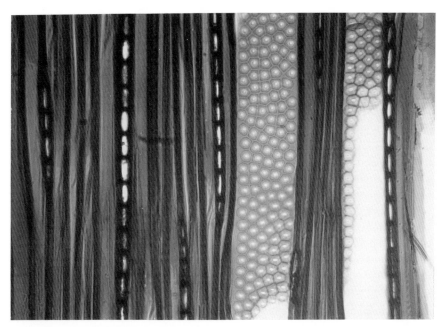

INTERVESSEL PITTING IN *POPULUS*
A tangential view of poplar showing the large alternate intervessel pitting and uniseriate rays. (250x)

RAYS IN *POPULUS*
As seen in radial view, ray cells in poplars are exclusively procumbent; therefore, the rays are homocellular. The ray-vessel pitting is large and conspicuous, especially in marginal rows of ray cells. (250x)

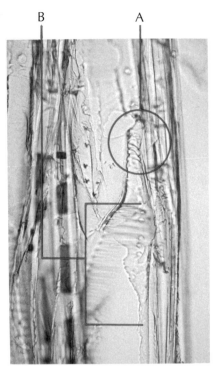

LIGULES IN SWEETGUM
A ligule with spiral thickenings (A) at the end of a vessel element in sweetgum. The scalariform perforation plate (B) is also visible. (250x)

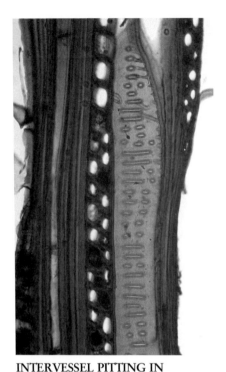

INTERVESSEL PITTING IN SWEETGUM
In sweetgum, the intervessel pitting commonly intergrades from scalariform to opposite, with up to three pits across the vessel. (250x)

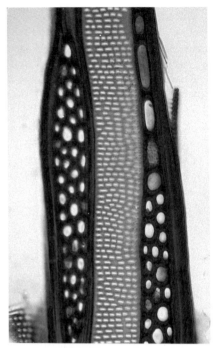

INTERVESSEL PITTING IN BLACK TUPELO
In black tupelo, intervessel pitting is opposite, with up to 6 or more pits across vessels. It only occasionally intergrades to scalariform. (250x)

as "sap-gum." The wood commonly has interlocked grain.

Neither the growth rings nor the rays are distinct to the eye, and neither is always clearly visible with a hand lens. The pores, which are extremely small and numerous, are solitary, in multiples, in short chains or in small irregular clusters. Some pores give the appearance of being channeled into a radial arrangement by pairs of closely spaced rays.

Sweetgum is much easier to identify when its sections are examined microscopically. The vessels have scalariform perforations with up to 20 or more bars. Intervessel pitting is opposite, up to three pits across, and regularly intergrades with scalariform pitting (see the photo at center above).

Spiral thickenings are present only in the ligules, the tail-like extensions of the vessel elements. Although of value in identification, spiral thickenings in the ligules are often difficult to find and see clearly (see the photo at left above). The

rays of sweetgum are heterocellular and 2-3 (mainly 2) seriate. In radial sections, the elongated thin-walled procumbent cells contrast clearly with the squarish, very thick-walled upright cells.

BLACK TUPELO

Black tupelo (*Nyssa sylvatica;* photo at far right on the facing page), also known as blackgum, is moderately dense, with an average specific gravity of 0.50. Its sapwood is pale grey, and its heartwood is a darker grey with a green or brown cast. The wood usually has interlocked grain.

In cross sections, neither growth rings nor rays are visible; they are indistinct even with a hand lens. Pores are very small and numerous. They are evenly distributed and are solitary, in radial multiples or in small clusters.

Microscopic examination shows scalariform perforation plates with up to 20 or more very thin bars. Intervessel pitting is opposite, with up to six pits per

row (see the photo above). The opposite pitting occasionally intergrades with scalariform pitting.

Spiral thickenings are present in the ligules at the ends of vessel elements. The rays are heterocellular; they range from 1-4 (mostly 2-3) seriate.

BUCKEYE, SWEETGUM AND TUPELO

Buckeye, sweetgum and tupelo are uniformly diffuse-porous with extremely fine texture. Growth rings and rays may not be distinct even with a hand lens. Confident identification depends on microscopic analysis. Buckeye has uniseriate rays and simple perforation plates. Sweetgum and tupelo have scalariform perforation plates; both also have scalariform intervessel pitting intergrading with opposite pitting. But opposite pitting is only up to 3 across in sweetgum, and up to 6 or more across in black tupelo.

YELLOW BUCKEYE
Aesculus octandra

Average Specific Gravity: 0.36
Heartwood Color: Creamy white to
 pale yellowish white
Pore Distribution: Diffuse-porous
Pores: Very small, solitary in multiples
 and small clusters
Ripple Marks: On tangential surface
 due to storied ray structure

Rays: Not clearly seen even with hand
 lens; uniseriate, heterocellular and
 homocellular; storied
Spiral Thickenings: Absent or sporadic
Perforation Plates: Simple
Intervessel Pitting: Alternate, medium
 to large

SWEETGUM
Liquidambar styraciflua

Average Specific Gravity: 0.52
Heartwood Color: Grey or reddish
 brown, sometimes with variegated
 pigment
Pore Distribution: Diffuse-porous
Pores: Very small, numerous, solitary
 and in multiples and small clusters,
 often in intermittent radial chains
Rays: Very fine, not distinct even with
 hand lens

Rays: 2-3 (mostly 2) seriate,
 heterocellular
Spiral Thickenings: Restricted to ligules
Perforation Plates: Scalariform with up
 to 20+ bars
Intervessel Pitting: Opposite
 intergrading to scalariform; 1-3 pits
 across vessel

BLACK TUPELO
Nyssa sylvatica

Average Specific Gravity: 0.50
Heartwood Color: Medium grey or
 grey with green or brown cast
 (wood usually has interlocked grain)
Pore Distribution: Diffuse-porous
Pores: Very small, numerous, solitary
 and in multiples and small clusters
Rays: Barely visible even with hand lens

Rays: 1-4 seriate, heterocellular
Spiral Thickenings: Sometimes present
 in ligules
Perforation Plates: Scalariform with up
 to 20+ bars
Intervessel Pitting: Opposite,
 occasionally intergrading to
 scalariform; up to 6-8 pits across
 vessel

TABLE 10.5: SUMMARY CHECKLIST OF FEATURES FOR RING-POROUS HARDWOODS IN SUBGROUP I-3

NAME	AVG. SPEC. GRAV.	HEARTWOOD COLOR		LATEWOOD (LW) CHARACTERISTICS
WHITE ASH (*Fraxinus* spp.)	0.56-0.61	Light brown or greyish brown (sapwood creamy white; may be very wide).	LW PORES THICK-WALLED	LW pores solitary and in radial multiples of 2-3, surrounded by vasicentric parenchyma or confluent parenchyma connecting pores into tangential lines, especially in outer LW.
BLACK ASH (*Fraxinus nigra*)	0.49	Greyish brown to medium or dark brown (sapwood creamy white; fairly narrow).		LW pores solitary and in radial multiples of 2-3, not numerous, rarely with aliform or confluent parenchyma in outer LW.
SASSAFRAS (*Sassafras albidum*)	0.45	Grey, greyish brown or brown.		LW pores solitary and in radial multiples of 2-3, grading to very small with pronounced aliform parenchyma in outer LW.
CATALPA (*Catalpa* spp.)	0.41	Greyish brown, sometimes with a faint lavender cast.	SPIRAL THICKENINGS COMMONLY PRESENT IN SMALLER LW VESSELS	LW pores variable, larger near earlywood, grading down to very small in outer LW, numerous in nests.
RED MULBERRY (*Morus rubra*)	0.66	Orange-yellow to golden or russet brown (turning deep russet on exposure to atmosphere).		LW pores in nestlike groups, merging laterally to form wavy or interrupted bands, especially in outer LW.
OSAGE-ORANGE (*Maclura pomifera*)	0.81	Golden yellow to orange or orange-brown (darkening on exposure).		LW pores small, in clusters, often merging or connected by confluent parenchyma into wavy or concentric bands in the outer LW.
BLACK LOCUST (*Robinia pseudoacacia*)	0.69	Olive or yellow-brown to dark golden yellow-brown (turning dark russet brown on exposure).		LW pores in nestlike groups, which merge into interrupted or somewhat continuous bands in the outer latewood.
KENTUCKY COFFEETREE (*Gymnocladus dioicus*)	0.60	Light red to reddish or orange-brown.		LW pores nested and sometimes merging into short irregular tangential bands in the outer LW.
HONEYLOCUST (*Gleditsia triacanthos*)	0.66	Light red to reddish brown or orange-brown.		LW pores solitary in radial multiples, and, especially in outer latewood, in nestlike groups connected into bands by confluent parenchyma.
STAGHORN SUMAC (*Rhus typhina*)	0.47	Olive green to greenish yellow, often with orange or russet brown in earlywood (R, T), variable color in latewood (X).		LW pores solitary or in radial multiples, but merging into interrupted or continuous bands in the outer LW.

EARLYWOOD (EW) CHARACTERISTICS	VESSEL OCCLUSIONS	RAY CHARACTERISTICS	REMARKS AND MISCELLANEOUS
EW 2-4 pores wide, pores surrounded by lighter tissue. Pores 150-250μm.	Tyloses fairly numerous (some vessels open).	Rays not distinct, barely visible to eye, but clearly visible with lens. 1-3 (mostly 1-2) seriate.	
EW 2-4 pores wide, pores surrounded by lighter tissue. Pores 200-300μm, easily visible to eye.	Tyloses fairly numerous (some vessels open).	Rays not distinct, barely visible to eye, but clearly visible with lens. 1-3 seriate.	Trees commonly slowly grown, growth rings narrow.
EW 3-8 pores wide. Pores 150-200μm.	Tyloses fairly abundant, with sparkle.	Rays barely visible to eye. 1-4 seriate.	Perforation plates occasionally scalariform in latewood. Rays contain scattered oil cells, appearing very large with bright yellow contents. Spicy odor and taste.
EW 3-8 pores wide. Pores 200-300μm.	Tyloses variably abundant, with sparkle.	Rays indistinct to eye but clearly visible with lens. 1-6 (mostly 2-3) seriate.	Faint but unique pungent/spicy odor. May be semi-ring-porous. Sapwood never more than 2 growth rings wide.
EW 2-8 pores wide. Pores 150-250μm.	Tyloses variably present in EW pores. Pores may also contain white deposits.	Rays plainly visible to eye. 1-8 (mostly 5-7) seriate.	Radial surfaces show conspicuous ray fleck.
EW 2-3 pores wide. Pores 150-250μm.	Tyloses abundant, packing vessels and making adjacent vessels indistinct.	Rays barely visible to eye. 1-6 (mostly 2-4) seriate.	Contains water-soluble dye. Shavings when soaked turn warm water yellow.
EW 2-3 pores wide. Pores 200-300μm.	Tyloses abundant, solidly packing vessels, with overall yellow cast, and with sparkle.	Rays barely visible to eye, but distinct with lens. 1-7 (mostly 3-5) seriate.	Earlywood appears as bright yellow band, contrasting with dark amber brown latewood. Sapwood never more than 2-3 rings wide. Bright yellow fluorescence under UV light.
EW 3-6 pores wide. Pores 250-300μm.	Tyloses absent. Reddish gum deposits only rarely present.	Rays uniformly fine, barely visible to eye. 1-7 (mostly 3-5) seriate. Tallest <1200μm high.	Deep yellow, bright fluorescence under UV light.
EW3-5+ pores wide. Pores 200-300μm.	Tyloses absent. Regularly contain reddish gum, especially evident on longitudinal surfaces.	Rays variable in size, largest conspicuous to eye. 1-14 (mostly 6-9) seriate. Tallest >1200μm high.	Bright yellow fluorescence under UV light.
EW 2-6 pores wide. Pores 150-250μm.	Tyloses variably abundant, with iridescent sparkle.	Rays faintly visible to eye, not conspicuous with lens. 1-3 (mostly 1-2) seriate.	Variable yellow to bluish fluorescence under UV light.

TABLE 10.6: SUMMARY CHECKLIST OF FEATURES FOR DIFFUSE-POROUS HARDWOODS IN SUBGROUP III-3

NAME	AVG. SPEC. GRAV.	HEARTWOOD COLOR	VESSEL FEATURES		
			PORE SIZE	PORE DISTRIBUTION	PERFORATION PLATES
HARD MAPLE SOFT MAPLE (*Acer* spp.)	0.46-0.63	Creamy white to light reddish brown. (Soft maple often has greyish cast or streaks.)	Small (50-100µm).	Solitary and in radial multiples of 2 to several.	Simple.
DOGWOOD (*Cornus* spp.)	0.73	Dark brown (sapwood creamy with flesh or pinkish cast).	Very small (<100µm).	Mostly solitary and occasionally in radial multiples of 2 to several.	Scalariform (with 20+ fine bars).
HOLLY (*Ilex opaca*, et al)	0.56	White, with grey or blue-grey cast or streaks.	Very small (50-75µm).	Mostly in multiples further grouped into chains, sometimes extending across growth-ring boundaries.	Scalariform (with 15+ bars).
CHERRY (*Prunus serotina*, et al)	0.50	Light to dark cinnamon or reddish brown.	Small (50-100µm).	Solitary and in radial multiples and small clusters of 2 to several (larger in a row at earlywood edge of ring).	Simple.
YELLOW-POPLAR (*Liriodendron tulipifera*)	0.42	Green, or yellow to tan with greenish cast (sapwood creamy-white).	Small-medium small (100-150µm).	Solitary and in multiples and small clusters of 2 to several.	Scalariform (with 2-16 stout bars).
CUCUMBER-TREE, MAGNOLIA (*Magnolia* spp.)	0.49	Yellow, or medium to dark greenish yellow, or dark green.	Small (50-100µm)	Solitary and in multiples and small clusters of 2 to several.	See remarks.
BASSWOOD (*Tilia americana*, et al)	0.37	Creamy white to light brown	Small (<100µm)	Numerous, solitary and in radial multiples and small clusters.	Simple.
BIRCH (*Betula* spp.)	0.55-0.65	Light brown to dark brown or reddish brown.	Small-medium (100-200µm)	Solitary and in radial multiples of 2 to several.	Scalariform (with 5-20+ bars).
EASTERN HOPHORN-BEAM (*Ostrya virginiana*)	0.70	Light brown to brown tinged with red.	Small (50-100µm)	Solitary and multiples of 2 to several, sometimes aggregated into irregular radial flamelike groups.	Simple. Occasional scalariform (with 3-6 bars).
CALIFORNIA-LAUREL (*Umbellularia californica*)	0.55	Light brown to greyish brown with darker streaks.	Small-medium (100-200µm)	Solitary, and in multiples of 2 to several, each pore or multiple surrounded by vasicentric parenchyma.	Simple.

(continued on pp. 140-141)

| INTERVESSEL PITTING | SPIRAL THICKENINGS | RAY FEATURES | | REMARKS AND MISCELLANEOUS |
		HOMOGENOUS OR HETEROGENEOUS	SERIATION	
Alternate, rounded to angular (5-10μm).	Present (fine, evenly and closely spaced).	Homogeneous.	Hard maple: 1-8 (mostly 5-7). Soft maple: 1-5 seriate.	Vessels may have gum deposits.
Alternate to opposite/oval, intergrading with scalariform.	Absent.	Heterogeneous and homogeneous.	2-8 seriate. Uniseriate (upright).	Uniseriate margins of multiseriate rays with up to several rows ("tails," T-view) of upright cells.
Opposite (sometimes linear), in low spiral rows (4-7μm).	Present.	Heterogeneous and homogeneous.	3-5 (mostly 5) seriate. Uniseriate (upright).	Spiral thickenings in fibers also.
Alternate (7-10μm).	Present (variable in thickness).	Heterogeneous and homogeneous.	1-6 (mostly 3-4) seriate.	Vessels with orange to ruby-red contents. Characteristic gum defects common.
Opposite (oval to rectangular).	Absent.	Heterogeneous and homogeneous.	1-5 (mostly 2-3) seriate. Noded.	Rings delineated by light line of marginal parenchyma.
Scalariform.	See remarks.	Heterogeneous and homogeneous.	1-5 seriate. Noded. See remarks.	Cucumber: Simple perforations, spirals absent, rays mostly 1-2 seriate. Magnolia: Mostly scalariform, 6-10 stout bars, usually with spirals. Line of marginal parenchyma distinct.
Alternate, usually angular/crowded (5-8μm).	Present (thick).	Homogeneous.	1-6 seriate. Noded.	Faint but characteristic musty odor. (T) Ray cells appear laterally compressed.
Alternate, usually with confluent apertures (2-4μm).	Absent.	Homogeneous.	1-5 seriate.	(X) Some pores (with lens) appear filled with whitish substance.
Alternate, round/oval to angular (6-10 μm).	Present.	Heterogeneous and homogeneous.	1-3 seriate.	(X) Often growth rings distinct by whitish spots along outer latewood. Banded parenchyma faintly visible with lens.
Alternate, round/angular (6-10μm).	Occasionally present.	Heterogeneous.	1-4 (mostly 2) seriate.	Spicy odor characteristic. (R) Oil cells present in rays. Fibers thin-walled, commonly septate.

TABLE 10.6 (CONTINUED):
SUMMARY CHECKLIST OF FEATURES FOR DIFFUSE-POROUS HARDWOODS IN SUBGROUP III-3

NAME	AVG. SPEC. GRAV.	HEARTWOOD COLOR	VESSEL FEATURES		
			PORE SIZE	PORE DISTRIBUTION	PERFORATION PLATES
WILLOW (*Salix* spp.)	0.39	Light brown to pale reddish or greyish brown.	Small-medium (50-150µm).	Numerous, solitary, and in multiples of 2 to several.	Simple.
ASPEN, POPLAR, COTTONWOOD (*Populus* spp.)	0.33-0.40	Aspen: creamy white to light greyish brown. Poplar/Cottonwood: greyish to light greyish brown.	Very small-medium. Aspen: 50-100µm Poplar/cottonwood: 50-150µm.	Solitary and in multiples (mostly radial) of 2 to several.	Simple.
BUCKEYE, HORSE-CHESTNUT (*Aesculus* spp.)	0.36-0.45	Creamy white to pale yellow white.	Very small (50-100µm).	Numerous, solitary, in multiples of 2 to several, and in small clusters.	Simple.
SWEETGUM (*Liquidambar styraciflua*)	0.52	Grey or reddish brown (sometimes with variegated pigment), sapwood light, often blue-stained.	Very small (50-100µm).	Numerous, solitary, in multiples of 2 to several (often in short intermittent chains and in small clusters.)	Scalariform (with 10-20+ thin bars).
BLACK TUPELO (*Nyssa sylvatica*)	0.50	Medium grey or grey with green or brown cast.	Very small (50-100µm).	Numerous, solitary, in radial multiples of 2 to several, and in small clusters.	Scalariform (with 20+ thin bars).

| INTERVESSEL PITTING | SPIRAL THICKENING | RAY FEATURES | SERIATION | REMARKS AND MISCELLANEOUS |
		HOMOGENOUS OR HETEROGENEOUS		
Alternate, round/oval to angular (6-10μm).	Absent.	Heterogeneous.	Uniseriate. Large pits in upright ray cells to vessels.	Wood often semi-diffuse-porous. Wood may have interlocked grain.
Alternate, round/oval to angular (8-12μm).	Absent.	Homogeneous.	Uniseriate. Large pits in marginal ray cells to vessels.	Aspen: Creamy white, no odor. Poplar/cottonwood: Foul odor when wet, never creamy white. Sometimes appears semi-diffuse-porous.
Alternate, round/oval to angular (5-10μm).	See remarks.	Heterogeneous and homogeneous.	Uniseriate.	Spiral thickenings present in vessels of horsechestnut; absent or sporadic in buckeye. Yellow buckeye has ripple marks, horsechestnut sometimes does.
Opposite (up to 3 across) intergrading with scalariform.	See remarks.	Heterogeneous and homogeneous.	2-3 (mostly 2) seriate. Uniseriate (upright).	Spiral thickenings in ligules of vessel elements only. Wood may have interlocked grain.
Opposite (up to 8 across) occasionally intergrading with scalariform.	See remarks.	Heterogeneous.	1-4 seriate.	Spiral thickenings, when present, in ligules of vessel elements only. Wood usually has interlocked grain.

CHAPTER ELEVEN

SOFTWOOD IDENTIFICATION

This chapter is devoted to the identification of the more common North American softwoods. Because the physical and anatomical characteristics of softwoods are quite different from those of hardwoods, the chapter is organized very differently from the previous one. For example, there are several choices as to how to proceed, depending on what is already known or what is yet to be determined about an unidentified wood. It will become clear that softwood identification involves greater reliance upon the microscope. Again, it is recommended that you read the chapter at least once to get a sense of its organization before attempting to identify an unknown softwood.

Initial Screening

Upon determining that an unknown wood is a softwood, note all the anatomical features that are visible with the unaided eye or hand lens and place the unknown softwood in one of the three major groups (see the photos on the facing page).

The automatic next question should be whether or not the sample has resin canals. About half the conifers do; the remainder do not.

Woods with resin canals can be separated into two groups. Those with large, numerous, evenly distributed resin canals form what we will call Group I, the pines. Those with small, less numerous, erratically distributed resin canals form Group II, the Douglas-fir/spruce/larch group. Woods that lack resin canals entirely form Group III: fir, hemlock, cedars, baldcypress,

sequoia and yew. Group III is tentatively separated into Group III-1 or 2 and Group III-3 or 4 on the basis of heartwood color, but this tentative initial breakdown is subject to revision, as some Type-2 cedars have nondescript heartwood and others have distinction coloration (see the chart on p. 144).

In most pines, the larger-sized resin canals can be detected immediately by eye on smoothly machined surfaces, either as dark lines on longitudinal surfaces or as tiny dark dots or specks on the end grain. In either case it is advisable to clean an area of end grain with a razor blade for hand-lens examination. This resolves any lingering doubt as to whether resin canals are present. If they are present, their size and distribution can be evaluated. It is also possible to assess such other features as ring width, percentage of earlywood and latewood, evenness of grain, transition from earlywood to latewood (abrupt or gradual) and sometimes the presence/abundance of parenchyma.

However, even when longitudinal resin canals are not visible on a transverse surface — possibly because the sample is too small or because the resin canals are too sparse — their presence can be confirmed by locating fusiform rays, which contain horizontal resin canals. On transverse surfaces, fusiform rays are conspicuously larger than the narrower rays where the plane of cut dissects a resin canal (see the photo on p. 23).

Where only a small sliver is available for examination or where a suitable end-grain surface cannot be exposed, an alternative or supplementary approach can be taken: direct micro-

INITIAL SCREENING OF SOFTWOODS BASED ON RESIN CANALS
An unknown softwood can be assigned initially to one of three groups based on whether resin canals are large, small or absent. (15x)

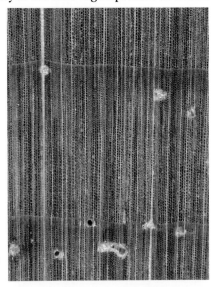
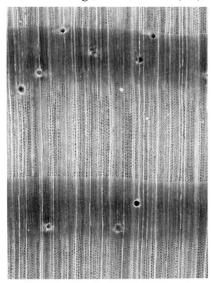

RESIN CANALS LARGE

I. PINES
(turn to p. 144)
Microscopic checks:
1. Fusiform rays present; thin-walled epithelial cells.
2. Cross-field pits windowlike or pinoid.
3. Ray tracheids always present, may be dentate.

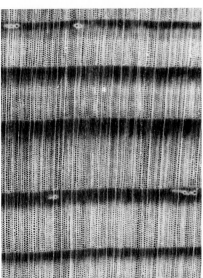

RESIN CANALS SMALL

II. DOUGLAS-FIR, SPRUCE, LARCH
(turn to p. 150)
Microscopic checks:
1. Fusiform rays present; thick-walled epithelial cells.
2. Cross-field pits piceoid.
3. Ray tracheids always present, never dentate.

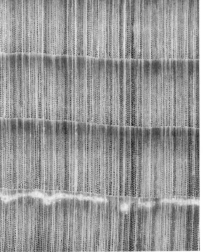
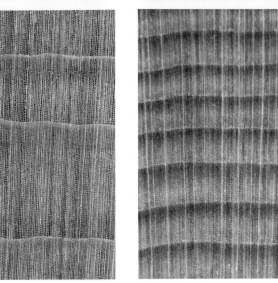

RESIN CANALS ABSENT

III. HEMLOCK, TRUE FIR, CEDARS, BALDCYPRESS, REDWOOD, YEW
(turn to p. 154)
Microscopic checks:
1. Fusiform rays absent.
2. Cross-field pits cupressoid or taxodioid.
3. Ray tracheids may be absent.

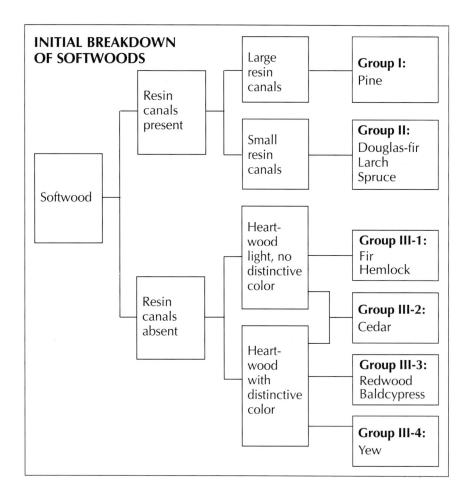

INITIAL BREAKDOWN OF SOFTWOODS

- Softwood
  - Resin canals present
    - Large resin canals → **Group I:** Pine
    - Small resin canals → **Group II:** Douglas-fir Larch Spruce
  - Resin canals absent
    - Heartwood light, no distinctive color
      - **Group III-1:** Fir Hemlock
      - **Group III-2:** Cedar
    - Heartwood with distinctive color
      - **Group III-3:** Redwood Baldcypress
      - **Group III-4:** Yew

scopic examination of longitudinal sections. In a tangential section, the occurrence of fusiform rays (see the photos on p. 23) indicates a species that has resin canals. If the epithelial cells are thin-walled (see the photos on p. 20), the wood is a pine in Group I; if the epithelial cells are thick-walled, the wood belongs to Group II, the Douglas-fir/spruce/larch group. If fusiform rays are not found, the wood belongs to Group III.

If only a radial section is available, the initial classification can be deduced by microscopic examination of ray cells:

1. If ray tracheids are absent,
the wood does not have resin canals and is in Group III. The cross-field pitting will be either taxodioid or cupressoid. If, in addition, longitudinal parenchyma are present in the body of the growth ring, the wood is in Group III-3 or 4. (Note: If you find spiral thickenings, proceed directly to yew.)

2. If ray tracheids are present,
A) and they are dentate, the wood is a hard pine in Group I (p. 147).

B) and they are smooth-walled
(i) and the cross-field pitting is windowlike, the wood is a soft pine in Group I (p. 145).

(ii) and the cross-field pitting is piceoid, the wood is in the Douglas-fir/spruce/larch group (Group II), but it could possibly also be hemlock in Group III-1. (Note: If spiral thickenings are present, go directly to Douglas-fir in Group II on p. 150.)

(iii) and the cross-field pitting is cupressoid, go to hemlock in Group III-1 (p. 156), or Alaska-cedar in Group III-3 (p. 160).

Radial or tangential sections usually provide all the information necessary for initial classification. Transverse sections can also be examined, but they are more difficult to prepare.

GROUP I: THE PINES

The pines, which comprise the genus *Pinus* in the family *Pinaceae*, include a very large and diverse group of species. They typically have large-diameter resin canals whose epithelial cells are thin-walled, in contrast to the smaller resin canals and thick-walled epithelial cells of the Douglas-fir/spruce/larch group (see the photos on p. 20).

The pines can be separated into two major subgroups, soft pines and hard pines. With some abnormal exceptions, such as slow growth rate and reaction wood, the qualities of the wood of these two groups can be summarized as follows:

SOFT PINES
Medium-low density
Fairly soft
Even-grained
Gradual earlywood/latewood transition
Nondentate ray tracheids
Windowlike cross-field pits

HARD PINES
Medium to high density
Moderately hard to hard
Uneven-grained
Abrupt earlywood/latewood transition
Dentate ray tracheids
Windowlike or pinoid cross-field pits

Subgroup I-1: The Soft Pines

Trees of this subgroup can be recognized by the fact their needles are borne along the twigs in groups of five. The three principal soft pine species are eastern white pine (*P. strobus*), western white pine (*P. monticola*) and sugar pine (*P. lambertiana*). Because of the general similarities among these woods, individual specimens cannot always be identified with certainty as to species. Table 11.1 on p. 146 summarizes and compares some of the major characteristics of these three soft pines.

As indicated, eastern white pine generally has the finest texture and the smallest resin canals. Sugar pine has the coarsest texture and largest resin canals. In western white pine they are intermediate in size. The resin canals in sugar pine are usually quite conspicuous, often appearing as prominent dark streaks along longitudinal surfaces. These streaks can still be seen in older samples, because the wood never ages to the dark shades that white pines commonly do.

The horizontal resin canals of sugar pine are the largest among the three species. These transverse canals in fusiform rays often show as distinct dark specks on tangential surfaces. However, the occasional piece of sugar pine may have surprisingly inconspicuous resin canals, just as the occasional piece of eastern white pine will display prominent resin canals.

It should be noted, too, that although the wood of all three soft pines is generally even-grained, reaction wood drastically changes the wood's appearance and usually produces an uneven grain due to the dense latewood formation.

Although texture and resin-canal size will usually provide a basis for separating a typical piece of eastern white pine from

THE THREE PRINCIPAL SOFT PINES

Woods in the soft pine group have large resin canals that are mostly solitary, numerous and evenly distributed. Woods are fairly even-grained, with a gradual transition from earlywood to latewood.

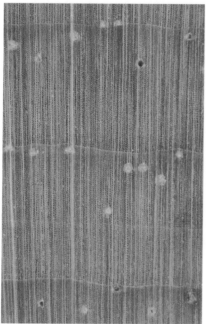 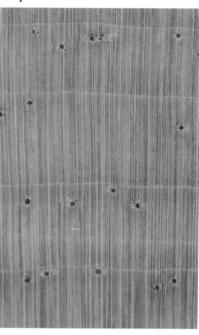 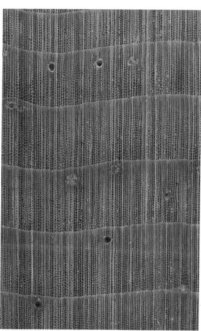

EASTERN WHITE PINE
P. strobus

*Average Specific Gravity: 0.35
Odor: Pleasant, piney
Heartwood: Distinct, darkening with
  age
Grain Appearance: Fairly even
Earlywood/Latewood Transition:
  Gradual
Resin Canals: large, numerous, mostly
  solitary, evenly distributed

**Cross-Field Pitting: Windowlike
Ray Tracheids: Smooth-walled
Epithelial Cells : Thin-walled
Longitudinal Parenchyma: Absent

WESTERN WHITE PINE
P. monticola

Average Specific Gravity: 0.38
Odor: Pleasant, piney
Heartwood: Distinct, darkening with
  age
Grain Appearance: Fairly even
Earlywood/Latewood Transition:
  Gradual
Resin Canals: Large, numerous, mostly
  solitary, evenly distributed

Cross-Field Pitting: Windowlike
Ray Tracheids: Smooth-walled
Epithelial Cells: Thin-walled
Longitudinal Parenchyma: Absent

SUGAR PINE
P. lambertiana

Average Specific Gravity: 0.36
Odor: Faint but distinct; resinous
Heartwood: Does not darken with age
Grain Appearance: Even
Earlywood/Latewood Transition:
  Gradual
Resin Canals: Very large, numerous,
  mostly solitary, evenly distributed

Cross-Field Pitting: Windowlike
Ray Tracheids: Smooth-walled
Epithelial Cells : Thin-walled
Longitudinal Parenchyma: Absent

* Throughout the chapter, macroscopic features are listed in the text block immediately under the species name.
** Microscopic features are below the macroscopic and separated from them by a line.

TABLE 11.1: COMPARATIVE RESIN-CANAL AND TRACHEID MEASUREMENTS OF THE THREE PRINCIPAL SPECIES OF THE SOFT PINE GROUP

		Eastern white pine (P. strobus)	Western white pine (P. monticola)	Sugar pine (P. lambertiana)
Resin canal diameter	Typical range	90-120 µm	135–150 µm	175-225 µm
	Maximum	150µm	200 µm	300 µm
Texture		Medium	Medium to medium-coarse	Medium to medium-coarse
Tracheid diameter	Typical range	25-35 µm	35-45 µm	40-50 µm
	Maximum	45 µm	60 µm	65 µm

a typical piece of sugar pine, the structure of either wood can intergrade with western white pine; microscopic anatomy provides a more reliable means of separation. The cross-field pitting of all three soft pines is classified as windowlike, but there is sufficient variation in this pitting to provide a basis to attempt separation of all three woods (see the photos below).

Examine cross fields in several early-wood areas to arrive at a general or average cross-field appearance; never make a judgment based solely on a small portion of a single ray. Note the number of windowlike pits, as well as the shape and percentage of the cross field occupied by the pits.

In eastern white pine, there are generally large, single pits that nearly fill the

WINDOWLIKE CROSS-FIELD PITTING IN SOFT PINES
Typical appearance of cross-field pitting in eastern white pine (left), western white pine (center) and sugar pine (right). (200x)

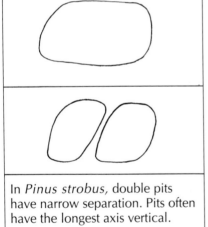

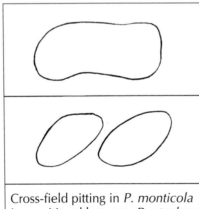

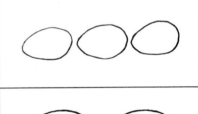

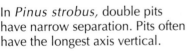
In *Pinus strobus*, double pits have narrow separation. Pits often have the longest axis vertical.

Cross-field pitting in *P. monticola* is transitional between *P. strobus* and *P. lambertiana*.

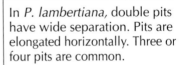
In *P. lambertiana,* double pits have wide separation. Pits are elongated horizontally. Three or four pits are common.

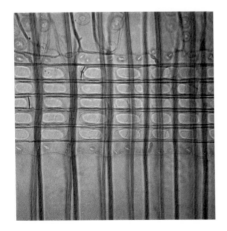

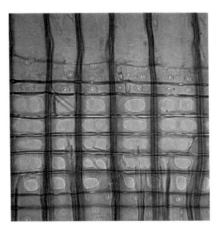

cross field. Where double pits are present, only a narrow strip of cell wall remains to separate them, and each of the paired pits is elongated slightly in the vertical direction or has about equal vertical and horizontal dimensions. More than two pits per cross field are not common. In sugar pine, the cross fields appear as longer rectangles, and multiple pits are more common. Where double pits occur, the individual pits are commonly lemon-shaped or oval, with the long axis horizontal and a conspicuously wide band of cell wall separating the pair. Cross fields with three pits are commonly found, and occasionally four or five pits are seen.

Western white pine is more or less intermediate. Its paired pitting is more common than in eastern white pine and more angular, less oval and more crowded than in sugar pine. However, eastern and western white pines are sometimes quite similar.

Two additional species of soft pine found in western North America are limber pine (*P. flexilis*) and whitebark pine (*P. albicaulis*). Whitebark pine can usually be distinguished visually from the other soft pines by the dimples on tangential surfaces; limber pine can be identified microscopically by its resin-canal crystals.

Subgroup I-2: The Hard Pines

This group includes trees whose needles are generally borne along their twigs in groups of two or three. Because the woods of the many species in this large group cannot be separated reliably, it is appropriate to consider them in the following groups:

THE SOUTHERN YELLOW PINES
loblolly pine (*P. taeda*)
shortleaf pine (*P. echinata*)
longleaf pine (*P. palustris*)
slash pine (*P. elliottii*)
pitch pine (*P. rigida*)

HARD PINES

Woods in the hard pine group have numerous resin canals that are usually large (the smallest are in red pine); they also show an abrupt transition from earlywood to latewood. Of the principal hard pines, southern yellow pines are typically the densest and most uneven-grained; ponderosa pine is usually most even-grained, with narrow growth rings. Both intergrade with red pine in appearance.

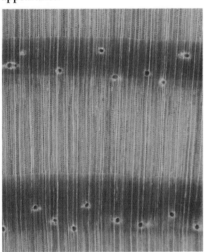

SOUTHERN YELLOW PINE
Pinus spp.

Average Specific Gravity: 0.51-0.61
Odor: Sharp, "pitchy" pine odor
Heartwood: Distinct
Grain Appearance: Uneven
Earlywood/Latewood Transition:
    Abrupt
Resin Canals: Large, numerous, mostly
    solitary, evenly distributed

Cross-Field Pitting: Pinoid
Ray Tracheids: Dentate
Epithelial Cells: Thin-walled
Longitudinal Parenchyma: Absent

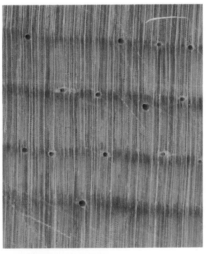

PONDEROSA PINE
P. ponderosa

Average Specific Gravity: 0.40
Odor: Mild, piney
Heartwood: Distinct, often reddish
Grain Appearance: Fairly even,
    especially if growth rings narrow
Earlywood/Latewood Transition:
    Fairly abrupt, latewood narrow
Resin Canals: Numerous, medium,
    mostly solitary, evenly distributed

Cross-Field Pitting: Pinoid
Ray Tracheids: Dentate
Epithelial Cells: Thin-walled
Longitudinal Parenchyma: Absent

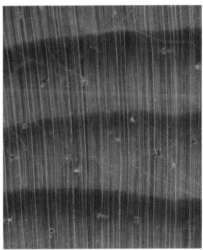

RED PINE
P. resinosa

Average Specific Gravity: 0.44
Odor: Harsh, piney
Heartwood: Distinct
Grain Appearance: Moderately uneven
Earlywood/Latewood Transition:
    Fairly abrupt
Resin Canals: Numerous, medium,
    mostly solitary, evenly distributed

Cross-Field Pitting: Windowlike
Ray Tracheids: Dentate
Epithelial Cells: Thin-walled
Longitudinal Parenchyma: Absent

RED PINE/SCOTS PINE
red pine *(P. resinosa)*
Scots pine *(P. sylvestris)*

THE WESTERN YELLOW PINES
ponderosa pine *(P. ponderosa)*
lodgepole pine *(P. contorta)*

There are eleven species of southern yellow pine, of which four — loblolly, shortleaf, longleaf, and slash — account for about 93% of the standing timber. These trees grow principally in the southeastern and south-central states. Pitch pine has a range extending from northern Georgia to southern Maine. Although it accounts for only 1% of the total volume of southern yellow pine, it is included here because its range overlaps with that of red pine; in the Northeast, the two woods are often confused, especially in period furniture.

Wood of the southern yellow pines is characterized by its heaviness and hardness; it is the densest and strongest among the commercially significant softwoods. Wood of average growth rate displays very uneven grain due to the conspicuous bands of very dense latewood. Slow-growth samples or juvenile wood may look more even-grained due to narrower or lighter latewood. South-

ern yellow pine is sometimes confused with tamarack or Douglas-fir but is distinguished from these species by its larger resin canals and distinctive "piney" odor.

The western yellow pines include two important species, ponderosa pine and lodgepole pine. These woods, especially old-growth ponderosa pine, commonly have very narrow growth rings and extremely narrow, almost imperceptible latewood. Samples of this kind appear very even-grained. However, when a transverse surface is viewed with a hand lens, the clear, abrupt transition from earlywood to latewood that characterizes the hard pines is evident. Even on longitudinal surfaces, careful examination of the growth rings reveals a sharp delineation between earlywood and latewood. The growth rate of western yellow pines ranges from slow to moderate; thus, some pieces have growth rings of medium width (they are seldom wide) that have more distinct latewood.

Generally, medium to wide growth rings and pronounced, very dense latewood in a hard pine indicate southern yellow pine. Wood samples that have medium-narrow to very narrow growth rings with narrow latewood and moderate density contrast are probably west-

ern yellow pine. However, there is considerable overlap in general appearance, and many samples cannot be identified with certainty.

Even with a microscope, there are no features that consistently and absolutely separate the southern and western yellow pines. Both groups have dentate ray tracheids and pinoid cross fields, the pits of which are variable among all species (see the photos below).

Intermediate to the southern yellow pines and western yellow pines is red pine *(P. resinosa)*, whose gross characteristics unfortunately range from those of southern yellow pine to those of western yellow pine. Microscopically, however, red pine is unique among native hard pines in that it has windowlike cross-field pitting in conjunction with dentate ray tracheids. This provides an absolute basis for separating it from all other native pines.

It should be noted here that the wood of red pine is very similar to that of Scots pine *(P. sylvestris)*, a principal conifer of Europe and Asia. (In the United States, Scots pine is sometimes incorrectly called Scotch pine.) Scots pine is similar to red pine in its gross features and in its windowlike cross-field pitting. Both woods are placed in a worldwide taxo-

RAY CELLS IN HARD PINES
Radial views of southern yellow pine (left), ponderosa pine (center) and red pine (right) show dentate ray tracheids (DRT) and ray parenchyma (RP). (275x)

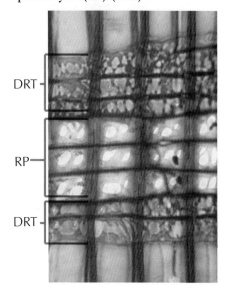
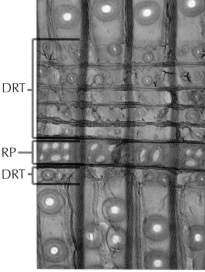
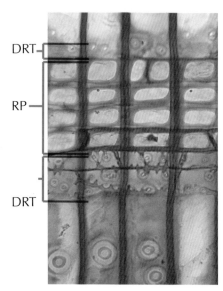

nomic subgroup called sylvestris pine; it has traditionally been assumed that the woods of the two species cannot be separated with certainty. This is a significant problem because distinguishing between the two woods is often important in the interpretation of early furniture, particularly in deciding whether the object in question has a North American or Eurasian origin.

Recent research by Sharon Zarifian at the University of Massachusetts at Amherst was directed at finding a basis for separating woods of the two species. After extensive anatomical study, the only discriminating feature discovered was the fusiform rays, which were found to be taller in Scots pine than in red pine. To test this feature, the height of 30 fusiform rays is measured with a calibrated eyepiece micrometer and averaged. If the average is greater than 340 microns, the wood is Scots pine; if it is less than 280 microns, the wood is red pine. In about one out of four samples of either species, the average will fall between 280 and 340 microns, indicating a common range of overlap and an inconclusive result.

In practice, however, an object such as a piece of furniture is usually an assemblage of many individual pieces of wood. Since it can usually be assumed that the woods in an object are all either one species or the other — not a mixture of both — there are usually some pieces that have average fusiform ray measurements above or below the overlap range. This enables them to be identified clearly as either Scots pine or red pine. We feel that this data constitutes an important breakthrough in separating the two species.

Not all microscopes have the eyepiece micrometer that allows this precise measurement to be made. However, I have found that with a little experience in examining tangential sections of known samples, the generally taller fusiform rays of Scots pine (photos at right) become familiar and the majority of unknown specimens of either Scots pine or red pine can be correctly identified by microscopic inspection.

Among the western yellow pines, several gross features help distinguish pon-

derosa pine from lodgepole pine. First, since ponderosa pines grow to a larger size than lodgepole, a growth-ring curvature that implies a stem diameter in excess of 2 ft. strongly suggests ponderosa pine. Since lodgepole pine commonly has small knots, large planks of clear wood also hint at ponderosa pine.

The heartwood of ponderosa pine is commonly dark or reddish in color, but

this is seldom true of lodgepole pine. The resin canals in ponderosa pine are conspicuously larger than they are in lodgepole pine — a good way to observe this difference is to compare the transverse surfaces of known samples. Finally, lodgepole pine often has pronounced dimpling (see the photo on p. 65), which is less common and less pronounced in ponderosa pine.

RAY HEIGHT IN RED PINE AND SCOTS PINE
In Scots pine (top photo), fusiform rays are relatively tall, usually having uniseriate 'tails' up to several cells long. In red pine (bottom photo), fusiform rays are shorter, tapering to blunt ends, commonly without uniseriate extensions. (150x)

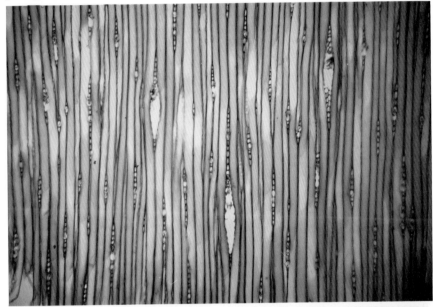

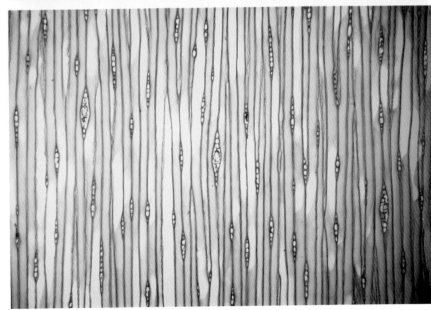

GROUP II: DOUGLAS-FIR, SPRUCE AND LARCH

Douglas-fir (*Pseudotsuga menziesii*), spruce (*Picea* spp.) and larch (*Larix* spp.), all members of the family *Pinaceae*, have resin canals that are relatively small, unevenly distributed either singly or in tangential groups, and sometimes relatively sparse. In fact, some growth rings seem not to have any resin canals at all. Other gross features, however, alone or in com-

bination, allow at least a tentative separation of the genera.

In samples of average growth rate, Douglas-fir and larch have a generally dense appearance, owing to the unevenness of grain and the dense bands of pronounced latewood, which occupies about a third of the growth ring. In both species the transition from earlywood to latewood is abrupt.

Average growth in spruce is typically much more even-grained, suggesting a softer wood. Although its rings are visually distinct, the earlywood makes a

gradual transition to fairly narrow but clear bands of latewood, which can look like white pine. But spruce can be quite variable. Some pieces have wider, denser latewood, with a more abrupt transition that more closely resembles fir or hemlock, especially when reaction wood is present.

Woods of Douglas-fir, larch and spruce can be confused when the growth rate is very slow. In wood from large, old, overmature trees — trees that have passed their prime and are beginning to stagnate — growth may slow down to the

---

DOUGLAS-FIR

Douglas-fir has small resin canals that are relatively small in number, irregularly distributed and sometimes in tangential groups. The transition from earlywood to latewood is abrupt. Wood of normal growth rate is typically uneven-grained; however, wood with extremely narrow rings appears more even-grained because of the inconspicuous latewood.

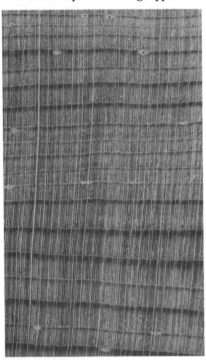  

DOUGLAS-FIR
Pseudotsuga menziesii

Average Specific Gravity: 0.48
Odor: Unique, resinous
Heartwood: Distinct reddish brown
Grain Appearance: Uneven
Earlywood/Latewood Transition: Abrupt
Resin Canals: Medium to small, relatively few and variable in distribution, solitary or up to several in tangential groups

---

Cross-Field Pitting: Piceoid
Ray Tracheids: Present
Epithelial Cells: Thick-walled
Longitudinal Parenchyma: Absent

rate of 70 to 80 rings per inch. In such slowly grown wood the latewood may be unusually narrow, and the nature of the earlywood/latewood transition cannot be evaluated meaningfully.

The heartwood and sapwood of most spruce species are light-colored and, consequently, inseparable. Therefore, if a piece of wood has a deep, reddish-brown color, it cannot be spruce. But be alert in interpreting a lighter color — it could be spruce, or it might be the sapwood of Douglas-fir or larch. The placement, number and curvature of the growth rings will help to suggest whether or not a sample is heartwood.

Additionally, Douglas-fir has a distinctive odor, whereas the others have no odor. Larch is somewhat waxy or greasy, whereas spruce and Douglas-fir feel typically dry. Spruce often has a sheen or luster, as compared to the normal dullness of the others. And spruce is the only wood that occasionally has dimpling.

Microscopic examination permits much more reliable separation. As al-

ready noted, all three genera have thick-walled epithelial cells in vertical and horizontal resin canals; they also have ray tracheids and piceoid cross-field pitting.

Douglas-fir is quickly and positively identified on the basis of its abundant spiral thickenings in longitudinal tra-

cheids (photo below left), in conjunction with evidence — fusiform rays, for example — that resin canals or ray tracheids are present. (Note: Yew also has spiral thickenings but lacks ray tracheids and resin canals; larch occasionally has just a few tracheids, usually in the late-

EASTERN SPRUCE AND SITKA SPRUCE

Spruces have small resin canals (they are larger in Sitka spruce) that are irregularly distributed and sometimes occur in tangential groups. The transition from earlywood to latewood is usually gradual.

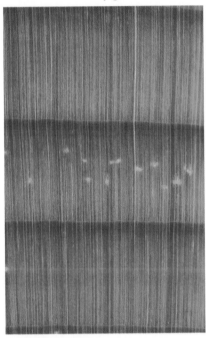

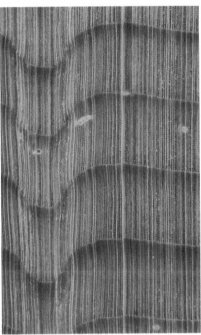

SPIRAL THICKENINGS IN LONGITUDINAL TRACHEIDS OF DOUGLAS-FIR

Spiral thickenings in longitudinal tracheids, together with resin canals (evidenced here by the fusiform ray), provide positive identification for Douglas-fir. (150x)

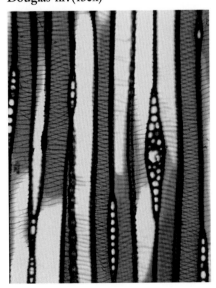

EASTERN SPRUCE
Picea spp.

Average Specific Gravity: 0.40
Odor: None
Heartwood: Light in color; indistinct from sapwood
Grain Appearance: Fairly even to moderately uneven
Earlywood/Latewood Transition: Gradual
Resin Canals: Small, relatively few and variable in distribution, solitary or up to several in tangential groups

Cross-Field Pitting: Piceoid
Ray Tracheids: With *Picea* Type I and II pits
Bordered Pits: Large bordered pits on radial walls of tracheids seldom paired
Epithelial Cells : Thick-walled
Longitudinal Parenchyma: Absent

SITKA SPRUCE
P. sitchensis

Average Specific Gravity: 0.40
Odor: None
Heartwood: Lighter sapwood grading to darker heartwood of light to medium pinkish brown
Grain Appearance: Fairly even to moderately uneven
Earlywood/Latewood Transition: Gradual
Resin Canals: Medium sized, sparse to numerous, variable in distribution, solitary or up to several in tangential groups

Cross-Field Pitting: Piceoid
Ray Tracheids: With *Picea* Type I and II pits
Bordered Pits: Large bordered pits on radial walls, commonly paired
Epithelial Cells : Thick-walled
Longitudinal Parenchyma: Absent

wood, that show spiral thickenings.) Conversely, the absence of spiral thickenings excludes Douglas-fir.

Separating larch and spruce is more challenging. It involves the examination of various macroscopic and microscopic features, none of which, alone, is always a reliable indicator. But when these features are considered collectively, they are quite dependable. Macroscopically, larch is distinguished by its greater overall density, more uneven grain, more abrupt transition from earlywood to latewood and darker heartwood. Microscopically, the following features should be analyzed on radial sections of these woods:

1. On radial walls of longitudinal tracheids, the large bordered pits are usually single in spruce (except Sitka spruce) but paired in larch (see the photos on p. 19). To check this feature, scan the first five to ten tracheids in the earlywood in several areas of the section. If paired pits are common, larch is indicated; if the pits are essentially single, with no more than occasional pairing, spruce is indicated.

2. The bordered pit pairs separating adjacent ray tracheids, as seen in radial view, can also be useful in the separation of spruce and larch (see the photos on the facing page). But this feature is harder to assess because the radial section must by chance have been cut to show clear detail of the bordered pit pair. It is usually necessary to scan considerable areas of a radial section to find a location where the pit-pair detail shows clearly. Moreover, because of the minuteness of this detail, high magnification — 400x or more — and high resolution are necessary to observe it clearly.

In larch, the pit borders are more slender, with smooth, rounded edges along a fairly wide aperture (opening). Appropriately, these are called *Larix* (larch)-type pits. In spruce, the pit borders are thick by comparison and form a very narrow aperture. If the borders are thick and typically angular, the pitting is called *Picea* (spruce) Type I; if the borders appear to have projections or "horns" on either or both sides of the aperture, the pitting is *Picea* Type II.

LARCH

Larch has small resin canals that are relatively few in number and sometimes occur in tangential groups. The transition from earlywood to latewood is abrupt.

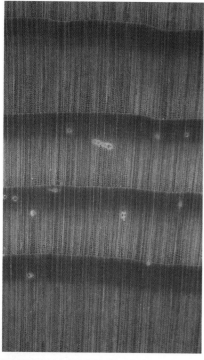

EASTERN LARCH (TAMARACK)
Larix laricina

Average Specific Gravity: 0.53
Odor: None
Heartwood: Distinct, yellowish to orange brown or russet brown
Grain Appearance: Uneven
Earlywood/Latewood Transition: Abrupt
Resin Canals: Small, relatively few and variable in distribution, solitary or up to several in tangential groups
Texture: Medium-fine

Cross-Field Pitting: Piceoid
Ray Tracheids: With *Larix*-type pits
Bordered Pits: On radial walls of earlywood tracheids commonly paired
Epithelial Cells: Thick-walled
Longitudinal Parenchyma: Absent

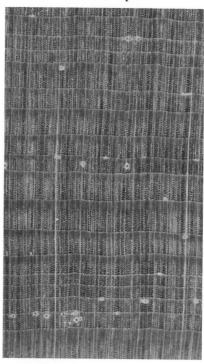

WESTERN LARCH
L. occidentalis

Average Specific Gravity: 0.51
Odor: None
Heartwood: Distinct, russet to reddish brown
Grain Appearance: Moderately uneven; rings generally narrow
Earlywood/Latewood Transition: Abrupt
Resin Canals: Small, relatively few and variable in distribution, solitary or up to several in tangential groups
Texture: Medium-coarse

Cross-Field Pitting: Piceoid
Ray Tracheids: With *Larix*-type pits
Bordered Pits: On radial walls of earlywood tracheids commonly paired
Epithelial Cells: Thick-walled
Longitudinal Parenchyma: Absent

LARCHES

Among the larches, eastern larch or tamarack (*Larix laricina*) has growth rings that are narrow to moderately wide, whereas western larch (*L. occidentalis*) typically has growth rings that are mostly narrow with less pronounced latewood. In eastern larch the heartwood is usually a yellowish to orange-brown or russet brown, whereas western larch heartwood varies from russet to reddish brown (but is not as red as in some Douglas-fir). The texture of eastern larch is fine to medium-fine; in western larch it is medium-coarse to coarse. An occasional biseriate narrow ray is more apt to occur in western than in eastern larch. European larch (*L. decidua*) is very similar to American larches.

SPRUCES

The woods of red spruce *(Picea rubens)*, white spruce *(P. glauca)* and black spruce *(P. mariana)* are usually designated simply as eastern spruce because they cannot be separated from one another. In fact, they are similar to Engelmann spruce *(P. engelmannii)*, which is often grouped with eastern spruce under the designation "transcontinental spruce." However, Engelmann spruce usually has a more even growth rate, and its narrow growth rings often show a more abrupt transition from earlywood to latewood. Norway spruce *(P. abies)* is indistinguishably similar to eastern spruces.

Sitka spruce *(P. sitchensis)* is different in several ways from the transcontinental spruces. Whereas the heartwood of other spruces is the same color as the sapwood, in Sitka spruce the heartwood is light to medium brown, typically with a pinkish hue, which is in distinct contrast to its nearly white sapwood. A moistened cross section shows a deeper, almost orange color, compared to the generally yellow color characteristic of transcontinental spruces.

Sitka spruce has a coarser texture and larger resin canals than the transcontinental spruces, which sometimes causes it to be confused with the pines. It may have indented rings that show up as "bear scratches" (see the photo on p. 65) on tangential surfaces. Dimpling on split tangential surfaces is more common than among the eastern spruces.

Microscopically, on tangential sections, the ray cells in Sitka spruce are rounded or squarish (isodiametric), whereas in other spruces the ray cells in tangential view appear oval or elongated in the grain direction. Sitka spruce has abundant yellowish-brown to amber deposits in ray cells. Large bordered pits are commonly paired on the radial walls of earlywood tracheids, in contrast to the usually single pitting in other spruces.

RAY TRACHEID BORDERED PITTING IN SPRUCE AND LARCH

In spruce, ray tracheid pitting in sectional view shows narrow apertures and thick, angular borders (left, *Picea* Type I), sometimes with hornlike projections on either side of the aperture (center, *Picea* Type II). In larch, pitting shows wide apertures and thinner, smoothly rounded borders (right, *Larix*-type). (550x)

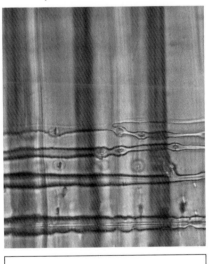 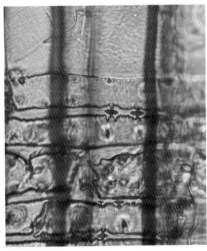 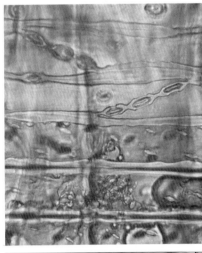

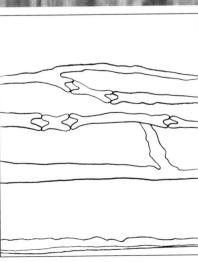 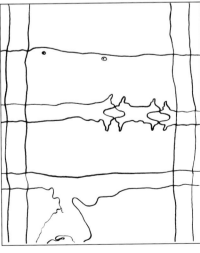 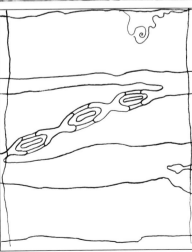

GROUP III: SOFTWOODS LACKING RESIN CANALS

The identification of woods in this group varies from easy to very difficult. A few species, such as yew, eastern redcedar and Alaska-cedar, can usually be recognized because of their distinctive color or smell. But separation of most of the other woods in this group is neither simple nor straightforward.

The separation sequence differs depending upon whether it is done on the basis of gross or microscopic features.

Wherever possible it is recommended that both approaches be used. To begin, follow one procedure or the other until it suggests a final species. Then consult the descriptive discussion associated with that particular species and check the results against Table 11.3 (pp. 164-165).

Beginning with an examination of gross features, carefully assess the following characteristics of the sample: density, heartwood color, odor, growth-ring characteristics (evenness of grain, ring width, percentage of latewood, nature of earlywood/latewood transition) and texture. Look for any sign of parenchyma — either darker tangential zones on transverse surfaces or very fine, dark

dotted lines running longitudinally along split surfaces, as viewed with a hand lens.

Keep in mind that if only gross features are used as the basis for identification, you rely heavily on such characteristics as heartwood color and odor, so the chance of correct identification is poor when the sample is small, aged by exposure, weathered or decayed. Be sure to consider the geographic origin of a sample, if known. It can be extremely helpful in narrowing the possibilities, because the ranges of most domestic conifers are polarized toward eastern or western North America. The map in the drawing below is a quick indica-

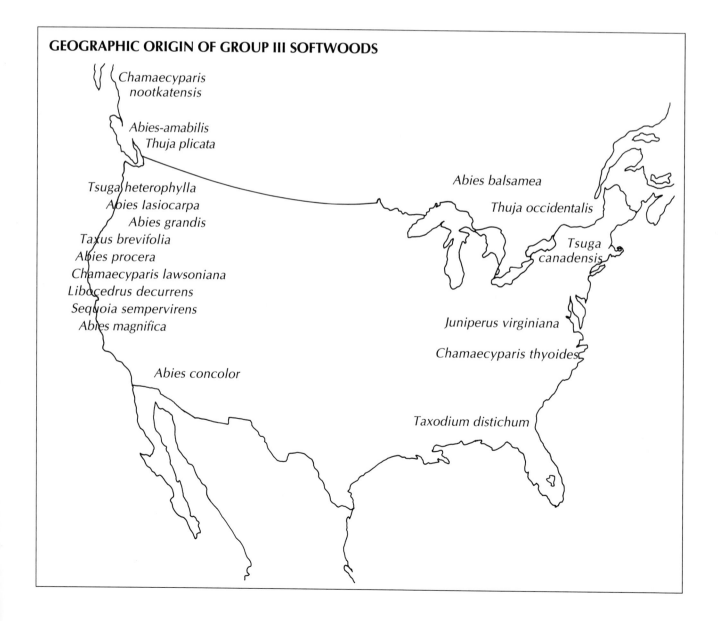

GEOGRAPHIC ORIGIN OF GROUP III SOFTWOODS

tor as to whether a species has an eastern or western origin.

Based on gross features, the separation scheme in the chart below can be used to place the wood in one of the four family subgroups. When a group is indicated, turn to the appropriate page, read the var-ious descriptions of the species included and complete the identification based on the best agreement of characteristics.

If a microscope is used, sections of wood tissue can be examined in any order. If preliminary examination of gross features has pointed to a certain species, consult the verbal description of that species for a list of key microscopic characteristics, then test your choice with a microscopic check for those features. As you scan the microscopic sections, compare your observations against Table 11.3 on pp. 164-165.

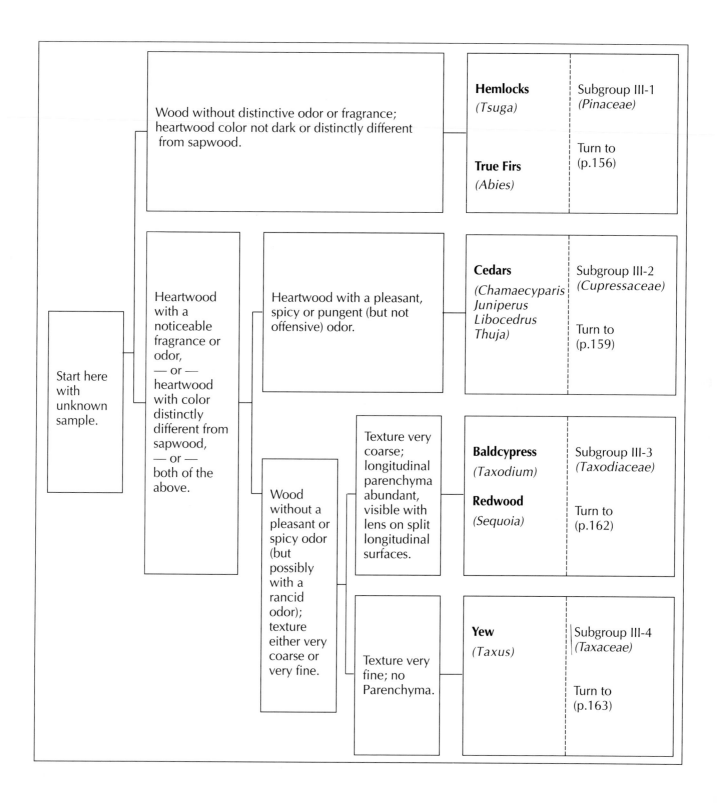

To use the table, lay a piece of paper with its edge along the dotted line just beneath the headings. As you find various features, write them on the edge of the paper under the proper heading. By sliding the paper up and down as you add new information and matching to the table, you get a good idea of which woods "look good" and which are being eliminated. A word of advice, however: Never eliminate a wood based on a single mismatch.

If, on the other hand, you do not have a particular plan or set of features to look for, try the following:

I. Examine a radial section:

1) If spiral thickenings are present in the longitudinal tracheids, go directly to yew on p. 163.

2) If ray tracheids are present in one or two rows along rays containing ray parenchyma, go directly to hemlock on p. 157.

3) If some rays are composed entirely of ray tracheids and other rays are entirely ray parenchyma, go directly to Alaska-cedar on p. 160.

4) If there are crystals in the ray cells, go directly to fir on p. 157.

5) If none of the above is found, continue to examine the radial section as suggested below and tally features against Table 11.3 on pp. 164-165, where applicable:

a) Examine the end walls of ray cells. Are they smooth or nodular?

b) Examine the cross-field pits. Are they cupressoid or taxodioid?

c) Note the color and abundance of ray cell contents.

d) Examine the earlywood tracheids. How many bordered pits are there across the tracheid walls?

II. Next, examine a transverse section:

1) Study the tracheids. If they are small — resulting in a fine texture — and rounded, and intercellular spaces are abundant, go directly to eastern redcedar on p. 159.

2) Look for longitudinal parenchyma, as evidenced by dark contents or, in latewood, by thin walls. Note the relative abundance of longitudinal parenchyma cells and the region of the growth ring where they seem most abundant. Sometimes they are numerous in the latewood, and sometimes bands of zonate parenchyma occur in the earlywood. If none can be found, check fir *(Abies* spp.) on p. 157 or cedar *(Thuja* spp.) on p. 159, particularly if other evidence also points to these species.

III. Finally, examine a tangential section:

1) Scan a host of rays, checking to see whether:

a) none seems taller than 12 cells;

b) you regularly find rays which exceed 30 cells in height.

2) Examine the ray seriation and note where rays are more than one cell wide.

3) Check whether ray cells are vertically oval or round.

4) Look for longitudinal parenchyma cells. If there are cell contents, note their color. Check also to see if the horizontal end walls are conspicuously nodular, faintly nodular or smooth.

Compare your collective observations against Table 11.3 on pp. 164-165 and find a "best fit." Then turn to the specific section on the wood or woods indicated to make sure that the additional features described there are also present in the sample. Double-check any other features that are in question.

Subgroup III-1: Hemlock and Fir in the Family *Pinaceae*

I would rate hemlocks *(Tsuga* spp.) and firs *(Abies* spp.) as the most difficult of all conifers to separate visually. They seem totally lacking in consistent, distinctive features. It is interesting that western hemlock and western firs are lumped into a single commercial grade designation: HEM-FIR (see the drawing below). The main reason, of course, is that their strength properties are similar enough that they can be used interchangeably in structural applications.

Both hemlocks and firs lack odor and heartwood color, characteristics that are common to the other species in Group III. I have concluded that samples showing a very gradual earlywood/latewood transition more often turn out to be firs, and those with an abrupt transition, hemlock, especially eastern hemlock, which often has reaction wood. If the transition is neither clearly gradual nor abrupt, consider other features.

The density range of fir is slightly lower than it is for hemlock; if a sample is extremely soft, it suggests fir. You might also try tasting the wood (at your own risk — its taste might best be described as lousy). If, by chance, the sample is a piece of fir rich in calcium oxalate crystals, the unpleasant, bitter taste (which can linger for some time) will let you know. The next step is to check the ver-

HEM-FIR GRADE STAMP
Structural lumber of western hemlock and western firs is sometimes marketed together under the commercial grade designation of HEM-FIR.

bal descriptions of these woods to see which set of features gives the best fit.

Fortunately, microscopic separation of hemlock and fir is more routine. Ray tracheids, mostly absent from fir, are present in hemlock (see the photo at the bottom of p. 24), usually as one or two rows along the edges of the rays. Because the ray tracheids have thinner walls and may be conspicuously narrower than the ray parenchyma cells — especially in western hemlock — they may appear very faint and are not easily seen at first glance. The presence of reasonable numbers of crystals in ray cells indicates western fir, but the absence of crystals is nondiagnostic. The cross-field pitting is variable in both genera, but it is more apt to appear taxodioid in fir (*Abies* spp.) and piceoid in hemlock (*Tsuga* spp.). But both intergrade with cupressoid.

In tangential view, ray height is too variable to be very useful, although a sample with ray height regularly in excess of 30 cells is more likely to be fir than hemlock. Firs rarely have longitudinal parenchyma, but hemlocks, especially western hemlock, may have scattered cells of marginal parenchyma along the growth-ring boundary.

HEMLOCK

Because of the wide separation of their geographic ranges, either of the two principal hemlocks, eastern hemlock (*Tsuga canadensis*) and western hemlock (*T. heterophylla*), can be distinguished if the origin is known. Otherwise, samples cannot always be separated with certainty, although a comparison of certain features may be helpful. Western hemlock usually has fairly narrow but uniform growth rings; in eastern hemlock they are variable, ranging from narrow to fairly wide. In eastern hemlock the earlywood/latewood transition tends to be abrupt and texture appears coarser. Of the two, eastern hemlock more commonly has paired pitting on the radial walls of earlywood tracheids.

FIR

The color of fir heartwood and sapwood is similar, ranging from very light buff to very light brown, sometimes with a tinge

of lavender in the darker latewood. Individual species are inseparable on the basis of macroscopic features; however, where the geographic origin is known it can assist in the separation process. Balsam fir (*Abies balsamea*) is the only major eastern species, but its range extends westward across Canada as far as Alberta. The remaining principal firs are western. The ranges of some overlap, but others occupy their own regions. Collectively, the western species of fir are called "white fir."

Microscopically, firs are characterized by a lack of resin canals and ray tracheids, and by the presence of nodular ray end walls and taxodioid cross-field pitting. The separation scheme in Table 11.2 on p. 158 is based on the occurrence of crystals and the color of the contents of ray cells.

The principal European species, Norway fir (*A. alba*), also called silver fir or whitewood, has rays consistently exceeding 30 cells in height. It is a low-density species similar to *A. balsamea*, and its general features are otherwise consistent with those described for North American firs.

HEMLOCK

Eastern hemlock is usually uneven-grained, with a fairly abrupt transition from earlywood to latewood and medium to medium-coarse texture. Western hemlock is typically more even-grained with narrower rings, and it has a more gradual transition from earlywood to latewood, as well as a finer texture.

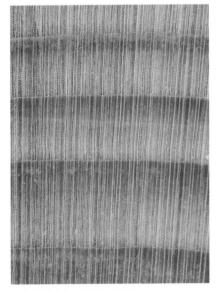

EASTERN HEMLOCK
Tsuga canadensis

Average Specific Gravity: 0.40
Odor: None
Heartwood: Indistinct, light in color
Grain Appearance: Fairly uneven
Earlywood/Latewood Transition:
    Fairly abrupt to gradual
Texture: Medium to medium-coarse

Cross-Field Pitting: Piceoid/cupressoid
Ray Tracheids: Often very narrow and
    inconspicuous
Ray Parenchyma End Walls: Nodular
Longitudinal Parenchyma: Absent (or
    marginal if present)

WESTERN HEMLOCK
Tsuga heterophylla

Average Specific Gravity: 0.44
Odor: None
Heartwood: Indistinct, light in color
Grain Appearance: Fairly uneven
Earlywood/Latewood Transition:
    Usually gradual
Texture: Medium

Cross-Field Pitting: Piceoid/cupressoid
Ray Tracheids: Usually very narrow
    and inconspicuous
Ray Parenchyma End Walls: Nodular
Longitudinal Parenchyma: Sparse
    (marginal if present) or absent

TABLE 11.2: GUIDE TO SEPARATION OF NORTH AMERICAN FIRS (*Abies* spp.)

FREQUENCY OF CRYSTALS IN RAY CELLS	SPECIES	COLOR OF RAY-CELL CONTENTS
Usually abundant	white fir *(A. concolor)*	Brownish
Regularly found; sparse to moderately abundant	grand fir *(A. grandis)* California red fir *(A. magnifica)* noble fir *(A. procera)*	
Very sparse or absent	Pacific silver fir *(A. amabilis)*	
	subalpine fir *(A. lasiocarpa)*	Pale yellow or colorless
Lacking	balsam fir *(A. balsamea)*	

TRUE FIR

The several species of fir are moderately even to slightly uneven-grained, usually with a gradual transition from earlywood to latewood and medium to medium-coarse texture. The examples shown below include one eastern fir (left) and two western firs (center and right).

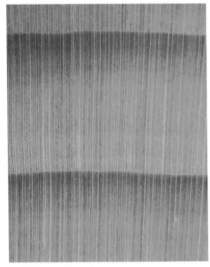

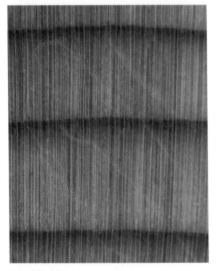

BALSAM FIR
Abies balsamea

Average Specific Gravity: 0.36
Odor: None
Heartwood: Indistinct, light in color
Grain Appearance: Moderately uneven to moderately even
Earlywood/Latewood Transition: Very gradual
Texture: Medium

Cross-Field Pitting: Taxodioid
Ray Tracheids: Absent
Ray Parenchyma End Walls: Nodular
Crystals: None in ray cells
Longitudinal Parenchyma: Usually absent

WHITE FIR
A. concolor

Average Specific Gravity: 0.34-0.40
Odor: None (*A. lasiocarpa* sometimes has foul odor); may have bitter taste
Heartwood: Indistinct, light in color
Grain Appearance: Moderately uneven to moderately even
Earlywood/Latewood Transition: Gradual
Texture: Coarse to medium-coarse

Cross-Field Pitting: Taxodioid
Ray Tracheids: Absent
Ray Parenchyma End Walls: Nodular
Crystals: Sometimes present in ray parenchyma
Longitudinal Parenchyma: Usually absent

GRAND FIR
A. grandis

Subgroup III-2: Cedars in the Family *Cupressaceae*

Distinctive odors—fragrant, spicy or otherwise—are important identification features of the cedars; most cedars have noteworthy heartwood color as well. In freshly cut pieces of average growth rate and reasonably large size, such as shingles or lumber, careful evaluation of the gross features, particularly texture, evenness of grain and visibility of parenchyma, will result in an accurate identification. But for atypical, smaller, aged or weathered specimens, microscopic confirmation is usually necessary.

EASTERN REDCEDAR

Freshly cut eastern redcedar (*Juniperus virginiana*) virtually identifies itself on the basis of its strong, distinctive "cedar-chest" aroma and its unmistakable purplish-red heartwood color. But the wood turns a flat reddish brown upon exposure and its scent may fade slightly, so it is worth noting some other distinctive gross features. These include the uniform firmness of its very fine-textured wood and its abundant longitudinal parenchyma, whose dark contents and zonate arrangement usually appear as "false rings." The dark contents of the ray cells produce a fine but distinct dark ray fleck.

To confirm redcedar microscopically, look for rounded tracheids with numerous intercellular spaces in transverse view, cupressoid pitting and nodular ray end walls in radial view, and nodular longitudinal parenchyma end walls in tangential view.

THE OTHER CEDARS

The remaining cedars are more difficult to separate, since each is confusingly similar to at least one other. Again, because of the wide separation in geographic range of the eastern and western groups, knowledge of origin is very helpful.

EASTERN CEDAR

Eastern cedars usually exhibit a gradual transition from earlywood to latewood. Eastern redcedar has a deep purplish-red heartwood color, a familiar "cedar-chest" aroma and very fine texture. White-cedars have fine texture, milder but distinct aroma and medium to light shades of heartwood color. Eastern redcedar and Atlantic white-cedar usually have visible zonate parenchyma.

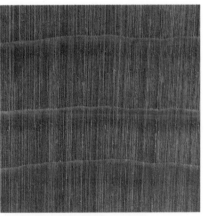

EASTERN REDCEDAR	ATLANTIC WHITE-CEDAR	NORTHERN WHITE-CEDAR
Juniperus virginiana	*Chamaecyparis thyoides*	*Thuja occidentalis*

EASTERN REDCEDAR
Juniperus virginiana

Average Specific Gravity: 0.47
Odor: Distinct "cedar-chest" odor
Heartwood: Distinct, deep purplish red, aging to reddish brown
Grain Appearance: Moderately uneven to fairly even; latewood narrow
Earlywood/Latewood Transition: Gradual
Texture: Very fine
Longitudinal Parenchyma: Abundant, with dark red contents; usually zonate in double ring

Cross-Field Pitting: Cupressoid
Ray Tracheids: Absent
Ray Parenchyma End Walls: Nodular
Longitudinal Tracheids: Rounded with intercellular spaces
Longitudinal Parenchyma: End walls nodular

ATLANTIC WHITE-CEDAR
Chamaecyparis thyoides

Average Specific Gravity: 0.32
Odor: "Sweet" and aromatic
Heartwood: Distinct, light brown with pinkish or flesh tones
Grain Appearance: Moderately even
Earlywood/Latewood Transition: Gradual
Texture: Fine to medium-fine
Resin Canals: Absent
Longitudinal Parenchyma: Abundant, with bright orange contents; usually zonate

Cross-Field Pitting: Cupressoid
Ray Tracheids: Absent
Ray Parenchyma End Walls: Smooth
Longitudinal Parenchyma: End walls smooth

NORTHERN WHITE-CEDAR
Thuja occidentalis

Average Specific Gravity: 0.31
Odor: Pungent but mild aroma
Heartwood: Faintly distinct to distinct, light to medium straw brown
Grain Appearance: Fairly even
Earlywood/Latewood Transition: Usually gradual
Texture: Fine to medium-fine
Longitudinal Parenchyma: Usually sparse, but occasionally abundant, zonate

Cross-Field Pitting: Taxodioid
Ray Tracheids: Absent
Ray Parenchyma End Walls: Smooth
Longitudinal Parenchyma: End walls nodular

WESTERN GROUP
Alaska-cedar
    (*Chamaecyparis nootkatensis*)
Port-Orford-cedar
    (*Chamaecyparis lawsoniana*)
western redcedar
    (*Thuja plicata*)
incense-cedar
    (*Libocedrus decurrens*)

EASTERN GROUP
Atlantic white-cedar
    (*Chamaecyparis thyoides*)
Northern white-cedar
    (*Thuja occidentalis*)

Separating the cedars that occur within the same region is in many cases a more common and more difficult problem than separating the woods within a genus. Woods within the same genus often come from different regions, and, particularly where origin is known, this facilitates separation. For example, western redcedar (*Thuja*) is more often confused with incense-cedar (*Libocedrus*) than with the other *Thuja*, northern white-cedar.

There are no definitive features common to all three *Chamaecyparis* species, other than cupressoid cross-field pitting and the fact that none has the darker

brown or deep reddish-brown heartwood of the other western species. The cedars discussed below have been paired for discussion because they are most commonly confused with each other.

ALASKA-CEDAR AND PORT-ORFORD-CEDAR
Freshly cut Alaska-cedar has clear yellow heartwood (it may darken to a dull yellow brown on prolonged exposure) and a distinctive odor, appropriately described as resembling raw potatoes, potato peelings or turnips. Port-Orford-cedar is also slightly aromatic, but with a pungent,

WESTERN CEDARS
Each western cedar has a distinctive heartwood color and a recognizable aroma. Western redcedar has an abrupt transition from earlywood to latewood; the transition is gradual in the other western cedars. Texture is fine in Alaska-cedar, medium in incense-cedar and medium to medium-coarse in Port-Orford-cedar and western redcedar. Zonate parenchyma is sometimes visible in Port-Orford-cedar and incense-cedar.

ALASKA-CEDAR
Chamaecyparis nootkatensis

Average Specific Gravity: 0.44
Odor: Distinct, like raw potatoes
Heartwood: Distinct, yellow to dark yellow
Grain Appearance: Fairly even
Earlywood/Latewood Transition: Gradual; rings usually very narrow
Texture: Fine to medium-fine

Cross-Field Pitting: Cupressoid
Ray Tracheids: Rays either entirely ray parenchyma or entirely ray tracheids
Ray Parenchyma End Walls: Smooth or faintly nodular
Longitudinal Parenchyma: Sparse to abundant; end walls nodular

PORT-ORFORD-CEDAR
C. lawsoniana

Average Specific Gravity: 0.42
Odor: Gingerlike
Heartwood: Distinct, yellowish or pinkish brown
Grain Appearance: Fairly uneven
Earlywood/Latewood Transition: Gradual
Texture: Medium to medium-coarse

Cross-Field Pitting: Cupressoid
Ray Tracheids: Absent
Ray Parenchyma End Walls: Smooth
Longitudinal Parenchyma: Abundant, sometimes zonate; end walls nodular

WESTERN REDCEDAR
Thuja plicata

Average Specific Gravity: 0.33
Odor: Pungent cedar odor; faint bitter taste
Heartwood: Distinct, medium to dark brown or reddish brown
Grain Appearance: Moderately uneven
Earlywood/Latewood Transition: Abrupt (or gradual in wide rings)
Texture: Medium to medium-coarse

Cross-Field Pitting: Taxodioid
Ray Tracheids: Usually absent
Ray Parenchyma End Walls: Smooth
Longitudinal Parenchyma: Absent, sparse or in some rings, abundant; end walls nodular

gingerlike odor. Its color varies from light yellowish to pale or pinkish brown.

Alaska-cedar appears even-grained because of its very narrow growth rings and thin, inconspicuous latewood. Its texture ranges from fine to medium-fine. Port-Orford-cedar has medium to medium-coarse texture, and wider, more distinct growth rings. Longitudinal parenchyma is fairly abundant, often showing zonate arrangement when viewed in a moist cross section with a hand lens. In Alaska-cedar the longitudinal parenchyma, though present, is usually inconspicuous and not visible with the hand lens.

INCENSE-CEDAR
Libocedrus decurrens

Average Specific Gravity: 0.37
Odor: Familiar "pencil-cedar" aroma; acrid taste
Heartwood: Distinct, medium reddish to purplish brown
Grain Appearance: Moderately even
Earlywood/Latewood Transition: Gradual
Texture: Medium

Cross-Field Pitting: Cupressoid
Ray Tracheids: Absent
Ray Parenchyma End Walls: Nodular
Longitudinal Parenchyma: Abundant, often zonate; end walls nodular

Microscopic separation is more positive, because Alaska-cedar has ray tracheids. These tracheids are the sole cell type of some rays; the other rays are wholly ray parenchyma whose end walls are smooth or faintly nodular. In Port-Orford-cedar the ray parenchyma end walls are smooth. Longitudinal parenchyma cells have nodular end walls in both species. The parenchyma is usually abundant in Port-Orford-cedar, but only sparse to moderately abundant in Alaska-cedar.

ATLANTIC WHITE-CEDAR AND NORTHERN WHITE-CEDAR

These two species are commonly confused, perhaps because of their eastern origin, their softness and their faint but pleasant "cedary" odor. In northern white-cedar, the odor is perhaps more pungent than the "sweet" scent of Atlantic white-cedar, but the difference may not be pronounced enough to be remembered easily. The heartwood of Atlantic white-cedar is light brown with a pinkish cast, more or less flesh-colored. Northern white-cedar heartwood is a soft pale or straw brown; it lacks the pink tinge of Atlantic white-cedar and is never as dark as the average shade of brown in western redcedar.

In both these eastern white-cedars the texture is fine to medium-fine, with moderately distinct growth rings and a gradual transition to latewood. In northern white-cedar, the parenchyma is usually sparse (although it can be abundant in the occasional ring) and therefore it is not usually visible with a hand lens. In Atlantic white-cedar, longitudinal parenchyma is fairly abundant, usually zonate and clearly visible as reddish tangential bands that may appear as "false rings" on a moistened cross section.

Microscopically, the ray end walls are smooth in both species, but are occasionally nodular near the growth-ring boundary in northern white-cedar. However, Atlantic white-cedar has cupressoid cross-field pitting and smooth longitudinal parenchyma end walls, whereas northern white-cedar has taxodioid cross-field pitting and nodular longitudinal parenchyma end walls.

WESTERN REDCEDAR AND INCENSE-CEDAR

The heartwood of western redcedar is medium dark reddish to pinkish brown when freshly cut, but it ages to a dull coffee brown. It has distinct growth rings due to a band of denser latewood with an abrupt earlywood/latewood transition. Along with its distinct rings, abrupt transition and darker color, its medium to medium-coarse texture distinguishes it from finer-textured northern white-cedar. (Microscopically, western redcedar commonly has paired bordered pits on the walls of earlywood tracheids; in northern white-cedar they are in a single row.) The odor of western redcedar is aromatic and is often described as sweet or chocolate-like. Longitudinal parenchyma cells are sometimes very sparse; when present in greater quantities, they are usually confined to the latewood, where they may be visible with a hand lens as a broken or irregular tangential line.

Incense-cedar is reddish brown to dull brown, sometimes with a purplish cast; its color range intergrades confusingly with western redcedar. It, too, has medium texture and distinct growth rings, but it is heavier and firmer than western redcedar. Parenchyma is usually very abundant and sometimes zonate and visible due to the dark contents of the cells.

Whereas western redcedar is fairly tasteless, incense-cedar has an acrid taste. Incense-cedar is also readily recognized by its "pencil-cedar" odor. (Note: Eastern redcedar was once the prime wood for pencilmaking. But it has largely been replaced by incense-cedar, and we now associate the odor of incense-cedar with pencils.)

Microscopic separation is more positive. Ray end walls are conspicuously nodular in incense-cedar, in contrast to the smooth end walls of western redcedar. Incense-cedar has ray cells with dark gummy contents and cupressoid cross-field pits, whereas the cross-field pits in western redcedar are taxodioid. In incense-cedar it is common to find rays that are biseriate in part, but rays in western redcedar are consistently uniseriate. Longitudinal parenchyma end walls are nodular in both species.

Subgroup III-3: Baldcypress and Redwood in the Family *Taxodiaceae*

Redwood (*Sequoia sempervirens*) and baldcypress (*Taxodium distichum*) are often confused for several reasons. First, because the heartwood of both is highly decay-resistant, both (along with cedars) are used for applications such as siding, decking, benches, planters and other outdoor uses. Anatomically they also have certain similarities. For example, the old-growth wood of both species can have narrow (but variable-width) rings of irregular or sinuous contour, with very narrow latewood. The latewood in both is denser and distinct because of the abrupt transition from the earlywood.

With a hand lens, both species appear very coarse-textured, and on split longitudinal surfaces minute dotted lines reflect the strands of longitudinal parenchyma

---

BALDCYPRESS AND REDWOOD

In both baldcypress and redwood, growth-ring width is variable from very narrow (even-grained) to moderately wide (fairly uneven-grained). The earlywood-latewood transition is generally abrupt. Both species have very coarse texture and longitudinal parenchyma that is visible with a lens on split longitudinal surfaces.

 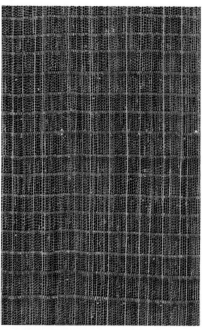 

BALDCYPRESS
Taxodium distichum

REDWOOD
Sequoia sempervirens

Average Specific Gravity: 0.46
Odor: Faint to moderately rancid
Heartwood: Usually distinct
Grain Appearance: Uneven
Earlywood/Latewood Transition:
    Abrupt; earlywood medium yellow
    brown, latewood amber to dark
    brown
Texture: Coarse to very coarse
	Average Specific Gravity: 0.40
	Odor: None
	Heartwood: Distinct, medium to deep reddish brown
	Grain Appearance: Fairly uneven
	Earlywood/Latewood Transition: Abrupt
	Texture: Coarse to very coarse

Cross-Field Pitting:
Cupressoid/taxodioid
Rays: Biseriate rays occasionally present
Ray Tracheids: Absent
Ray Parenchyma End Walls: Smooth
Bordered Pits: On radial walls of
    earlywood tracheids occasionally 3+
    across
Longitudinal Parenchyma: Abundant,
    diffuse; end walls strongly nodular

Cross-Field Pitting: Taxodioid (large)
Rays: Commonly 2-3 seriate in some samples
Ray Tracheids: Absent
Ray Parenchyma End Walls: Usually smooth
Bordered Pits: On radial walls of earlywood tracheids occasionally 3+ across
Longitudinal Parenchyma: Abundant, diffuse; end walls usually smooth

cells with dark reddish contents (see the photo on p. 21).

The two can be separated either by gross or macroscopic features. Redwood is odorless, whereas baldcypress may have a sour, rancid smell that some find offensive. Of the two, redwood feels drier and softer to the touch, whereas baldcypress usually has a greasy or even tacky feel. When a moistened transverse surface of baldcypress is viewed with a hand lens, a two-tone color effect is often observed — the medium yellow to straw color of the earlywood contrasts with the amber to dark brown of the latewood. In redwood, there is a definite reddish hue, the contrast between earlywood and latewood being a matter of different shades of the same color. On split radial surfaces, the longitudinal parenchyma specks of baldcypress have greater contrast, often appearing cherry red against the generally yellowish background; in redwood, the dots of parenchyma more closely match the reddish background color.

When necessary, the two can be reliably separated with a microscope. In tangential sections of baldcypress, the end walls of longitudinal parenchyma cells have conspicuous nodules (see the photo on p. 22); in redwood, they are smooth or only slightly irregular. On radial sections, cross-field pitting in redwood is usually large and "textbook" taxodioid, whereas in baldcypress it intergrades confusingly with cupressoid.

The ray cells of redwood and baldcypress, when viewed in tangential sections, appear to be more or less square (or isodiametric), in contrast to the ray cells of other conifers in the group, which appear to be oval or elongated in the grain direction. Redwood has the widest uniseriate rays among conifers.

Redwood can be separated from incense-cedar by observing the end walls of longitudinal and ray parenchyma. In redwood they are smooth or only faintly nodular, whereas they are distinctly nodular in both locations in incense-cedar. Incense-cedar has its "pencil-cedar" aroma; redwood is odorless. Redwood has the coarser texture. Redwood can be separated from western redcedar by color: western redcedar usually ages to a greyer, coffee-brown shade, in contrast to the reddish-brown hue retained by redwood.

Microscopically, the nodular end walls of longitudinal parenchyma in western redcedar distinguish them from the smooth walls in redwood. If this feature is difficult to assess because repeated sectioning fails to produce any parenchyma, this may itself be significant. Such sparseness of parenchyma points toward western redcedar, since redwood always has abundant parenchyma.

The giant sequoia (*Sequoiadendron giganteum*) has wood very similar to redwood. However, the giant sequoia exists principally in protected groves and has not been extensively harvested since the early days of logging.

Subgroup III-4: Yew in the Family *Taxaceae*

Very little needs to be said about yew (*Taxus* spp.) since it can be confirmed so readily. Microscopically, it has spiral thickenings but lacks ray tracheids or resin canals; this combination separates it from all other woods in the book. In fact, it stands alone even after an examination of gross features. Its extremely fine texture, orange to russet heartwood (as distinct from the purple-red of the comparatively fine-textured *Juniperus* spp. and relative hardness are definitive. However, the wood of Pacific yew (*Taxus brevifolia*) of the Pacific Northwest is indistinguishable from that of European or common yew (*T. baccata*) of the British Isles.

YEW
Yew is moderately even-grained, with a gradual transition from earlywood to latewood. It is the densest of the conifers, and has a distinctive reddish-orange to russet heartwood color.

PACIFIC YEW
Taxus brevifolia

Average Specific Gravity: 0.62
Odor: None
Heartwood: Distinct, orange to russet
Grain Appearance: Fairly uneven
Earlywood/Latewood Transition: Gradual
Texture: Very fine

Cross-Field Pitting: Cupressoid
Ray Tracheids: Absent
Ray Parenchyma End Walls: Smooth
Longitudinal Parenchyma: Absent
Longitudinal Tracheids: With spiral thickenings

TABLE 11.3: SUMMARY CHECKLIST FOR FEATURES OF GROUP III SOFTWOODS

SPECIES	GROSS AND MACROSCOPIC FEATURES				
	AVG. SPEC. GRAV.	HEARTWOOD COLOR	DISTINCTIVE ODOR	EARLYWOOD/ LATEWOOD TRANSITION	TEXTURE
EASTERN HEMLOCK (*Tsuga canadensis*)	0.40	Near white to pale brown, sometimes with roseate cast.	None.	Typically abrupt (or gradual).	Medium to medium-coarse
WESTERN HEMLOCK (*Tsuga heterophylla*)	0.44	Near white to pale brown.	None.	Typically gradual (or abrupt).	Medium.
(TRUE) FIR (*Abies* spp.)	0.34-0.43	Near white to pale brown, some with lavender tinge in latewood.	None. (Samples of subalpine fir [*A. lasiocarpa*] may have a foul odor.)	Gradual (more abrupt in some).	Medium to medium-coarse.
EASTERN REDCEDAR (*Juniperus virginiana*)	0.47	Deep purplish red.	Familiar cedar-chest aroma.	Typically gradual (or abrupt).	Very fine to fine.
ALASKA-CEDAR (*Chamaecyparis nootkatensis*)	0.44	Yellow to dark yellow.	Distinct, like raw potatoes.	Gradual.	Fine to medium-fine.
PORT-ORFORD CEDAR (*Chamaecyparis lawsoniana*)	0.42	Yellowish or pinkish brown.	Distinct, gingerlike scent.	Gradual.	Medium to medium-coarse
ATLANTIC WHITE-CEDAR (*Chamaecyparis thyoides*)	0.32	Light brown with pinkish or flesh-colored cast.	Moderately strong sweet cedar aroma.	Gradual.	Fine to medium-fine.
NORTHERN WHITE-CEDAR (*Thuja occidentalis*)	0.31	Light to medium or straw brown.	Mild but somewhat pungent odor.	Usually gradual.	Fine to medium-fine.
WESTERN REDCEDAR (*Thuja plicata*)	0.33	Medium to dark brown or reddish brown.	Distinct pungent cedar odor.	Typically abrupt (or gradual, especially in wide rings).	Medium to medium-coarse.
INCENSE-CEDAR (*Libocedrus decurrens*)	0.37	Medium reddish to purplish brown.	Familiar pencil-cedar aroma.	Gradual.	Medium.
BALDCYPRESS (*Taxodium distichum*)	0.46	Earlywood medium yellow brown, latewood amber to dark brown.	Faint to moderate rancid odor.	Abrupt.	Coarse to very coarse.
REDWOOD (*Sequoia sempervirens*)	0.40	Medium to deep reddish brown.	None.	Abrupt.	Coarse to very coarse.
YEW (*Taxus* spp.)	0.62-0.64	Orange to russet.	None.	Gradual.	Very fine.

| MICROSCOPIC FEATURES | | | | REMARKS AND MISCELLANEOUS |
| LONGITUDINAL PARENCHYMA | | RAY PARENCHYMA | | |
PRESENCE	END WALL VIEWED TANGENTIALLY	CROSS-FIELD TYPE	END WALL VIEWED RADIALLY	
Absent (or sparse, marginal).	Nodular, if present.	Piceoid/ cupressoid.	Nodular.	Heartwood/sapwood color not distinguishable. Ray tracheids present.
Sparse, marginal or absent.	Nodular, if present.	Piceoid/ cupressoid.	Nodular.	Heartwood/sapwood color not distinguishable. Ray tracheids present (usually very narrow).
Usually absent (or very sparse).	Nodular if present.	Taxodioid.	Nodular.	Heartwood/sapwood not distinguishable. Crystals in ray cells of some species. May have very bitter taste. Most samples with a few rays of 30+ cells in height.
Abundant (usually zonate).	Nodular.	Cupressoid.	Nodular.	In transverse view, tracheids rounded with intercellular spaces.
Sparse to moderately abundant.	Nodular.	Cupressoid.	Smooth (or faintly nodular).	Faint bitter-spicy taste. Growth rings usually very narrow. Some rays entirely ray tracheids.
Abundant (often zonate).	Nodular.	Cupressoid.	Smooth.	Faint bitter-spicy taste.
Abundant, zonate (occasionally sparse).	Smooth.	Cupressoid.	Smooth.	Faint bitter taste. Longitudinal surfaces may feel slightly oily.
Sparse (abundant in occasional rings).	Nodular.	Taxodioid.	Smooth.	Faint bitter taste. Ray end walls occasionally nodular in latewood portions of rays.
Sparse/absent (sometimes abundant).	Nodular.	Taxodioid.	Smooth.	Faint bitter taste. Scattered ray tracheids occasionally present.
Abundant, often zonate.	Nodular.	Cupressoid.	Nodular.	Acrid taste. Biseriate rays occasionally present.
Abundant, diffuse.	Nodular.	Cupressoid/ Taxodioid.	Usually smooth (occasionally nodular).	Greasy or tacky to the touch. 3+ bordered pits common across radial walls of earlywood tracheids. Biseriate rays occasionally present. Occasional rays 40-50+ cells in height.
Abundant, diffuse.	Smooth (or faintly nodular).	Taxodioid.	Smooth (or faintly nodular).	3+ bordered pits common across radial walls of earlywood tracheids. 2-3-seriate rays common in some samples.
Absent.	N.A.	Cupressoid.	Smooth.	Longitudinal tracheids with spiral thickenings.

CHAPTER TWELVE

IDENTIFICATION OF TROPICAL WOODS

No book on the identification of woods used in North America would be complete without the inclusion of some tropical hardwoods. As early as the 1600s, tropical woods were used in American furniture. During the early 18th century, mahogany became one of the woods of choice for furniture and cabinetry, and Spanish-cedar, satinwood, rosewood and ebony were also used. Through subsequent years, tropical woods such as teak have enjoyed periods of fashion.

In today's marketplace, we find a number of tropical species that are as familiar to us as many of our domestic woods. For example, species of the *Dipterocarpaceae*, including lauan, seraya, meranti, tangile and almon, are common among products such as lumber and plywood. We can hardly visit a home center without seeing tool handles of ramin. A cabinetmaker shopping for lumber and veneer finds tropical species such as sapele, korina, avodire, mahogany, mansonia and afrormosia among the alternatives to the traditional domestic favorites.

The subject of tropical woods is a veritable treasure trove — the list of important examples seems endless. This chapter, then, is a compromise. Rather than attempt to survey all the im-

portant tropical woods, I have chosen a representative selection. The woods included are among the more familiar ones. You will likely recognize most, although it is probable that one or more of your favorites is missing.

Tropical woods are commonly diffuse-porous, whether coarse-textured as in African mahogany or fine-textured as in mansonia. In this chapter, the woods presented illustrate the extreme characteristics found in tropical woods of the world. They are arranged with anatomically similar woods placed next to one another to permit easier comparisons. For example, the specific gravity of lignum vitae is ten times that of balsa. The nearly black color of ebony stands in stark contrast to the pale cream of obeche. Nevertheless, anatomical uniqueness is represented by the precise aliform parenchyma of ramin, the white chalky contents of American mahogany and the distinctive radial-pore multiples of jelutong.

This chapter comprises macroscopic transverse views and brief summaries of the properties and characteristics of 30 tropical woods. References listed in the Bibliography on pp. 212-215 provide further information on the anatomy and the identification of tropical woods.

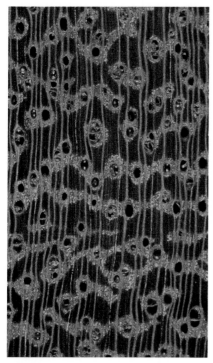

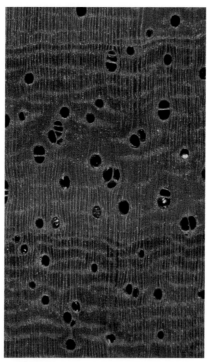

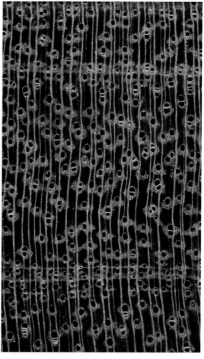

IROKO
(Chlorophora excelsa)

*Average Specific Gravity: 0.62
Heartwood: Light olive brown, frequently with darker streaks, darkening on exposure
Grain: Usually interlocked
Pore distribution: Diffuse-porous
Pores: Large, distinct without lens, mostly solitary and in radial pairs; vessels may contain tyloses and white deposits; vessel lines conspicuous on longitudinal surfaces
Parenchyma: Mostly aliform, sometimes confluent among pores; marginal parenchyma may delineate rings of growth
Rays: Barely visible on cross section

**Rays: 1-6 (mostly 3-5) seriate
Crystals: Ray cells and longitudinal parenchyma cells may have solitary crystals
Intervessel Pitting: Alternate, large

PADAUK
(Pterocarpus spp.)*

Average Specific Gravity: 0.67
Heartwood: Yellowish brown to red brown or deep red, even-colored or with darker streaks
Grain: Usually roey, producing a ribbon figure
Pore Distribution: Diffuse-porous or sometimes ring-porous; growth rings delineated by terminal parenchyma
Pores: Distinct without lens, large, unevenly distributed, solitary and in radial multiples of 2 to several
Parenchyma: Aliform with short to long wings, confluent forming wavy bands; also irregular patches of apotracheal
Rays: Indistinct without lens on cross section
Ripple Marks: Present; not usually visible, but distinct with lens

---

Rays: Uniseriate (rarely biseriate), mostly 10 cells or less high
Crystals: Parenchyma cells with crystals in chains
Intervessel Pitting: Alternate, medium

PURPLEHEART
(Peltogyne spp.)*

Average Specific Gravity: 0.83
Heartwood: Brown when freshly cut, becoming deep violet purple with exposure to light
Grain: Varies from straight to roey
Pore Distribution: Diffuse-porous; growth rings usually indistinct
Pores: Small to medium, barely visible without lens, numerous, evenly distributed in solitary and radial multiples of 2-4
Parenchyma: Mostly aliform, intergrading to confluent, and occasionally as bands of terminal parenchyma
Rays: Distinct on radial surfaces, but barely visible without lens on cross section

---

Rays: 1-6 (mostly 2-4) seriate, homogeneous
Crystals: Parenchyma occasionally with crystals
Gum: Vessels with red gum
Perforations: Simple
Intervessel Pitting: Alternate, quite small

---

* Throughout the chapter, macroscopic features are listed in the text block immediately under the species name.
** Microscopic features are below the macroscopic and separated from them by a line.

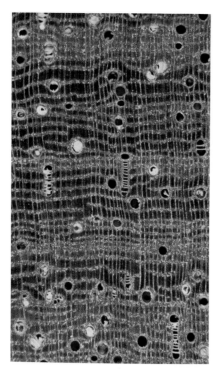

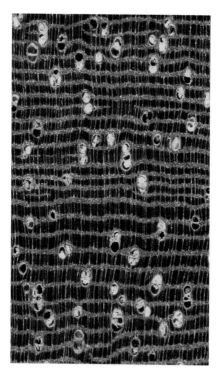

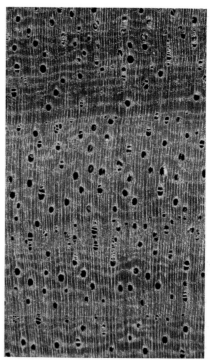

BASRALOCUS
(Dicorynia guianensis)

Average Specific Gravity: 0.72
Heartwood: Russet when freshly cut, becoming lustrous brown with reddish or purplish cast
Grain: Usually straight, occasionally mildly interlocked
Pore Distribution: Diffuse-porous
Pores: Large, visible without lens, not numerous, evenly distributed, solitary and in radial multiples of 2-6
Parenchyma: Distinct without lens, mostly aliform with short to long wings; confluent forming tangential bands; also irregular patches of apotracheal
Rays: Not distinct without lens on cross section
Ray Fleck: Distinct ray fleck, darker than background
Ripple Marks: Distinct and regular, all elements storied

Rays: 1-4 (mostly 3) seriate
Gum: Vessel and parenchyma lumens with brown gum deposits
Intervessel Pitting: Alternate, medium

EKKI
(Lophira alata)

Average Specific Gravity: 1.00
Heartwood: Dark reddish to chocolate brown with conspicuous yellowish solid deposits in vessels
Grain: Usually roey or irregular
Growth Rings: Not distinct
Pores: Distinct without lens, medium to large, solitary or in close radial arrangement of 2-4
Parenchyma: Distinct without lens, apotracheal in closely spaced wavy tangential bands
Rays: Not visible without lens

Rays: 1-3 (mostly 2) seriate
Crystals: Longitudinal parenchyma contains crystals
Gum: Vessels with conspicuous yellow deposits
Intervessel Pitting: Alternate, small

INDIAN ROSEWOOD
(Dalbergia latifolia)

Average Specific Gravity: 0.76
Heartwood: Light to dark violet brown to purple with near-black pigment layering that resembles growth rings
Grain: Irregular to roey
Growth Rings: Indistinct
Pores: Variable in size, large to small, irregularly distributed, solitary and in radial multiples
Parenchyma: Aliform with short to very long wings; confluent forming wavy tangential bands; also apotrachael forming broken tangential lines or in smaller patches
Rays: Not visible without lens
Ripple Marks: Distinct and regular; all elements storied

Rays: 1-3 (mostly 2-3) seriate; up to 15 cells high
Intervessel Pitting: Alternate, medium-large

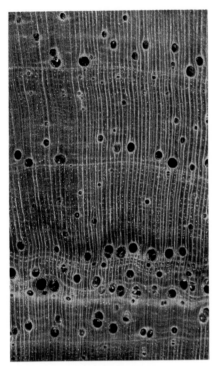

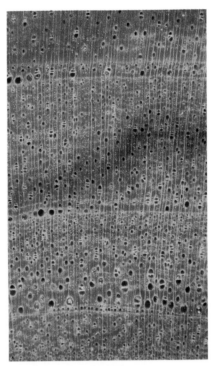

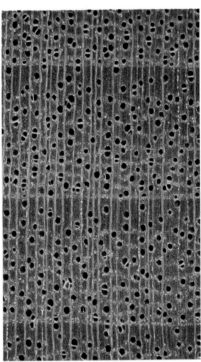

HONDURAS ROSEWOOD
(Dalbergia stevensonii)

Average Specific Gravity: 0.90
Heartwood: Light brown with purple
  or purplish-brown streaks
Grain: Straight to only slightly roey
Growth Rings: Often distinct due to
  concentric lines of parenchyma
Pores: Variable, small to large, not
  numerous, unevenly distributed,
  mostly solitary, some in radial
  multiples of 2-3
Parenchyma: Terminal, vasicentric
  (aliform and concentric rare) and
  apotracheal in closely spaced
  concentric lines
Rays: Not distinct on cross section
  without lens
Ripple Marks: Distinct and regular; all
  elements storied

Rays: 1-3 seriate; mostly 6-8 cells high
Crystals: Longitudinal parenchyma cells
  with crystals
Intervessel Pitting: Alternate,
  medium-large

BRAZILIAN TULIPWOOD
(Dalbergia fructescens, var. *tomentosa)*

Average Specific Gravity: 0.86
Heartwood: Alternating stripes
  varying from bright yellow to light
  red or violet
Grain: Straight to slightly interlocked
Growth Rings: Distinct, in some cases
  due to ring-porous distribution, in
  others due to terminal parenchyma
Pores: Medium to very small, unevenly
  distributed, solitary and in radial
  groups of 2-3
Parenchyma: Terminal, vasicentric,
  aliform with short wings (rarely
  confluent); also apotracheal in
  scattered tangential lines
Rays: Not distinct without lens
Ripple Marks: Distinct and regular; all
  elements storied

Rays: 1-3 (mostly 1-2) seriate, mostly 10
  or fewer cells high
Crystals: Parenchyma cells may contain
  crystals in chains
Intervessel Pitting: Alternate, medium
  large

SAPELE
(Entandrophragma cylindricum)

Average Specific Gravity: 0.60
Heartwood: Light red to dark reddish
  brown with a purplish cast
Odor: Cedarlike and aromatic
Grain: Straight to roey
Growth Rings: Distinct due to lines of
  terminal parenchyma
Pores: Small to medium-sized, visible
  without lens, rather numerous,
  evenly distributed, solitary or in
  short radial multiples
Parenchyma: Terminal, vasicentric,
  aliform with short wings and
  confluent, connecting 2-3 pores and
  forming short tangential bands
Rays: Visible as fine white lines on cross
  section
Ray Fleck: Dark against lighter
  background on radial surfaces
Ripple Marks: Distinct and regular

Rays: 1-5 (mostly 3-5) seriate, 15-20 cells
  high; storied
Crystals: Some marginal ray cells
  contain crystals; longitudinal
  parenchyma may have crystals in
  chains
Gum: Some vessel lumens with red gum
  and white deposits; vessel lines conspicuous due to vessel contents
Intervessel Pitting: Alternate, very small

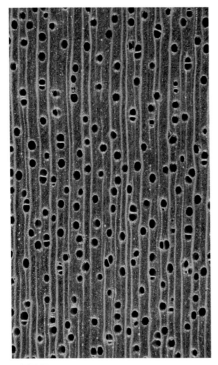

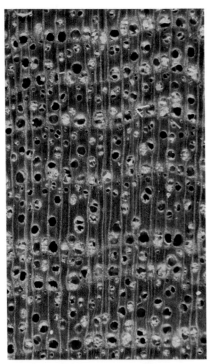

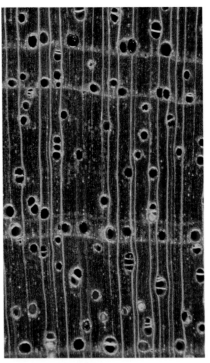

UTILE
(Entandrophragma utile)

Average Specific Gravity: 0.59
Heartwood: Uniformly red brown or
    purplish brown
Grain: Irregularly interlocked
Growth Rings: Distinct due to
    tangential lines of terminal
    parenchyma
Pores: Medium to large, visible without
    lens, fairly numerous, evenly
    distributed, solitary and in radial
    multiples of 2-4
Parenchyma: Terminal, vasicentric,
    aliform with short wings, confluent
    forming wavy bands; also
    apotracheal in wavy bands
Rays: Barely visible on cross section
Ray Fleck: Dark on radial surfaces
Ripple Marks: Absent or irregularly
    distributed

Rays: 1-4 (mostly 2-3) cells wide and
    15-40 cells high
Crystals: Longitudinal parenchyma
    may contain crystals
Gum: Some pores have red gum or
    white deposits
Intervessel Pitting: Alternate, small

TEAK
(Tectona grandis)

Average Specific Gravity: 0.57
Heartwood: Dark golden yellow
    turning dark brown or nearly black
    with age; waxy feel
Grain: Straight
Odor: Characteristic spicy odor
Growth Rings: Distinct; wood usually
    ring-porous
Pores: Earlywood pores very large,
    solitary and in radial groups of 2-3;
    latewood pores smaller, not
    numerous, evenly distributed;
    vessels with tyloses or with yellowish
    or whitish deposits
Parenchyma: Terminal and vasicentric
Rays: Distinct without lens on cross
    section

Rays: 1-5 (mostly 3-4) seriate
Gum: Ray and longitudinal
    parenchyma with brown gum
    contents
Intervessel Pitting: Alternate, medium

SPANISH-CEDAR
(Cedrela spp.)

Average Specific Gravity: 0.44
Heartwood: Pale, medium or dark
    brown with orange, reddish or
    purple tinge
Odor: Characteristic spicy or cedarlike
    odor
Grain: Straight to roey
Growth Rings: Usually distinct,
    delineated by thin band of terminal
    parenchyma in diffuse-porous
    specimens, and by predominance of
    largest pores along earlywood in
    ring-porous specimens
Pores: Medium-large to medium-small,
    not numerous, evenly distributed,
    solitary and in radial multiples
Parenchyma: Terminal and vasicentric
Rays: Visible without lens on cross
    section
Ray Fleck: On radial sections, ray fleck
    is darker than background
Gum: Vessels with red gum; vessel lines
    distinct as reddish scratches

Rays: 1-5 (mostly 3-4) seriate
Crystals: Marginal cells with crystals;
    longitudinal parenchyma
    occasionally with crystals
Intervessel Pitting: Alternate, medium
    large

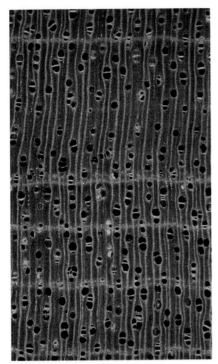

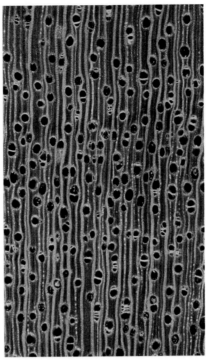

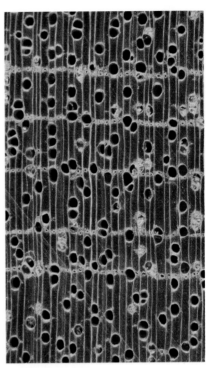

CENTRAL AMERICAN MAHOGANY
(Swietenia spp.)

Average Specific Gravity: 0.58
Heartwood: Pale brown or pink to dark reddish brown
Grain: Straight or roey
Growth Rings: Distinct due to concentric lines of terminal parenchyma easily visible without a lens
Pores: Medium-large, visible without lens, rather numerous, evenly distributed, solitary and in radial groups of 2-10
Gum: Some pores with red gum contents or whitish deposits
Parenchyma: Terminal and vasicentric
Rays: Barely visible without lens on cross section
Ray Fleck: On radial surfaces, ray fleck darker than background
Ripple Marks: Usually distinct and regular; occasionally irregular, all elements storied

Rays: 1-6 (mostly 3-4) seriate
Crystals: Crystals in some marginal ray cells; longitudinal parenchyma occasionally contains crystals
Intervessel Pitting: Alternate, very small

AFRICAN MAHOGANY
(Khaya spp.)

Average Specific Gravity: 0.63
Heartwood: Pale rosy red to dark reddish brown, often with purplish cast
Grain: Typically interlocked, producing even stripe figure
Growth Rings: Usually indistinct, but sometimes distinct due to increased fiber density in outer latewood or weakly defined terminal parenchyma
Pores: Medium to medium-large, visible without lens, relatively few to numerous, evenly distributed, solitary and in radial groups of 2-8
Gum: Some pores with red gum contents (but usually not white)
Parenchyma: Usually not distinct without lens; terminal parenchyma occasionally present, poorly defined; vasicentric parenchyma narrow
Rays: Distinct on cross section without lens
Ray Fleck: On radial surfaces, ray fleck darker than background
Ripple Marks: Usually not present; if present, indistinct and irregular

Rays: Of two sizes: smaller mostly 2-3 seriate, larger 4-7 seriate, with larger cells on flanks of rays (tangential view)
Crystals: Marginal cells and longitudinal parenchyma may contain crystals
Intervessel Pitting: Alternate, very small

WHITE LAUAN
(Shorea spp.)

Average Specific Gravity: 0.46
Heartwood: Pale greyish or yellowish brown with pinkish cast and silvery sheen
Grain: Roey
Growth Rings: Not distinct
Pores: Medium-large to very large, distinct without lens, evenly distributed to echelon (diagonal lines), solitary and in radial groups of 2-3, some with tyloses
Gum Ducts: In long tangential lines, embedded in bands of parenchyma; free of contents
Parenchyma: Vasicentric and aliform with short wings
Rays: Barely visible on cross section without lens

Rays: 1-5 (mostly 3-4) seriate
Intervessel Pitting: Alternate, medium-large

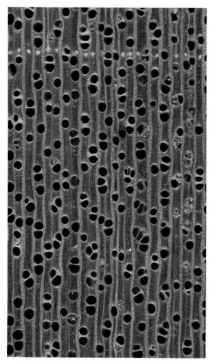

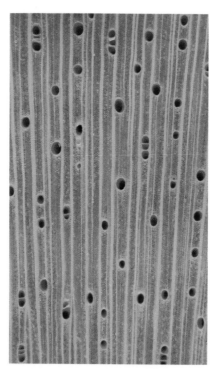

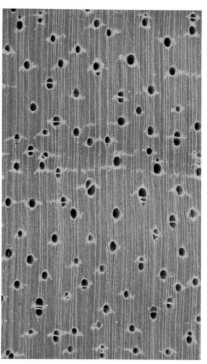

YELLOW MERANTI
(*Shorea* spp.)

Average Specific Gravity: 0.50
Heartwood: Light yellowish brown
    with cinnamon cast and golden
    luster
Grain: Roey
Growth Rings: Indistinct
Pores: Medium to large, visible without
    lens, evenly distributed, solitary and
    in radial groups of 2 to several, some
    with tyloses
Gum Ducts: In tangential lines or arcs,
    embedded in tangential bands of
    parenchyma, with whitish contents
Parenchyma: Not distinct with lens,
    vasicentric and aliform with very
    short wings; occasionally
    diffuse-in-aggregates
Rays: Barely visible on cross section
    without lens

Rays: 1-5 (mostly 3-4) seriate
Crystals: Some ray cells with crystals
Intervessel Pitting: Alternate, medium
    to medium large

BALSA
(*Ochroma pyramidale*)

Average Specific Gravity: 0.16
Heartwood: White, cream or pale
    brown, sometimes with pinkish tinge
Grain: Straight
Growth Rings: Not apparent
Pores: Medium-large, distinct without
    lens, numerous to moderately few,
    solitary and in radial multiples of 2-3
Parenchyma: Usually not distinct with
    lens
Rays: Conspicuous on all surfaces

Rays: 1-3 (mostly 5-6) seriate
Intervessel Pitting: Alternate, medium
    diameter

KORINA
(*Terminalia superba*)

Average Specific Gravity: 0.49
Heartwood: Wood in most commercial
    grades uniformly cream or pale
    yellow; may also be greyish brown
    or with streaks of dark brown or
    black
Grain: Straight to wavy or irregular
Growth Rings: Usually distinct because
    of increased fiber density
Pores: Medium-sized, distinct without
    lens, not numerous, evenly
    distributed, solitary or in radial
    multiples of 2-3
Parenchyma: Vasicentric, aliform with
    short to long wings and confluent
    forming occasional wavy tangential
    bands or lines
Rays: Not visible without lens, fine

Rays: Uniseriate (rarely biseriate),
    homogeneous
Crystals: Elongated crystals in
    longitudinal parenchyma
Intervessel Pitting: Alternate, large

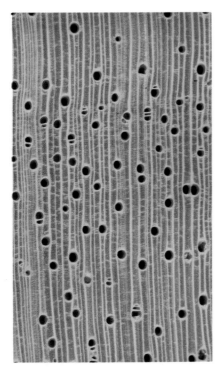

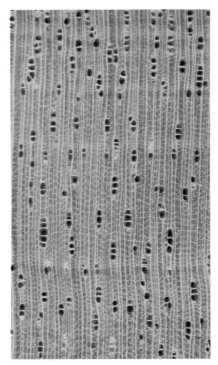

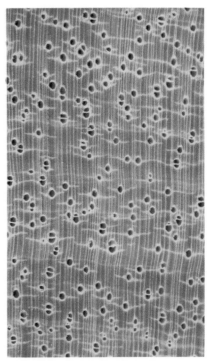

OBECHE
(Triplochiton scleroxylon)

Average Specific Gravity: 0.34
Heartwood: Heartwood and sapwood
  not distinguishable; wood uniformly
  cream to pale yellowish brown
Grain: Straight to interlocked
Growth Rings: Faintly distinct due to
  increased fiber density
Pores: Medium-large, distinct without
  lens, not numerous and irregularly
  distributed
Parenchyma: Visible with lens,
  apotracheal in fine tangential lines
  between rays
Rays: Visible as distinct light lines on
  cross sections
Ripple Marks: Distinct and regular

Rays: 1-8 (mostly 5-6) seriate
Crystals: Ray cells with crystals;
  longitudinal parenchyma with
  crystals
Intervessel Pitting: Alternate, medium
  diameter

JELUTONG
(Dyera costulata)

Average Specific Gravity: 0.40
Heartwood: Not distinct; wood pale
  straw to creamy white, its clear
  appearance interrupted by
  conspicuous latex slits at regular
  intervals of about 3 ft. on
  longitudinal surfaces
Grain: Straight
Growth Rings: Visible
Pores: Medium-sized, visible without
  lens, mostly in radial groups of 2 to
  several, or solitary
Parenchyma: In fine tangential lines,
  forming reticulate pattern with rays
Rays: Fine, faintly visible on cross
  section

Rays: 1-3 (mostly 2) seriate, many rays
  with horizontal latex tubes
Intervessel Pitting: Alternate, small
  diameter
Fibers: Thin-walled

RAMIN
(Gonystylus macrophyllum)

Average Specific Gravity: 0.59
Heartwood: Not distinct; wood whitish
  to pale yellow
Grain: Straight or shallowly interlocked
Growth Rings: May be faintly distinct
  due to fiber tracheids
Pores: Medium-large, evenly
  distributed, solitary or in radial
  multiples
Parenchyma: Aliform and confluent,
  fine bands connecting up to several
  pores
Rays: Fine, barely visible on cross
  section

Rays: Uniseriate
Intervessel Pitting: Alternate, small
  diameter

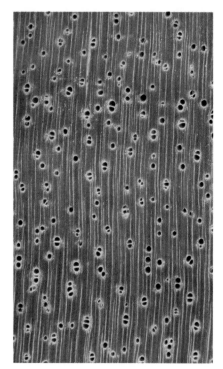

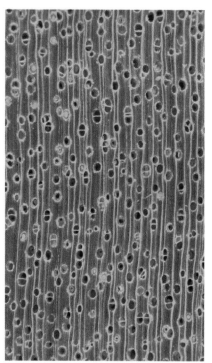

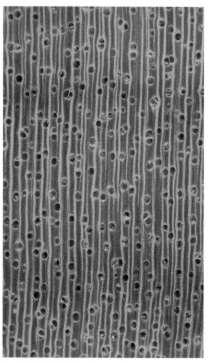

KOA
(Acacia koa)

Average Specific Gravity: 0.60
Heartwood: Light to dark brown with
    golden luster, often with irregular
    darker streaks
Growth Rings: Not distinct
Pores: Fairly small, not distinct without
    lens, numerous, evenly distributed,
    solitary and in radial multiples
Parenchyma: Vasicentric, visible with
    lens
Rays: Very fine, not visible without lens

Rays: Uniseriate or occasionally
    biseriate
Intervessel Pitting: Alternate,
    medium-small diameter

AVODIRE
(Turraeanthus africanus)

Average Specific Gravity: 0.52
Heartwood: Pale yellow to nearly white
    with high satiny luster
Grain: Usually roey
Growth Rings: Indistinct
Pores: Medium-sized, barely visible
    without lens, numerous, evenly
    distributed, solitary or in radial
    multiples
Parenchyma: Not visible with lens
    (paratracheal, scanty)
Rays: Indistinct on cross section
    without lens

Rays: 1-2 (mostly 2) seriate
Crystals: Parenchyma cells with
    crystals in chains
Intervessel Pitting: Alternate, small

PRIMA VERA
(Tabebuia donnell-smithii)

Average Specific Gravity: 0.41
Heartwood: Pale yellowish to light
    yellowish brown, often with darker
    streaks
Grain: Usually roey
Growth Rings: Distinct due to greater
    fiber density and terminal
    parenchyma
Pores: Medium-sized, distinct without
    lens, numerous, evenly distributed,
    solitary or in radial multiples or
    small clusters with moderately
    abundant tyloses
Parenchyma: Terminal and vasicentric
Rays: Distinct on cross section without
    lens
Ripple Marks: Fairly distinct to
    indistinct and irregular

Rays: 1-8 (mostly 3-6) seriate
Intervessel Pitting: Alternate,
    medium-small diameter

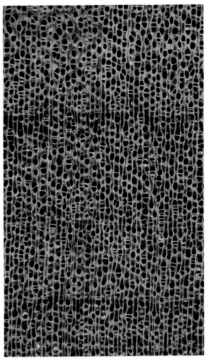

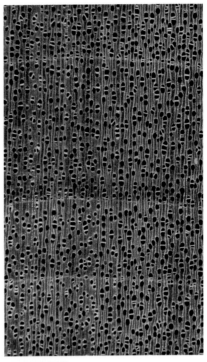

IMBUIA
(Phoebe porosa)

Average Specific Gravity: 0.57
Heartwood: Yellowish brown to chocolate brown, often somewhat variegated
Odor: Characteristically spicy
Grain: Straight to roey
Growth Rings: Distinct due to darker areas of denser fiber
Pores: Medium-small, barely visible, numerous, evenly distributed, solitary and in radial groups of 2-5, some with tyloses
Parenchyma: Not visible with lens (paratracheal scanty, vasicentric, apotracheal diffuse)
Rays: Barely visible on cross section

Rays: 1-3 (mostly 2) seriate
Oil Cells: Longitudinal parenchyma and rays with scattered oil cells
Intervessel Pitting: Alternate, medium large diameter

AFRORMOSIA
(Pericopsis elata)

Average Specific Gravity: 0.62
Heartwood: Golden brown, usually with darker brown streaks resembling growth rings
Grain: Straight to roey
Growth Rings: Faintly visible, delineated by terminal parenchyma
Pores: Medium-sized, barely visible, evenly distributed, solitary and in radial multiples of 2-4
Parenchyma: Terminal, vasicentric, aliform to confluent
Rays: Not distinct without lens

Rays: 2-3 (mostly 3) seriate
Crystals: Longitudinal parenchyma with crystals
Intervessel Pitting: Alternate, medium large diameter

MANSONIA
(Mansonia altissima)

Average Specific Gravity: 0.59
Heartwood: Yellowish to greyish to dark chocolate brown, sometimes with darker banding
Grain: Usually straight, but sometimes mildly interlocked
Growth Rings: Faintly visible due to indistinct lines of terminal parenchyma
Pores: Small to medium-small, not distinct without lens, numerous and evenly distributed, solitary and in radial multiples of 2-8
Parenchyma: Terminal and apotracheal diffuse-in-aggregates; visible with hand lens as short, fine tangential lines between rays
Rays: Not distinct without lens on cross sections

Rays: 1-4 (mostly 2-3) seriate
Crystals: Marginal ray cells with crystals
Intervessel Pitting: Alternate, small

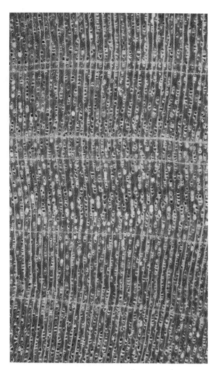

CEYLON SATINWOOD
(*Chloroxylon swietenia*)

Average Specific Gravity: 0.87
Heartwood: Light or golden yellow
Grain: Usually interlocked, producing a
    narrow stripe figure, and commonly
    with additional wavy grain,
    producing bee-wing figure
Growth Rings: Distinct due to lines of
    terminal parenchyma
Pores: Small, not visible, but distinct
    with lens, numerous and evenly
    distributed, mostly in radial
    multiples of 2-6 (some solitary)
Parenchyma: Terminal, also may be
    visible with lens as short tangential
    patches or lines
Rays: Not distinct without lens on
    cross sections
Ripple Marks: Distinct and regular

Rays: 1-5 (mostly 3-4) seriate
Crystals: Crystals in marginal ray cells;
    longitudinal parenchyma with
    crystals in chains
Intervessel Pitting: Alternate, small
    diameter

LIGNUM VITAE
(*Guaiacum* spp.)

Average Specific Gravity: 1.14
Heartwood: Medium to dark olive
    brown, typically with yellowish,
    greenish or darker streaking; oily or
    waxy feel
Odor: Characteristically pleasant and
    spicy
Grain: Usually finely interlocked or
    irregular
Growth rings: Usually indistinct
Pores: Small, visible only with lens, few
    and irregularly distributed, typically
    mostly solitary but in some
    specimens predominantly in
    radial rows
Parenchyma: Not distinct with lens; if
    visible, as irregular fine tangential
    lines
Rays: Very fine, faintly visible with lens
    on cross section
Ripple Marks: Regular but finely
    spaced, visible only with lens

Rays: Uniseriate, mostly 4-5 cells high;
    storied
Intervessel Pitting: Alternate, small
    diameter

EBONY
(*Diospyros* spp.)

Average Specific Gravity: 0.90
Heartwood: Uniformly black, or with
    lighter streaks; metallic luster
Grain: Straight to irregular or wavy
Growth Rings: Indistinct
Pores: Very small, barely visible with
    lens, not numerous, evenly
    distributed, solitary or in radial
    multiples of 2-3
Parenchyma: Not evident with lens
Rays: Very fine, visible only with lens

Rays: Uniseriate (occasionally biseriate)
Crystals: Numerous in rays
Intervessel Pitting: Alternate, small
    diameter

CHAPTER THIRTEEN

WOOD IDENTIFICATION IN THE REAL WORLD

As a student majoring in wood technology, I accepted my study of wood anatomy and identification as just one more of the many academic requirements for professional competence. I understood the importance of wood-identification skills to the many phases of wood technology, but gave little thought to ever using this expertise outside my chosen field. During subsequent years, however, I have been fascinated by the parade of wood-identification problems that have come my way from all walks of life. Of the calls and letters I receive asking for assistance in identifying wood, only the occasional inquiry is directly related to my own profession and usually involves some routine problem in lumber sales or manufacturing technology. Most of the requests come from the unrelated fields of science, commerce and law. In some cases, the identity of the wood is the only matter of concern; in others, the identification of one or more wood samples is but a small piece of a much larger and more complex problem.

The anecdotes in this chapter are offered as a sampling of the surprising breadth of applications of wood identification in the real world. They also serve to illustrate how the principles, techniques and details presented in this book are used to solve specific wood-identification problems.

Commercial Lumber Questions

As might be expected, disputes arise from time to time between vendors and customers concerning the species of hardwood or softwood lumber. If I were to single out the most frequent problem in this category, it would be the question of whether soft maple has been substituted for hard maple in a shipment.

Typically, the customer suspects that the lumber is not hard maple because an unusually large number of pith flecks are evident on the board surfaces after they are dressed. Pith flecks are found regularly in soft maples; however, they are occasionally numerous in hard maple as well. This is particularly true of black maple. Therefore, hard and soft maple are separated more reliably on the basis of the seriation of the largest rays.

In one instance, I examined a total of 12 tangential sections from three boards, and the largest rays were 4 and 5 seriate. Only two rays were 6 seriate. The boards also showed traces of greyish mineral streak. Therefore I concluded that the lumber was indeed soft maple, as claimed by the customer. In all other instances of this hard vs. soft maple controversy, however, I was able to find many rays that counted 8+ seriate in every tangential section sampled, indicating that the lumber was hard maple, as claimed by the supplier.

Another commercial-shipment question stands out in my mind because of the personal embarrassment it caused me. In the midst of a busy day I received a call from an engineering firm that was participating in the renovation of a large warehouse. Douglas-fir had been specified for the structural posts. Upon receiving the shipment, however, the firm suspected that another species had been supplied.

The project was on a tight construction schedule, and the contractor wanted confirmation that the timbers were Douglas-fir before going ahead. I assured the caller that checking for Douglas-fir was a simple matter and that I would be happy to do so as soon as they could send samples to me. Unfortunately, all I said was "samples," without specifying how large they should be. For the next two days I awaited delivery, but none came. Finally, on the third day, a trucker appeared at my office with a dolly laden with 20-in. lengths of 12x12s. I felt myself flush with embarrassment as I realized the unnecessary time and cost of shipping such large chunks when splinters in a first-class envelope would have been sufficient.

In examining the pieces, the reason for concern became obvious. The wood didn't look much like Douglas-fir. Some pieces were so slowly grown — there were 80 rings per inch in a few portions — that they appeared even-grained, lacking the usual distinct, uneven-grained rings so characteristic of Douglas-fir. The heartwood color was more yellowish brown than the familiar reddish brown of Douglas-fir heartwood. And some of the pieces had only a trace of the characteristic Douglas-fir odor.

Nevertheless, tangential sections examined microscopically revealed that each piece had both fusiform rays and abundant spiral thickenings in the earlywood tracheids, confirming that the wood was indeed Douglas-fir. In reporting the results, I assured the firm that the wood was the correct species, but urged that the material be checked to determine whether the structural grade requirements had been met. (As shown in the photos on pp. 10-11, extremely slow-growth Douglas-fir, as is typical in suppressed or overmature trees, can result in abnormally low wood density.) Since this incident, I have been very careful to give clear instructions regarding the size of samples to be submitted for identification.

Structural Restoration

A home owner called to ask if I would help settle an argument involving replacement ceiling beams that were being installed in her 250-year-old house. The builder had assured the woman's husband that the recycled hand-hewn barn timbers were chestnut. "They always used chestnut beams in those old barns," he said. But a friend had advised the couple that the beams were something else.

At my suggestion, she brought in a sample. It was immediately clear that this was not a ring-porous hardwood — or any kind of hardwood for that matter. Instead, it had the typical characteristics of a fairly uneven-grained softwood, but without any apparent resin canals. Checking tangential and radial sections with a microscope, I found no fusiform rays, so there were no resin canals. Nor were there any longitudinal parenchyma. But ray tracheids were present, cross-field pits were cupressoid and the ray parenchyma end walls were nodular. The "chestnut" beams were eastern hemlock.

Historical Archaeology

A more noteworthy case involved restoration work in Christ Church — more commonly known as Old North Church — in the North End district of Boston. This is the church made famous during the American Revolution as the vantage point chosen by Paul Revere to signal by lantern that the British were coming, "one if by land, two if by sea." In 1744, a set of eight bells was cast in England; the bells were installed in the belfry the following year. They were carried on a frame, specified in the original cutting list of 1744 as "all sound oak timber." Records show that the frame was rebuilt in 1847 of hard pine.

Now the frame was again in need of rebuilding. Someone reported that the bell frame timber was Douglas-fir, and therefore must be of relatively recent vintage because Douglas-fir must have come from the West. I was asked to help settle the question. This time I indicated that sound splinters would suffice for identification samples.

When the bell-frame samples arrived, their large resin canals as viewed with a hand lens initially suggested that they were not Douglas-fir. Radial sections viewed microscopically failed to show any spiral thickenings; rather, they had deeply dentate ray tracheids and pinoid cross-field pitting. This confirmed hard pine, probably the timber installed in the mid-19th century. The restoration was completed using white oak timbers as in the original frame.

Civil Engineering

Urban-redevelopment projects often uncover interesting wood-identification puzzles. For example, in Boston, a core boring assessing the integrity of abutments of the Charles River Bridge ran through some concealed timbers, fragments of which were sent to me. Though waterlogged and stained, the wood was apparently sound. A tangential microscopic section quickly revealed fusiform rays with thick-walled epithelial cells, but no spiral thickenings in the tracheids. This narrowed the choice to larch or spruce. A radial section showed no paired pits on the tracheid walls, and among the ray tracheids, bordered pits were *Picea* Type I and *Picea* Type II. Conclusion: the wood was eastern spruce.

Structural Engineering

One common wood-identification problem involves extensive sampling of the structural timbers of old buildings to enable an engineering analysis of their structural integrity. The examination of older buildings in the Northeast often suggests that only a relatively few species were used for framing — oak, chestnut and hard pine are perhaps the most common. However, repairs, later modifications, and additions may contain surprises.

The typical batch of samples is all one species. In one instance I received 24 samples from an old factory building. All 24 were southern yellow pine, as evidenced by their obvious high density, uneven grain, wide bands of latewood and distinct turpentine odor. Nevertheless, it was requested that this be verified microscopically. The repetitive checking of dentate ray tracheids and pinoid pitting became rather uninteresting by the time the last slide was finally on the microscope stage.

In another batch of 15 samples from an old mill building in eastern Massachusetts, however, ten samples were eastern spruce, two were southern yellow pine, one was eastern white pine, one was fir and one was hemlock. The hemlock was especially difficult to identify because it consisted of very severe reaction wood, with nearly the entire growth ring composed of abnormal tracheids with deep spiral checking. Only in a few of the earlywood tracheids could ray cross fields be found that were normal enough to assess the cupressoid pitting.

Decorative Arts

Compared to identifying a single sample of wood — or even a series of 20 or 30 samples — checking all the woods in a major furniture collection is a challenging task. Such an assignment presented itself when I was invited to assist in the identification of over 200 pieces of case furniture in the Garvan Collection and related collections at the Yale University Art Gallery.

Here the task had an added challenge: the samples had to be taken inconspicuously and with a minimum of damage to the objects. I had to read as much as possible from the surface characteristics of the wood. Assessing such physical features as weight, color, evenness of grain and prominence of rays became very important. Fortunately in woods such as beech or oak, which have conspicuous rays, old stain or paint can actually help highlight ray size and distribution.

In sampling primary woods (the visible exterior woods in a piece of furniture), small fragments could be removed from an inconspicuous spot, such as under a glide caster on the bottom of a foot or under a drawer lock at the edge of the original mortise.

Where tentative visual identification was possible, it suggested which specific microscopic view was needed for confirmation. It was often possible to remove inconspicuously the necessary section directly from the piece at a point of wear or minor damage, and it was sometimes possible to take tiny sections directly from the inside faces of shrinkage checks, which usually occur precisely along a radial plane.

The routine in surveying a piece of furniture is first to decide visually which components are of the same wood, then to establish a sampling plan to verify microscopically a representative number of samples of each apparently different wood type. Although microscopic checking most often simply confirms the initial visual identification, occasional surprises do turn up.

For example, in one chest the side panels looked at a glance like hard pine because of obvious uneven grain. I reasoned then that a radial microscopic section should be taken for confirmation. Expecting to see dentate ray tracheids and pinoid cross-field pits, I was startled to find myself staring at smooth-walled ray tracheids and cupressoid cross-field pits. I realized I had followed my intuition and had failed to check for resin canals — which of course were not present. The panels were, in fact,

hemlock. Catching these occasional surprises is a sobering reminder that visual impressions alone can be quite deceptive and that microscopic follow-up is a comforting safety net.

The Garvan Collection experience made me especially alert when identifying woods whose surface features are obscured by old finish, stain or accumulated dust and dirt. A painted Windsor chair is the ultimate test in wood identification. The layers of earlywood pores in ring-porous species such as oak and ash are usually detectable, and the conspicuous rays of oak and beech will often show through even the muddiest of finishes. The diffuse-porous hardwoods are especially deceptive, if not impossible.

Maple — particularly soft maple — was perhaps the most commonly used wood for turnings, so traditionally it has been assumed that legs and similar turnings are maple. But a surprising number are not. A tangential section usually puts the question to rest. For example, the stout turned legs of 17th-century chairs are often found to be aspen, as quickly revealed by their uniseriate rays and confirmed in radial sections that show the rays to be homogeneous. In some New England furniture, birch can be recognized by its scalariform perforation plates and minute alternate intervessel pitting.

Perhaps the greatest single surprise in the Garvan Collection case furniture was a chest that had been labeled butternut. It certainly looked like butternut in surface color and figure. The routine microscopic sampling paid off, as the sections revealed gash-like pitting on the radial walls of the vessels and large crystals in many of the longitudinal parenchyma cells. These features reliably confirm black walnut.

Since this initial project on case furniture, I have also completed the identification of woods in tables and looking glasses in the Garvan Collection. More recently, I have had the opportunity to identify woods in furniture of several other collections, including the Bybee Collection at the Dallas Museum of Art, the Milwaukee Art Museum furniture collection, the Bayou Bend Collection in Houston, the furniture collection in the Langdon House in Portsmouth, New Hampshire (maintained by the Society for the Preservation Of New England Antiquities) and the furniture collection in the Diplomatic Reception Rooms at the United States Department of State in Washington, D.C. These marathon identification projects have taught me the value of repetitive observation in developing a familiarity with identification features. They have also helped me to appreciate the range of normal variation in the anatomical characteristics of wood. I have especially learned the wisdom of microscopically checking even those woods I feel sure I know by sight.

Fine Arts

Other identification problems have led me onto very uncertain ground. A college museum once asked me to identify the wood in a religious icon of presumed Russian origin. The wood of the painted panel was said to be "either fruitwood or European pine" — quite a choice.

I managed to remove a portion of the panel support to expose the panel, from which I took a piece of wood large enough to provide a radial section. One glance at the strange cross-field pits was all it took for me to realize that I had never seen such wood before. Using Greguss's book (see the Bibliography on pp. 212-215), I finally traced it down to a species of *Podocarpaceae,* whose members are found mostly in the southern hemisphere, with a couple of species extending up into northern Africa and the Mediterranean region. I had no known material to check it against, and to this day I cannot be absolutely sure I was correct.

A local gunsmith once asked me to check out the wood in an old European military baton that had a dagger concealed within it. The sheath for the dagger blade was rosewood (*Dalbergia* spp.), but I knew immediately that I had never seen the wood of the handle portion before. Using Kribs's manual (see the Bibliography) as my guide, I painstakingly keyed it out as a wood called dika (*Irvingia gabonensis),* of African origin. I have not seen or heard of this wood since, and probably never will. We had no sample in our collection against which I could check it, but the features agreed perfectly with the description in Kribs. Still, I'll never really know if I had that one right either.

Prehistoric Archaeology

Archaeological work also turns up challenging problems in wood identification. The wood has usually been unearthed and is often discolored. In one case, I received two pieces of wood taken from the remains of a coffin excavated from a slave cemetery in Frederick County, Maryland. Although badly decayed, the splinters had areas sound enough to surface with a razor blade. It was evident that both were ring-porous. In one, the broad rays, radial patches with numerous indistinct latewood pores and abundant tyloses indicated white oak. In the other, only uniseriate rays were present, which indicated chestnut.

Because wood is biodegradable, a more common form of wood recovered intact from archaeological sites is charcoal. I received a sample from an excavation of a Paleo-Indian site in Washington, Connecticut. Other material from the site had been carbon-14-dated at 10,190 +/-300 years before the present. The sample comprised many small fragments of charcoal, each measuring only a millimeter or two. By examining fracture surfaces of the fragments, two species were evident. One was a ring-porous hardwood with both broad and narrow rays and distinct latewood pores. But it had no tyloses and was therefore presumed to be red oak. The other species was a softwood, with fine texture and no evidence of resin canals. Its rays were up to only 8 cells high, with most between 3 and 6 cells high. My identification was somewhat uncertain, but it was probably *Juniperus* spp.

Wildlife Biology

Wood identification has drawn me into other areas of science, including wildlife biology, fisheries biology and entomology. In one study, analysis of stomach contents from raccoons trapped in Rhode Island produced a variety of unidentified fragments. I was asked if it was possible to determine whether or not they were wood, and if so, what species of wood.

I found a considerable amount of wood among the fragments. Softwood fragments, upon microscopic examination, were found to contain dentate ray tracheids. These were sometimes accompanied by windowlike pits, which indicated red pine (*P. resinosa).* In other cases, the pits were pinoid, indicating pitch pine (*P. rigida).* A third species, soft maple, could also be discerned based on vessel perforations, pits and spirals, and ray seriation and type.

In another case, fish mortality in New Hampshire was investigated, and it was found that the dead fish had their gills clogged with microscopic debris. Careful examination of the debris revealed that it was principally fragments of softwood tracheids with piceoid pitting. Samples of water taken at the onset of the fish kill contained physical pollutants, primarily softwood fiber, also with piceoid pitting, in amounts more than 1,000 times greater than the quantity of wood fiber that normally occurs as a result of natural attrition of driftwood.

The microscopic evaluation narrowed the possibilities to woods that were all species being used by an upstream paper mill. The mill was under investigation for illegal dumping of pulp residues. The information produced through wood identification was used in an effort to enforce corrective actions to clean up the pollution.

In evaluating the nest-building habits of the white-faced hornet, entomologists wanted to know what species of wood were being chewed free from partially rotted logs to form the intricate paper nests that housed the hornet colony. Bits of the hornets' nest "paper" examined with a microscope showed obvious hardwood vessel elements, elaborate alternate intervessel pitting and spiral thickenings, indicating maple. Fibers with bordered pits boasting long, deep, extended apertures also indicated a hardwood origin. However, the large bordered pits of tracheids that were found indicated that softwoods were also being used by the hornets; the cross-field pitting, which was piceoid or cupressoid, suggested spruce or hemlock.

Law

I have been involved as a consultant and as an expert witness in lawsuits where wood or wood products were involved and wood identification was in some way critical to the outcome. In my very first case, a home owner had purchased a stockade-style cedar enclosure for an outdoor swimming pool. After three years, both the posts and the fence sections had rotted to the point of collapse — hardly the performance expected of cedar.

Eight samples were brought to me for identification. Of the eight, five were balsam fir and three were eastern spruce. The home owner promptly sued the supplier. When the case came to trial I was summoned to appear in court as a witness. The plaintiff's attorney asked me routine questions about the identity of the woods, and, surprisingly, the defendant's lawyer offered no cross-examination. However, the judge questioned me thoroughly as to whether or not spruce, fir and cedar were in fact "different kinds" of wood. She asked specifically how I could be sure, beyond a doubt, that the fencing was not cedar.

I found myself giving a wood-anatomy lecture in court, and I was glad I had had many "rehearsals" in the classroom. Apparently my responses were to the judge's satisfaction. The home owner won the case.

Other cases have been more tragically serious. In one incident, a car veered from the highway, killing the driver and passenger. Upon examining the wreckage, a fragment of wood was found lodged in a shattered area of the fiberglass car body. One hypothesis was that the car had hit something wooden in the road, perhaps a tree limb, causing the driver to lose control and swerve to the side.

The fragment of wood was sent to me for examination. It was a walnut-sized chunk gouged from a larger piece. I identified it as cherry, based on the fine texture, diffuse-porous distribution with earlywood pores aligned at the growth-ring boundary, distinct rays, spiral thickenings, etc. I also noted that the nugget had a clean, smooth cambial surface, free of any weathering or discoloration, suggesting that the piece had come from green wood during the growing season when the bark readily separates. The curvature of the cambium suggested a stem diameter of 5 in. to 6 in.

We visited the accident site some six years after the accident. Several feet to one side of the presumed path of the car we found a 6-in.-dia. cherry tree with a gouge on one side. Although callous tissue and six years of added diameter had begun to close the wound, the depression clearly showed gouge marks that perfectly matched the specimen from the car's fender.

The wood analysis had clearly established that the car left the road first, then hit the tree. It also helped pinpoint the path of the car as it crashed, and the height of the scar on the tree indicated that the car became airborne after it left the road. Wood identification provided a small but important piece of evidence in this case.

The most difficult single problem that I ever encountered resulted from an accident where a window washer fell when his ladder suddenly broke. The man suffered head injuries that left him permanently incapacitated.

The ladder was sold as having hemlock rails. I identified one rail of the ladder as western hemlock, an acceptable species for ladder rails. The other was apparently fir, individual species of which are usually considered indistinguishable on the basis of wood tissue alone. Confusingly, the ladder code allows noble fir (Abies procera) but not other species, so it became critical to know which fir species it was.

Crystals in the ray parenchyma cells were extremely sparse. Fortunately, I remembered a journal article reporting work done at Forintek Laboratory in Vancouver, British Columbia,

that established a correlation between the ray-parenchyma crystal count and various fir species. I made crystal counts, then consulted the paper. The low number suggested that the wood was not A. procera, but probably A. amabilis or A. lasiocarpa.

As a check of my own work, I submitted a sample of wood to Forintek Laboratory. Their findings were similar. Next, I tried a color reaction test using Ehrlich's reagent. This reagent gives a positive reaction — a purple coloration — with a known sample of subalpine fir (A. lasiocarpa), but a negative reaction with Pacific silver fir (A. amabilis). The wood sample from the ladder gave no reaction. Ray-cell contents are reported to be clear or pale yellow in A. balsamea and A. lasiocarpa, but dark brown in other western firs. The contents of ray cells in the questionable ladder rail were dark brown.

I concluded that the ladder rail was probably Pacific silver fir, and that its extremely low density (specific gravity 0.25) and weakness were principal contributing factors in the ladder's failure.

Just for Fun

Wood identification need not always be serious or important. For a change of pace, I sometimes find myself identifying wood just for fun. This is not to say that the task is always successful or easy.

A friend once dropped off a small sack of assorted woods to "check out when you get a minute." When time permitted, I laid them out on my bench. I didn't recognize a single one. With a razor blade I cleaned up an end-grain surface on each for a closer look with a hand lens. They were all hardwoods, but strangers every one. A few looked like dipterocarps, perhaps lauan or meranti. I called my friend to ask him the source of such an exotic assortment. The reply was that they were "crating boards from a Japanese motorcycle." I threw in the towel.

APPENDIX I

ATYPICAL WOOD

Although this book is about the identification of wood, any precise definition of the term "wood" has been carefully avoided. The wood with which we are most familiar is mainly derived from the merchantable central stem tissue of tree-sized coniferous gymnosperms (softwoods) and dicotyledonous angiosperms (hardwoods); therefore the book has dealt primarily with timber species of commercial importance in North America.

There are, however, a surprising number of other plants — especially in other parts of the world — that yield useful materials that can be broadly classified as wood and whose applications are comparable to our familiar woods. Some are closely related botanically to our common species and therefore recognizably similar. Others represent unrelated plant groups with uniquely different anatomical features. It is fitting that at least brief mention be made of some of the more important and unusual ones. This appendix does not attempt to serve as a detailed identification guide to these other woody plants, but it should help you recognize some of the major groups.

Without a doubt, our perception of the domain of wood has been dictated at least in part by the commercial timber industry in our temperate region — an industry based on tree-sized softwoods and hardwoods. It is easy to forget that for some applications such as certain types of basket-making, shrubs and twigs are the most appropriate form of wood. A close look at American antiques turns up a variety of examples of the use of shrubs, twigs and vines. If we look at other regions of the world, we find that monocots are a principal source of functional material. It is not surprising, then, that in the modern American department store we see increasing numbers of imported items manufactured from monocots — mainly rattan and bamboo — especially in the home-furnishing department.

For convenience, I will consider these atypical or peripheral woody materials under two headings: unconventional tree parts or tree forms, and woody monocotyledonous angiosperms (monocots).

Unconventional Forms

In an old volume on frontier survival, I found instructions on how to build a turkey call. The principal component was a mountain-laurel peg. For other items, I have heard of sumac, lilac, hawthorn, hazelnut, spicebush and barberry being used. These woods are certainly recognizable as hardwoods, but one would be hard-pressed to find them described in wood-identification books.

Sometimes the unfamiliar-looking twigs used in various applications belong to species or genera we recognize quite readily in board form.

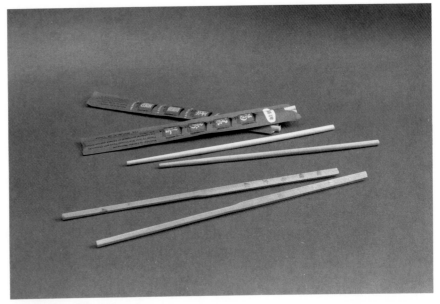

BAMBOO CHOPSTICKS
An increasing number of home furnishings and accessories, such as baskets, wicker furniture, and the chopsticks shown here, are sold in American stores and are made from monocots, principally bamboo and rattan.

The osiers, or flexible twigs, used as withes for weaving willow-ware or split-willow baskets *(Salix viminalis,* twig willow, for example) may not be recognized in twig diameters as willow. Certain tribes of American Indians used fine roots for weaving baskets; although we might be familiar with a tree's wood, its root wood can be strangely different (for more information on this interesting subject, see Schweingruber's *Microscopic Wood Anatomy).*

The bark of trees can also be used to make products. Basswood, for example, has strong sinewy fibers or "bast" in its inner bark. The bark can be used in strips for weaving baskets or chair seats; the fibrous strands can be woven into cord and rope. Strips of inner hickory bark, when green, are as pliable as wet leather, but they are hard and resilient when dry. Hickory bark also makes excellent chair-seat material. Although its cell structure is different from that of hickory wood, it is recognizable by the streaks in the surfaces of the strips (see the photo below right). These characteristic streaks correspond to the crevices between plates in the rough outer bark.

The best-known tree bark is cork. Commercial cork is the product of the cork oak *(Quercus suber),* a native of the Mediterranean — the world supply comes primarily from Portugal, Spain, Algeria and Morocco. Cork is the outer bark of the tree. Layers can be stripped away from the tree every 8 to 10 years during the tree's productive lifetime of 150 years or more.

Cork consists of physiologically dead cells that are thin-walled and essentially watertight, giving the cork its light weight. It has been used in its natural form by simply cutting it to shape to make such items as bottle stoppers, fly-rod handles and shoe soles. In addition to being used in solid pieces, however, cork is now commonly broken into fragments, mixed with a binder and pressed into sheets or blocks to make products such as gaskets, bottle-cap liners, tack boards and waterfowl decoy bodies.

Cork has faintly visible tangential growth layers, with dark streaks called *lenticels* running radially (see the top left photo on the facing page).

Microscopically, cork cells are nearly isodiametric, looking like a mass of tightly packed air-filled balloons (see the bottom left photo on the facing page).

Vines are also incorporated into various artifacts and craft objects. Grape vines are easily identified by their ring-porous structure, very large vessels and wide, closely spaced rays (see the top right photo on the facing page). I have a set of coasters that are 3-in. dia. cross-sectional wafers of wisteria vine, which has a ring-porous structure somewhat similar to that of grape.

The ginkgo or maidenhair tree, a native of the Orient, is the single species *(Ginkgo biloba)* in the gymnosperm order *Ginkgoales* (see the bottom right photo on the facing page). The tree has some similarity to hardwoods, with its fan-shaped deciduous leaves, but its cell structure more closely resembles that of the softwoods in the order *Coniferales.* Because there are only limited quantities and its wood is quite brittle, ginkgo is not a commercially important timber species.

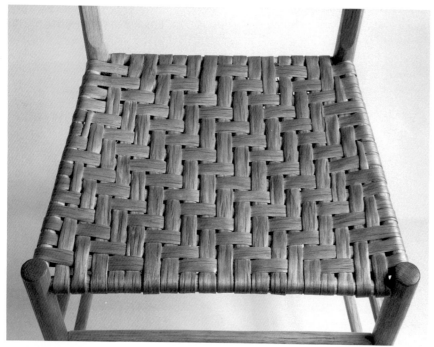

HICKORY INNER BARK
This chair seat is woven from strips of the inner bark of shagbark hickory.

LENTICELS IN CORK
This piece of cork oak bark shows growth layers and several dark, radially oriented lenticels.

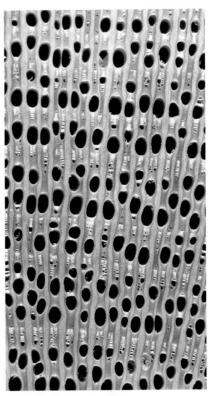

GRAPE STEM WOOD
Grape-vine wood is ring-porous, with numerous large rays. (15x)

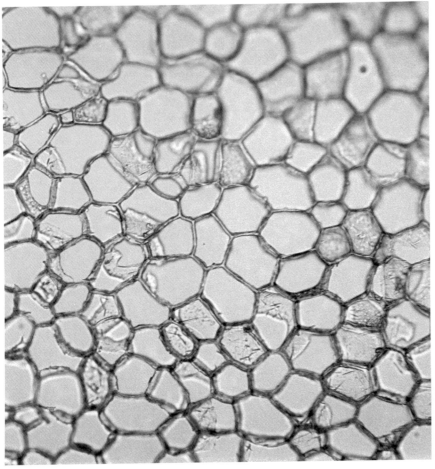

CORK CELLS
Cork cells are thin-walled and roughly spherical. (400x)

GINKGO
Wood of the ginkgo tree resembles a conifer without resin canals. (15x)

Monocots

In the United States, the tree forms of angiosperms are principally dicotyledonous, but in the southern and coastal states there are several tree-size monocots that yield woody stem material. The palms are the most familiar, but there are also a few bamboos, cacti and yucca. In other areas of the world, however, monocots are the dominant tree-form flora; in some cases the local economy depends on them.

BAMBOO

It is easy to argue that bamboo is the world's single most important plant. In some regions it provides many of life's necessities. The greatest concentrations of bamboo and its greatest use are in the eastern hemisphere from India to China along the Asian mainland, and from Japan to Java among the islands. The known bamboo flora encompass more than 700 species in some 50 genera, nearly half of which are in China.

The uses of bamboo are many. The large bamboos are used in stem form or as slats for construction; smaller stems are fabricated into such items as furniture and implements; split strips are fashioned into baskets, screens, utensils and novelties. We commonly see bamboo floral sticks and chopsticks.

While most bamboo genera are obscure, a few, including *Arundinaria*, *Bambusa*, *Dendrocalamus*, *Phyllostachys* and *Guadua*, are noteworthy. Tonkin cane from China, prized in fly-rod manufacture, is *Arundinaria amabilis*, but two-thirds of the bamboo China produces is mao chu *(Phyllostachys pubescens)*. In South America, *Guadua augustifolia* is a favorite for construction and furniture.

Bamboo is the largest member of the *Gramineae* or grass family. It has a slender stem or culm, which is hollow and jointed at regularly spaced nodes (see the photo above right).

A transverse section through the culm wall reveals typical monocotyledonous structure. The ground mass tissue is composed of nondescript parenchyma

cells that contrast sharply with the vascular bundles. Each bundle consists of two large vessels, thin-walled sieve tubes (a series of cells connnected by sievelike plates at the end walls), companion cells (special parenchyma cells associated with sieve tubes) and, usually, fibers. These groupings can take on the appearance of amusing faces (see the bottom photo below). The bundles are larger and farther apart toward the inner edge of the culm, but smaller, more numerous and more crowded near the denser periphery.

PALM

The *Palmae* or palm family comprises over 2,500 species, mainly in the tropics. The palms lack clear classification, although at least 13 distinct subgroups have been recognized. The most economically valuable are in the genus *Cocos* (coconut palms) and the genus *Phoenix* (date palms). These are especially important where they constitute the only woody vegetation. The few palms native to the United States include the California washingtonia *(Washingtonia filifera)*,

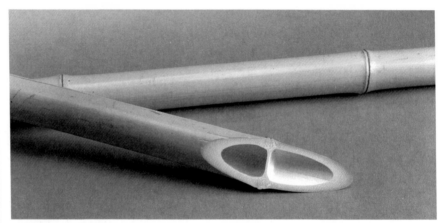

BAMBOO CULM
A cutaway view of this bamboo culm shows a cross wall at the node.

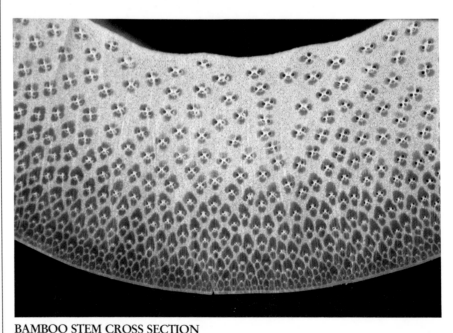

BAMBOO STEM CROSS SECTION
A transverse surface of the wall of a bamboo culm shows the fibrovascular bundles embedded in softer parenchymatous tissue. (15x)

**BAMBOO FLY-ROD
CROSS SECTION**
Six precisely milled segments of Tonkin cane were bonded together to form this fly-rod blank . (8x)

the Florida royal palm *(Roystonea elata)* and the cabbage palmetto *(Sabal palmetto).*

The anatomical stem structure among the various palms is surprisingly uniform, so it offers little taxonomic information. Identification of individual species relies on other plant parts, such as leaves and fruits. The outer portion of mature stems consists of fibrovascular bundles embedded in a ground mass of parenchyma tissue. Toward the center, the bundles are fewer and may disappear altogether; at the outer surface, they are smaller and more crowded.

Species with the densest outer layers of small bundles yield woody material that can be made into such items as walking sticks, umbrella handles and fishing rods that are beautifully figured and take a high polish. The material

farther away from the dense surface cannot be worked to a smooth surface because of its hornlike bundles in a soft parenchyma matrix. In fact, it is often extremely difficult to hand-section a transverse surface for study because of the variation in density between the bundles and the parenchyma mass (see the top photo below).

Longitudinal surfaces of palm wood get their characteristic appearance from the slightly irregular alignment of the bundles in the parenchyma mass; the surface roughness is traceable to the density variation between the bundles and the parenchyma (see the bottom photo below).

In addition to novelty items, we find palm wood in a variety of imported products such as crates, planter boxes and plant stakes.

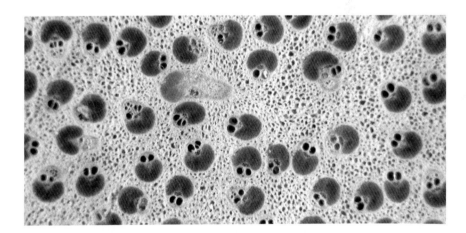

**CROSS-SECTIONAL SURFACE
OF PALM**
A transverse surface of the inner portion of a palm stem showing the fibrovascular bundles embedded in soft parenchymatous tissue. (15x)

PALM WOOD
As seen in these pieces of palm wood, the fibrovascular bundles are conspicuous on all surfaces.

RATTAN

A specialized group of the palm family known as the *Lepidocaryoid* palms include vinelike climbing forms commonly referred to as rattans. One principal genus is *Calamus,* embracing 250 to 300 species. Its stems are extremely long, slender and uniform in diameter. Pulled down from their host trees, these vines are scraped free of their outer covering, dried and fashioned into a host of products. Larger stems are used for walking sticks, furniture frames and broom handles. They are also used for hampers, baskets, baby carriages, chairs and other light furniture known as "wickerwork," although this can also be fashioned from other suitable materials such as willow and bamboo. The main framing in wicker is made of the stouter vines, the ribbing and weavers of slender rattan.

Rattan splits readily into very uniform strips. The dense outer shell of the vine provides split rattan cane for caned chair seats or for more exacting basketry such as the famous Nantucket Lightship baskets (see the top photos below).

The soft inner portion of the stem is split into reed, which is either left flat or pulled through round dies. This material is more flexible and capable of being bent into the intricate and ornate designs that are the hallmark of wicker furniture. Flat strips of split rattan with rough surfaces are also used to imitate split ash for baskets and basket-weave seats.

When viewed in cross section with a hand lens, rattan is easily recognized by the uniform size and spacing of the bundles, each of which contains a single large vessel or pair of vessels (see the photos at the bottom of this page).

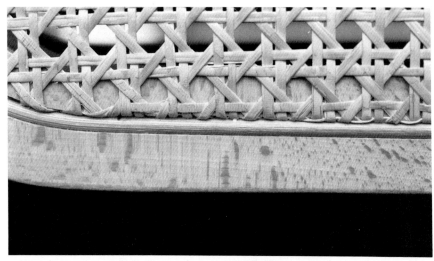

Susan Kahn

CANE CHAIR SEAT AND NANTUCKET LIGHTSHIP BASKET
The cane used to weave the classic octagonal seat design at left is stripped from the dense outer surface of the rattan stem. The same type of cane was used to make the basket at right in the Nantucket Lightship style.

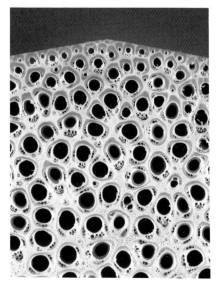

RATTAN STEM
The large single vessels of the vascular bundles are visible on the transverse surface of a rattan stem (near right). As seen through a hand lens (far right), the large-pored bundles help identify it as rattan.

CACTUS AND YUCCA

Some cacti, including the saguaro (*Cereus giganteus*) of the southwestern United States, grow to a substantial size. After the cactus dies and the fleshy tissue deteriorates, a woody "skeleton" remains. Such dried stems, especially of the chollas (*Opuntia* spp.), can be fashioned into such novelties as paperweights and lamp bases. They are usually recognized by the geometric pattern of the remaining woody tissue.

Of the eleven native species of yucca that reach tree size, the Joshua-tree (*Yucca brevifolia*) is perhaps the best known. Although its stem is occasionally used for structural posts or poles, its anatomical structure consists of coarse stringy fibers, and pieces cut from the stem cannot be smoothed in the manner of conventional wood. Yucca is recognized by the loose, fibrous nature of exposed surfaces (see the bottom photo at right).

Because yucca is now considered a threatened plant in some areas, attempts to collect it or use its woody tissue are discouraged.

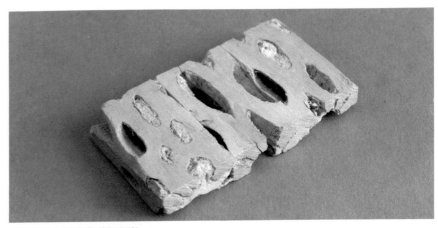

CHOLLA 'SKELETONS'
After the soft tissue of various cholla cacti disintegrates, characteristic 'skeleton' stems of woody tissue remain.

JOSHUA-TREE, AMERICAN SOUTHWEST
The Joshua-tree (*Yucca brevifolia*) is a tall yucca with a tree-like form and is locally common in the Mohave Desert and Joshua Tree National Monument in Southern California.

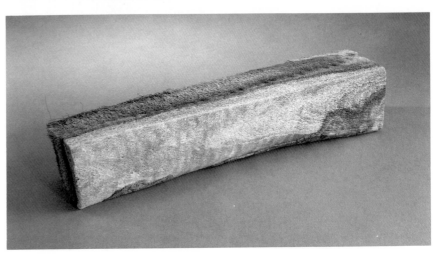

JOSHUA-TREE STEM SECTION
Yucca stems consist of stringy fibrous tissue that does not conform to our usual idea of tree 'wood.'

APPENDIX II

IDENTIFYING DECAYED WOOD AND CHARCOAL

It is sometimes necessary to identify decayed wood or charcoal — archaeological specimens, for example. In some instances, it is preferable to sample a decayed part of a historic object rather than deface a portion that is still in good condition. Identification of decayed wood or charcoal requires slightly modified techniques, and although the results are not as consistent as with sound wood, at least some success is possible if the nature of the deterioration is understood.

Decay in Wood

Fungi that inhabit wood tissue are broadly classified as either stain fungi or decay fungi. Stain fungi simply invade the cell structure and cause discoloration, primarily in the sapwood, but they do not significantly soften or weaken the wood. Microscopically, the thread-like hyphae may be seen in the cell lumens or passing through the pit pair's connecting cell. Stain fungi utilize the contents of parenchyma cells for sustenance. The individual hyphae appear translucent brown under a microscope, although sapstained wood usually appears bluish or grey to the eye. Wood cell structure and anatomical detail are otherwise intact, and stained wood can usually be sectioned and identified as normal wood.

Decay fungi cause eventual breakdown of cell structure. As decay fungi first begin to spread their hyphae through wood tissue, there may be a slight discoloration, but there is little or no perceptible softening of the wood. This is the **incipient decay** stage. As

hyphae spread and multiply, however, cell walls break down and the wood shows obvious deterioration, first by softening and then eventually by crumbling in what is called **advanced** or **typical decay.**

Decay fungi are classified as **brown rots** or **white rots** according to their appearance in the advanced stage (see the top photos on the facing page). Brown-rot fungi mainly attack the cellulosic portions of the cell walls, leaving a brown residue. The deterioration of cell-wall material causes a shrinkage of the wood and formation of cross-checks. This produces a cubical surface somewhat resembling charred wood. White-rot fungi attack both the cellulosic and lignin portions of the wood, leaving a whitish residue. An overall punky or stringy consistency results.

In only a very few instances, a unique form of decay may serve as a positive identification feature. Examples are the pocket rots in baldcypress and incense-cedar. Usually, however, decay is non-specific; rather than being helpful, it serves only to alter and obscure the useful normal features and to complicate preparation of the specimen.

Advanced decay produces drastic changes in many gross features. Useful characteristics such as color, evenness of grain, ray fleck and parenchyma arrangement may lose their original macroscopic appearance. Features such as resin canals and rays, although present, may be more difficult to detect. Therefore it is usually necessary to rely more on microscopic features.

However, three types of problems are often encountered in microscopic examination. First, decayed wood is usually very soft when wet and crumbly or powdery when dry; it is therefore quite difficult to section. When sections are finally cut, they are hard to handle without breaking. Second, the hyphae may confuse the appearance of the cell structure. Finally, the wood's structural detail may be altered by the decay. For example, pit borders and membranes may be eroded away, or the pits may become enlarged by bore holes. In brown-rotted wood, the cell walls retain their original thickness, but the numerous bore holes can be confused with pits; in white-rotted wood, eroded cell walls often appear thinner than normal or of variable thickness (see the bottom photos on the facing page).

If these modifications of gross features and cell structure are anticipated and understood, the cellular features of decayed wood can usually be evaluated well enough to allow identification. Examining samples of decayed wood of known species is valuable in learning how anatomical features appear in rotted wood.

Special Methods for Examining Decayed Wood

Decayed wood is usually quite variable in its degree of deterioration. Some areas will be in much better condition than others. Before attempting to identify a piece of decayed wood, search for those areas with the greatest integrity by probing the specimen with the point of a knife. If the wood is dry and crumbles when you attempt to cut it, don't persist. Instead, try breaking the piece. Then examine all flat, cleanly broken cross-sectional surfaces with a hand lens for macroscopic detail. Note the orientation of radial and tangential planes. If the wood is wet or damp, do not allow it to dry out. In old, waterlogged material, the cell walls may have already broken down somewhat — if the wood dries out, it will break up badly. Instead, store waterlogged material in water or in sealable plastic bags until it can be examined.

If the decayed wood is dry and crumbles when cut, soak it and then heat it in boiling water. Cut sections while the wood is still hot and soft. The fragile nature of the wood will make good sections hard to obtain. It is best to use fresh double-edged razor blades, taking extremely small, thin cuts with a sliding, slicing stroke. The best way of handling the sections is by suspending them in a drop or small puddle of water. This way the sections can be transferred to slides while they are still in the puddle. The slide should be topped with a cover glass in the usual manner.

CHARCOAL

Charcoal is common among archaeological specimens because unlike other forms of wood (except petrified wood, of course), charcoal is little affected by fungi and other wood-destroying organisms and may survive in the ground indefinitely. Officials investigating the remains of burned buildings are also interested in charcoal analysis. Charcoal identification is possible because the anatomical features of the wood remain essentially intact through the carbonizing process, although some of the fine detail — pit membranes, pit borders and perforation bars — may disintegrate. Because charcoal identification is more difficult than identification of normal wood, it is best to have a good working familiarity with wood structure and some experience with the identification procedure for normal wood before attempting to analyze charcoal.

BROWN ROT AND WHITE ROT
Typical brown rots soften and darken the wood, as shown at top. When it dries, it breaks up into cubical sections. White rots leave the wood light in color, but with a soft, sometimes stringy consistency, as shown above.

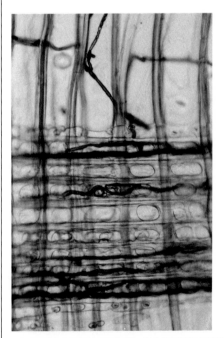
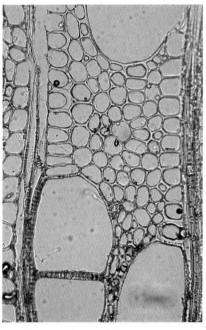

MICROSCOPIC VIEWS OF ADVANCED BROWN ROT AND WHITE ROT
As the advanced stage of brown rot develops in wood, the cell structure is invaded by hyphae (left). Advanced white rot results in erosion of cell walls (right). (300x)

Sample Preparation and Viewing

Due to the brittleness of charcoal, attempts to surface or section the wood using ordinary cutting techniques will fail. No matter how sharp the cutting edge, the cell walls will shatter into microscopic fragments. A better approach is to snap the piece in half again and again, attempting to create planes that approximate as closely as possible the true cross-sectional, radial and tangential surfaces. A smooth, continuous surface is easiest to obtain along a transverse plane. Radial and tangential surfaces tend to be somewhat irregular, though small areas or platelets usually cleave accurately along these planes when the piece is broken.

When charcoal is viewed under a bright light with a hand lens, a great deal of macroscopic detail is visible. However, the opaque, black, lustrous surface makes these details appear quite different from those of normal wood. As with decayed features, it takes experience to interpret carbonized features. In hardwoods, routine details such as ring-porosity, pore arrangement, parenchyma arrangement and ray widths are readily determined, so species such as oaks, ash, elm, hickory, chestnut and beech are most easily recognized (see the photo at right). With experience, you will easily recognize other species too.

In softwood charcoal, texture, evenness of grain, earlywood/latewood transition and the size and number of resin canals can be assessed, and an unknown wood can at least be narrowed to a group of possibilities. Since thin sections cannot be cut, microscopic examination is limited to available surfaces. To mount an irregular piece of charcoal, place a small dab of florist's clay on a standard glass slide, then gently push the fragment into the clay (see the photo at right). Because the clay is soft and plastic, the charcoal can be repositioned as needed to orient the desired surface parallel to the slide.

Because charcoal is dark in color, an intense light is needed to illuminate it.

A slide projector with an improvised baffle makes an effective, if cumbersome, light source. It will usually take a considerable amount of experimentation and adjustment of the angle and brightness of the light to bring the detail into proper focus and contrast. Once this is accomplished, however, such details as ray seriation and heterogeneity, spiral thickenings and pitting of vessels should be discernible.

More advanced methods of softening, embedding and sectioning of charcoal for microscopic examination have been developed, but they are well beyond the scope of this book.

OAK CHARCOAL
The fractured transverse plane of this charcoal fragment clearly shows anatomical features of red oak.

CHARCOAL FRAGMENT MOUNTED ON A GLASS SLIDE
To examine a fragment of charcoal, press it into a dab of florist's clay on a blank slide. The fragment can easily be pushed into the proper position to view the desired surfaces with a microscope using incident light.

APPENDIX III

IDENTIFYING WOOD FIBERS

Wood identification most commonly involves solid pieces of wood. Even if a sample has only the thickness of veneer or the size of a splinter, it is a cellular aggregate and its identification entails the usual macroscopic and microscopic examination. Occasionally identification must be based solely on individual wood cells. Examples are stream pollutant analysis for environmental studies and fiber analysis in forensic work. Without a doubt, however, fiber identification is most commonly involved in the analysis of paper.

In the paper industry, any and all wood cells used in the paper production process are called "fibers;" it is common, therefore, to speak of wood-cell identification as fiber identification. Although a complete treatment of fiber identification is impossible here, the following summary indicates what special considerations are involved and what kinds of results can be expected, especially as compared to the identification of solid wood.

Cell Characteristics

The relative arrangements of cells and the resulting tissue formations, so valuable in identifying solid wood, are entirely lost in fiber work. This includes macroscopic features such as resin canals, earlywood/latewood transition, parenchyma arrangement and ray size, and such basic gross features as density, color and figure. The analyst must rely totally upon recognizing the various cell types and their individual characteristics.

Cell length is the initial consideration and probably the only macroscopic criterion available. The average lengths of longitudinal tracheids of softwoods range from 2mm to 7mm; among hardwoods, the longest fibers are less than 2mm long. Therefore, wood cells longer than 2mm typically indicate a softwood species. Other than cell length, however, fiber analysis is basically a matter of microscopic evaluation.

For a quick introduction to paper fiber examination, prepare a microscope slide. This can be done in either of two ways. The faster method is simply to tear apart a piece of paper ¼ in. or smaller; its shape is not important. It is best first to moisten and then to flex the area to be torn. Next, carefully pry the fragment away from the sheet so the fibers are pulled loose rather than broken. The result is a fragment whose frayed edges and loose fiber ends will show the best detail under the microscope (see the photos below and on p. 194). The fragment can then be mounted on a slide with a drop of water and a cover glass, just as with razor-blade-cut sections of solid wood tissue.

PAPER FIBERS
To examine paper fibers, soak a piece of paper and tease apart a small piece, leaving the edges as frayed as possible. Mount the fragment on a slide with a drop of water and cover glass.

A second method is to soak a small piece of paper in warm or hot water for several minutes. Then tease it apart and place the fragments in a small vial or test tube with a small amount of water and several glass beads. Shake the contents vigorously for several minutes. This causes the paper to separate into individual cells. Allow the mixture to settle and pour off most of the water. Then transfer a drop of water with suspended fibers to a clean microscope slide. Spread the fibers with a needle and add a cover glass.

In commercial papermaking, pulp fibers are separated from wood by either mechanical or chemical methods. In mechanical pulping, the wood is abraded with a rough stone surface that rips the cells apart. A high pulp yield results, but individual cells are usually damaged. In chemical processing, wood chips are "cooked" in chemicals that dissolve the intercellular layer, causing the cells to disperse. Although this causes less damage to individual cells, smaller-sized cells may be washed away and lost when the cooking liquor is rinsed from the pulp.

To allow wood fibers to develop a tighter, more compact mat when the pulp sheet is formed into paper, the fibers are beaten. It is not surprising, then, that wood fibers in paper are often broken, split or frayed. Thin-walled cell types such as earlywood coniferous tracheids, thin-walled hardwood fibers and hardwood vessel elements are flattened into ribbons. But thicker-walled tracheids and hardwood fibers may retain their rounded or rectangular shape. It is important to note dimensions and characteristic shapes of cells, especially of cell ends. But the most valuable information comes from the cell wall itself: thickness, pitting, perforations and thickenings, primarily of the longitudinal cell types.

Softwood Fibers

In softwoods, earlywood tracheids are the principal source of information. Spiral thickenings, where present, are important. Ray-contact areas are especially meaningful, because they show the type of cross-field pitting, the number of pits per cross field and the angle of inclination of the pit aperture. Other information — such as ray height and whether the rays are homocellular or heterocellular — can be deduced. It is also important to note the number of large intertracheid bordered pits across the radial wall of the cell (see the photos below).

Occasionally, ray cells may contain valuable detail, such as dentate ray tracheids, which suggest that the cell belongs to the hard pine group.

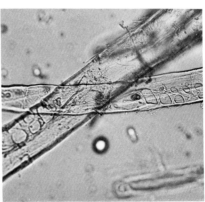

PHOTOMICROGRAPH OF SOFTWOOD PULP
At the edge of a frayed piece of paper (left), details of individual fibers are visible. The small photo above shows the ray-contact areas of two tracheids. Cross-field pitting is obviously windowlike, narrowing the possibilities to the soft pines or red pine, if the paper is domestic in origin.

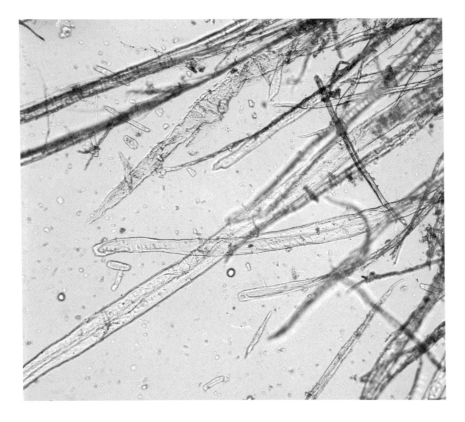

Hardwood Fibers

Among various hardwood pulps (see the photos below), vessel elements are the most informative cell types. Dimensional data, the shape of the ends and the size of the ligule can all be diagnostic. The type of end-wall perforation and intervessel pitting are especially valuable; other features such as spiral thickenings and ray-vessel pitting help narrow the possibilities.

Hardwood fibers are abundant in hardwood pulp, but they offer fewer distinctive and useful features than do vessel elements. Nonetheless, it is helpful to note whether the pits are bordered or simple and whether spiral thickenings are present. The presence of vasicentric tracheids is also diagnostic.

Limitations of Wood-Fiber Identification

It is important to recognize the inherent limitations of fiber identification. The lack of supporting gross and macroscopic features means that fibers cannot be identified as precisely as solid wood. Unless some uniquely definitive feature is present — the closely spaced spiral thickenings of Douglas-fir, for example — the identification can only be narrowed to a group of possibilities. In these cases, information as to the source of the wood becomes exceedingly important.

In examining pieces of solid wood, a small sample can be taken anywhere from the larger piece; if its features are definitive, it serves as evidence for the entire piece. However, paper can be a blend of pulps of more than one species. This problem is compounded when the paper is recycled. Therefore, diagnosis of a species based on a single fiber's characteristics indicates only the presence of that particular species in the pulp; it does not preclude the presence of others. A statistical sampling scheme must be employed to determine the probability of a single species being present or the rough percentages of component species in a mixture.

Further Reading

By now it should be clear just how important an understanding of the detailed nature of solid wood structure is to fiber identification. Although many of the solid wood features described in this book can be routinely applied to fiber-identification work, there is no substitute for special fiber photomicrographs of individual fibers for visual comparison.

More specific identification keys for fiber identification can be found in the following books, which are listed in the bibliography: Côté, *Papermaking Fibers;* Parham and Gray, *The Practical Identification of Wood Pulp Fibers;* Strelis and Kennedy, *Identification of North American Commercial Pulpwoods and Pulp Fibres;* Panshin and deZeeuw, *Textbook of Wood Technology* (Chapter 12).

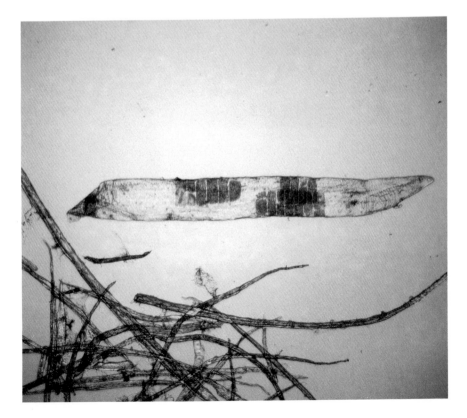

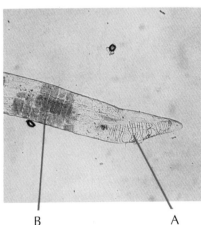

B   A

DISCERNIBLE IDENTIFICATION FEATURES IN HARDWOOD PULP FIBERS
Along the torn edge of a paper sample (left), a vessel element is clearly discernible among countless fibers. In the photo above, a scalariform perforation plate (A) and minute ray-vessel pitting (B) identify the vessel element as birch.

APPENDIX IV

CHEMICAL METHODS OF WOOD IDENTIFICATION

Chapter 5 introduced some of the chemical methods used to separate and identify woods. At right is a list of additional techniques and the species that can be separated by them. For each separation, details of the methods are discussed in the references given. The numbers in the list correspond to the references below.

REFERENCES

1: Panshin, A.J., and Carl de Zeeuw, 1980. *Textbook of Wood Technology.* McGraw-Hill.

2: Stearns, J.L., and C. Hartley, 1952. Physico-chemical methods for wood diagnosis. *Journal of the Forest Products Research Society* 2:4:58-61.

3: Pearson, T.W., G.S. Kriz and R.J. Taylor, 1977. Absolute identification of hydroxystilbenes, chemical markers in Engelmann spruce. *Wood Science* 10:2:93-98.

4: Swan, E.P. 1966. Chemical methods of differentiating the wood of several western conifers. *Forest Products Journal* 16:1:51-54.

5: Barton, G.M. 1973. Chemical color tests for Canadian woods. *Canadian Forest Industries* February: 57-62.

6: Kutscha, N.P., J.T. Lomerson and M.V. Dyer, 1978. Separation of eastern spruce and balsam fir by chemical methods. *Wood Science and Technology* 12:293-308.

WOODS TO BE SEPARATED	CHEMICAL TECHNIQUE	REFERENCE
Lodgepole pine *(Pinus contorta)* Jack pine *(Pinus banksiana)*	Chemical analysis of terpenes	1: p. 443
Ponderosa pine *(Pinus ponderosa)* Jeffrey pine *(Pinus jeffreyi)*	Chemical analysis of essential oils	1: p. 448
Red oaks *(Quercus* spp.) White oaks *(Quercus* spp.)	Color reaction to chemical reagent	1: p. 644
Red maple *(Acer rubrum)* Sugar maple *(Acer saccharum)*	Color reaction to chemical reagent	1: p. 644
Boxelder *(Acer negundo)* Maples *(Acer* spp.)	Color reaction to chemical reagent	2
Sweetgum *(Liquidambar styraciflua)* Black tupelo *(Nyssa sylvatica)*	Color reaction to chemical reagent	2
Pond pine *(Pinus serotina)* Sand pine *(Pinus clausa)*	Color reaction to chemical reagent	2
Slippery elm *(Ulmus rubra)* Rock elm *(Ulmus thomasii)* American elm *(Ulmus americana)*	Color reaction to chemical reagent	2
Colorado blue spruce *(Picea pungens)* Engelmann spruce *(Picea engelmannii)*	Chromatography/ Chemical analysis	3
Eastern white pine *(Pinus strobus)* Western white pine *(Pinus monticola)*	Paper chromatography	1: p. 439
Western hemlock *(Tsuga heterophylla)* Pacific silver fir *(Abies amabilis)*	Thin-layer chromatography	4
Subalpine fir *(Abies lasiocarpa)* Pacific silver fir *(Abies amabilis)*	Color reaction to chemical reagent	5
Spruces *(Picea* spp.) Pines *(Pinus* spp.)	Color reaction to chemical reagent	5
Eastern spruce *(Picea* spp.) Balsam fir *(Abies balsamea)*	Color reaction to chemical reagent	6

APPENDIX V

ORGANIZATIONS, INFORMATION AND MATERIALS

ORGANIZATIONS

International Wood Collectors Society
The IWCS is a nonprofit organization with some 1,100 members in 27 countries. It was founded in 1947 to promote wood collecting, the exchange of specimens among members and wood craftsmanship. It also promotes a standardized sample size (½ in. x 3 in. x 6 in.) and assists in the identification of wood. It holds regular international and national meetings and publishes the monthly bulletin *World of Wood*, which contains articles on all aspects of wood and collecting. The publication also lists woods and wood kits available for sale or exchange. The only requirement for membership is an interest in wood.
Write:
Secretary/Treasurer
International Wood Collectors Society
2913 Third Street
Trenton, MI 48183

International Association of Wood Anatomists
The IAWA was founded in 1931 as a professional society to advance the knowledge of wood anatomy. The *IAWA Bulletin*, published quarterly, includes papers on wood anatomy and identification, reviews of current literature, international news, IAWA meeting information and abstracts. Its membership directory is issued annually. Several major publications related to wood identification are available for purchase. Write:
Executive Secretary
International Association of Wood
  Anatomists
Institute of Systematic Botany
P.O. Box 80.102
3508 TC Utrecht
THE NETHERLANDS

IDENTIFICATION SERVICES

U.S. Forest Products Laboratory
The United States Department of Agriculture's Forest Products Laboratory in Madison, Wisconsin, will identify up to five wood samples free of charge for those with "a clear need for wood identification." Samples at least 1 in. x 3 in. x 6 in. are recommended, although any size large enough to be hand-held during sectioning is acceptable. Samples should be labeled individually and accompanied by any information on origin or popular name.
Send samples to:
Center for Wood Anatomy Research
U.S. Forest Products Laboratory
1 Gifford Pinchot Drive
Madison, WI 53705-2398

Agencies Offering Wood Identification on a Fee Basis
Write or call these agencies before sending samples.

Ethnobotanical Research Services
Department of Anthropology
Oregon State University
Corvallis, OR 97331
(503) 737-0123

Program Manager
Tropical Timber Identification Center
College of Environmental Science and
  Forestry
Syracuse, NY 13210
(315) 473-8788

WOOD SAMPLE KITS

Wood sample kits are available from these suppliers:

Bruce T. Forness
International Wood Collectors Society
Drawer B, Main Street
Chaumont, NY 13622

Norman Jones
Colonial Hardwoods, Inc.
212 North West Street
Falls Church, VA 22046

Veneer sample kits are available from:

Albert Constantine and Sons, Inc.
2050 Eastchester Road
Bronx, NY 10461

Wood Shed
1807 Elmwood Avenue
Buffalo, NY 14207

PREPARED MICROSCOPE SLIDES

Permanently stained and mounted wood-tissue sections or wood fibers are available in sets or individually from these suppliers. An asterisk indicates that the company also sells microscopes and magnifiers.

Ripon Microslides, Inc.
P.O. Box 262
Ripon, WI 54971
(414) 748-7770

*Triarch, Inc.
P.O. Box 98
Ripon, WI 54971
(414) 748-5125

Turtox, Inc.
P.O. Box 266
Palos Heights, IL 60463
(800) 826-6164

SUPPLIES, EQUIPMENT AND BOOKS

The following are sources for (A) microscopes and accessories, (B) hand magnifiers, (C) ultraviolet lamps and (D) books.

(A,B,C)   Arthur H. Thomas Co.
P.O. Box 779
Philadelphia, PA 19105

(A,B,C)   Carolina Biological
Supply Co.
Burlington, NC 21725

(A,B,C)   Edmund Scientific Co.
101 East Gloucester Pike
Barrington, NJ 08007

(A,B,C)   VWR Scientific Inc.
P.O. Box 232
Boston, MA 02101

(A,B,D)   McCrone
Accessories/Components
850 Pasquinelli Drive
Westmont, IL 60559

(A,B)   Forestry Suppliers, Inc.
P.O. Box 8397
Jackson, MS 39204

(B)   Brookstone Co.
120 Vose Road
Peterboro, NH 03458

(D)   Books About Wood
RR3
Owen Sound, Ontario
N41C 5N5, CANADA

(D)   R. Sorsky Bookseller
3845 North Blackstone
Fresno, CA 93726

(D)   Stobart & Son Ltd.
67/73 Worship Street
London EC2A 2EL
ENGLAND

(D)   Woodworking Books
Bark Service Co.
Box 637
Troutman, NC 28166

SOURCES OF DOMESTIC AND FOREIGN WOODS

All these sources supply domestic woods. Those with asterisks also furnish foreign woods.

*Robert M. Albrecht
18701 Parthenia Street
Northridge, CA 91324
(213) 349-6500

*Austin Hardwoods, Inc.
Mail Order Department
2125 Goodrich
Austin, TX 78704
(512) 442-4001

Cabell Forestry Services
2649 Jefferson Park Circle
Charlottesville, VA 22903
(804) 973-2938

*Cards of Wood
1271 House Road
Belmont, MI 49306
(616) 887-8257

Dan Cassens
600 N., 400 W.
West Lafayette, IN 47907
(317) 434-4691

*Chico Hardwoods
565 East Lindo Avenue
Chico, CA 95926
(916) 891-0570

Colonial Hardwoods, Inc.
(also sample sets)
7953 Cameron Brown Court
Springfield, VA 22153
(800) 466-5451

*Maurice L. Condon Co., Ir
250 Ferris Avenue
White Plains, NY 10603
(914) 946-4111

*Albert Constantine and Sons, Inc.
2050 Eastchester Road
Bronx, NY 10461
(212) 792-1600

Eastern Forest Products Laboratory
800 Montreal Road
Ottawa K1A 0W5
CANADA
(613) 767-1207

*Exotic Woodshed
65 North York Road
Warminster, PA 18974
(212) 672-2257

Falcon Rule
Auburn, ME 04210
(207) 784-4041

*A.N. Green
117 Fairfield
Bastrop, LA 71220
(318) 281-0610

M & M Hardwood
5344 Vineland Avenue
North Hollywood, CA 91601
(213) 766-8325

*MacBeath Hardwood Company
930 Ashby Avenue
Berkeley, CA 94710
(415) 843-4390

*J.H. Monteath Company
2500 Park Avenue
Bronx, NY 10451
(212) 292-9333

*Frank Paxton Lumber Company
5701 West 66th Street
Chicago, IL 60638
(312) 767-1207

Real Wood Veneers
107 Trumbull Street
Elizabeth, NJ 07206
(201) 351-1991

Sawdust and Shavings, Inc.
3518 Chicago Avenue
Riverside, CA 92507
(714) 781-0564

Schlesinger's for Tools, Ltd.
1261 Utica Avenue
Brooklyn, NY 11203
(212) 451-2556

*Herbert Sulerud
Fertile, MN 56540
(218) 945-6672

*Wood Carvers Supply Company
3112 West 28th Street
Minneapolis, MN 55416
(612) 927-7491

Woodcraft Supply Corporation
313 Montvale Avenue
Woburn, MA 01801
(617) 935-5860

For large quantities see:
Imported Wood Purchasing Guide
International Wood Trade Publications, Inc.
P.O. Box 34908
Memphis, TN 38134
(901) 372-8280

GLOSSARY

**ADULT WOOD
(MATURE WOOD, OUTER
WOOD, STEM-FORMED WOOD)**
Wood with mature characteristics that is
produced after cambial cells have
attained maximum dimensions. *See also*
Juvenile wood.

ADVANCED DECAY
The later stage of decay in which the
wood has become punky, soft and
spongy, stringy, pitted or crumbly.

ADVENTITIOUS BUD
A bud that emerges from somewhere
other than a stem or twig tip or leaf axil.
Adventitious buds sometimes develop
along a stem at a point of injury.

AGGREGATE RAY
A broad ray with longitudinal cells
passing through it. In tangential section
under magnification, it appears as a
composite of many smaller rays.

ALBURNUM
Same as sapwood.

ALIFORM PARENCHYMA
An arrangement of parenchyma cells, as
seen in cross section, that are grouped
closely around pores and form winglike
lateral extensions.

ALTERNATE PITTING
Pitting in diagonal rows on vessel
elements or tracheid walls; sometimes
so crowded that it gives the pit outlines
a hexagonal shape. *See also* Opposite
pitting.

ANGIOSPERM
A member of the seed plant class whose
seeds are enclosed in ovaries. The
subclass dicotyledons includes all
hardwood trees.

**ANGSTROM
(ABBREVIATION: Å)**
A unit used to measure electromagnetic
wavelengths. One angstrom = one
ten-billionth of a meter (or one-tenth of
a nanometer).

ANISOTROPIC
Not having the same properties in all
directions. (Wood is an anisotropic
material, since it has different properties
in the longitudinal, tangential and
radial directions.)

**ANNUAL RING (ANNUAL
GROWTH RING, ANNUAL
INCREMENT)**
See Growth ring.

APERTURE
Same as pit aperture.

APOTRACHEAL PARENCHYMA
Longitudinal parenchyma that is not in
contact with pores as seen in cross
section. *See also* Diffuse parenchyma.

ARM
The part of a microscope frame that
joins the body tube and stage, and
usually serves as a carrying handle.

AXIAL
Same as longitudinal.

AXIAL PARENCHYMA
Same as longitudinal parenchyma.

BANDED PARENCHYMA
Parenchyma cells, as viewed in cross
section, that collectively appear as thin
tangential lines, as in hickory.

BARK
The tree tissue outside the cambium. It
includes inner bark, which is living, and
outer bark, which is dead.

BARK POCKETS
Small isolated portions of bark encased
in wood tissue that commonly result
where wounds or branch stubs
are overgrown.

BARS
Same as perforation bars.

BASTARD GRAIN
Used in reference to lumber with
growth rings that form angles of 30° to
60° with the wide surfaces of the lumber.

BEAR SCRATCHES
The lines, or striations, on the tangential
surface of wood resulting from
indented rings.

BEE'S WING
A mottled figure, as in Ceylon
satinwood.

BINOMIAL
The scientific name of a plant or animal
species; a two-word term consisting of
the genus (the generic noun) and the
specific epithet (an adjective) denoting
the species.

BIRD'S-EYES
Small, localized areas in wood where
the fibers are indented and contorted to
form either a few or many small circular
or elliptical figures that resemble birds'
eyes on tangential surfaces. Bird's-eye
figure is common in sugar maple and is
used for decorative purposes, but it is
rare in other hardwood species and does
not occur in softwoods.

BISERIATE RAY
A ray two cells wide tangentially.

BLACK KNOT
Same as encased knot.

"BLACK LIGHT"
See Ultraviolet light.

BLISTER FIGURE
A figure on smooth, plainsawn, rotary-cut surfaces that appears to consist of small, rounded, more or less widely spaced, elevated or depressed areas.

BLUE STAIN (SAPSTAIN)
A bluish or greyish discoloration of the sapwood caused by the growth of certain dark-colored fungi on the surface and in the interior of the wood.

BODY TUBE
The cylindrical portion of a compound microscope with an objective lens at its lower end and an eyepiece lens at its upper end.

BOLE
The term used for the stem or trunk of a tree if it is large enough to convert into saw timber or veneer.

BORDER
Same as pit border.

BORDERED PIT
A pit in which the pit membrane is, in part, overarched by the secondary cell wall.

BORDERED PIT PAIR
An intercellular pairing of two bordered pits that share the same pit membrane.

BRASHNESS
Brittleness in wood characterized by abrupt rather than splintering failure. Causes include reaction wood, juvenile wood, compression failure, high temperature and extremes of growth rate.

BROKEN STRIPE
Ribbon figure in which the stripe effect is intermittent.

BROWN ROT
Decay caused by a type of fungus that attacks cellulose rather than lignin, leaving a brownish residue. *See also* White rot.

BURL (BURR)
A hard, woody, more or less rounded outgrowth on a tree usually resulting from the entwined growth of a cluster of adventitious buds. Such burls are the source of the highly figured veneers used for ornamental purposes.

CAMBIUM
The thin layer of living meristematic (reproductive) cells separating the bark and the wood, which, by cell division, forms additional bark and wood cells.

CAVITY
Same as cell cavity.

CELL (ELEMENT)
The basic structural unit of wood and other plant tissue. It consists of an outer cell wall surrounding a central cavity or lumen. Wood cell types include tracheids, vessel elements, fibers, ray cells, etc.

CELL CAVITY (CAVITY, LUMEN)
The inner space of a cell enclosed by the cell wall.

CELL WALL
The outer structural layer of the cell, as opposed to the inner cavity or lumen.

CELL-WALL CHECKING
See Spiral checking.

CHECK
Separation of wood cells along the grain due to uneven shrinkage. Checking is most common on end-grain surfaces of lumber.

CHEMOTAXONOMY
Classification of plants by analysis of their constituent chemical compounds.

CLASS
A unit or level of classification ranking below divisions and above orders.

CLOSE GRAIN (DENSE GRAIN, FINE GRAIN, NARROW GRAIN)
Used in reference to wood that is slowly grown and has narrow, usually inconspicuous growth rings, in contrast to coarse grain.

CLOSED GRAIN
A characteristic of wood with small and closely spaced pores; used especially in reference to a wood's finishing characteristics. However, fine-textured is a preferable term.

COARSE GRAIN
Used in reference to wood that has wide, conspicuous growth rings, in contrast to close grain. It is sometimes used synonymously with coarse texture in reference to wood with relatively large cell size.

COMB GRAIN
See Rift grain.

COMPOUND MICROSCOPE
An instrument in which an image magnified by a first lens system (the objective) is further magnified by a second lens system (the eyepiece or ocular). *See also* Microscope.

COMPOUND MIDDLE LAMELLA
A composite layer composed of the primary walls of two adjacent cells and the intercellular layer between; usually appears as a single layer under the light microscope.

COMPRESSION WOOD
See Reaction wood.

CONDENSER
A lens system beneath the stage of a microscope that focuses the light upward through the stage.

CONFLUENT PARENCHYMA
Parenchyma cells, as seen in cross section, that are so grouped as to form a more or less tangential band connecting two or more pores. Sometimes called zonate parenchyma.

CONIFEROUS WOODS (CONIFERS)
Same as softwoods.

CORE WOOD
Same as juvenile wood.

COVER GLASS (COVER SLIP)
The thin glass cover placed over the specimen when it is mounted on a standard microscope slide.

CRASSULAE
Arclike markings above and below bordered pits of conifer tracheids that appear to form parentheses around the pit when viewed in radial section.

CROSS FIELD (RAY CROSSING)
As seen in radial section, the rectangular area of cell-wall contact between a longitudinal tracheid and a ray cell in conifers. Cross-field pitting in the earlywood is a valuable identification feature. In fiber identification, the term ray crossing includes all individual cross fields of a given ray.

CROSS-FIELD PITTING
See Cupressoid pitting, Piceoid pitting, Pinoid pitting, Taxodioid pitting, Windowlike pitting.

CROSS GRAIN
Deviation of grain direction from the longitudinal axis of a piece of wood.

CROSS SECTION (TRANSVERSE SECTION)
A section cut perpendicular to the grain; also the surface exposed by such a cut.

CROTCH GRAIN
Figure produced by cutting centrally through a tree crotch in the common plane of both branches.

CROWN-FORMED WOOD
Same as juvenile wood.

CRYSTALLIFEROUS CELL
A cell with one or more crystals in its cell cavity.

CUPRESSOID PITTING
A type of cross-field pitting in the earlywood of some conifers (e.g., *Chamaecyparis, Libocedrus, Juniperus*) in which the pits are oval in outline with included apertures that are oval but narrower than the width of the border on either side.

CURL
The figure produced by the irregular cell structure at the junction of a branch and stem.

CURLY FIGURE (FIDDLEBACK GRAIN, TIGER GRAIN)
The figure produced on surfaces, particularly radial surfaces, of wood with wavy grain.

CURLY GRAIN
Same as wavy grain.

CYTOPLASM (CYTOPLAST)
The protoplasm (*see* protoplast) of a cell, excluding the nucleus.

DEAL
A British term loosely applied to softwoods used in construction or more specifically to *Pinus sylvestris* imported from Scandinavia or Russia. It is also used in reference to softwood lumber in board sizes.

DECAY (ROT)
The decomposition of wood by fungi.

DECIDUOUS TREES
Trees whose current year's leaves normally drop after the yearly growth period is over.

DEFECTS
Any irregularities or abnormalities in wood that lower its strength, grade, value or utility.

DELIQUESCENT GROWTH
Growth form in which the trunk branches and rebranches into a network of smaller limbs, as is typical of most hardwoods. *See also* Excurrent growth.

DENDRITIC GROWTH
Same as deliquescent growth.

DENDROCHRONOLOGY
The dating of wood by analyzing in sequence the year-to-year variation in the width of its growth rings and comparing this pattern of variation to an already dated master scale of year-to-year ring width. This study also provides information on past climatic and other environmental conditions.

DENDROLOGY
The branch of botany involving the study of trees and shrubs.

DENSE GRAIN
Same as close grain.

DENSITY
The weight of a body or substance per unit volume. See also specific gravity.

DENTATE RAY TRACHEIDS
Ray tracheids that appear to have uneven cell walls whose toothlike projections reach into the cell cavity when viewed in radial section; found in all species of the hard pine group.

DICHOTOMOUS KEY
An identification scheme based on a sequence of decisions, each requiring the choice of either of two alternatives.

DICOTS (DICOTYLEDONS)
A class of plants within the angiosperms characterized by having two cotyledons or seed leaves when sprouting. All hardwood trees are dicots. *See also* Monocots.

DIFFUSE-IN-AGGREGATES PARENCHYMA
Apotracheal parenchyma arranged in short tangential lines, often extending from ray to ray, as viewed in cross section. *See also* Reticulate parenchyma.

DIFFUSE PARENCHYMA
Longitudinal parenchyma occuring singly as viewed in cross section. In hardwoods they are independent of pores and apotracheal-diffuse; in softwoods, they are scattered throughout the growth ring and are sometimes termed metatracheal parenchyma.

DIFFUSE-POROUS WOOD
A hardwood in which the pores are of approximately uniform size and distributed fairly evenly throughout the growth ring. *See also* Ring-porous wood.

DIMPLES
Numerous small depressions in the growth rings, most obvious on split tangential surfaces. Dimpling occurs in certain conifers, notably ponderosa and lodgepole pines.

DISCONTINUOUS RING
A growth ring that does not extend around the entire circumference of the stem.

DIVISION
A major taxonomic group of the plant kingdom. This term is preferred to phylum, which is used in reference to divisions of the animal kingdom.

DURABILITY
A general term indicating resistance to deterioration, although it frequently refers specifically to decay resistance.

DURAMEN
Same as heartwood.

EARLYWOOD (SPRINGWOOD)
The first-formed portion of the growth ring, often characterized by larger cells and lower density.

ECHELON
Diagonal arrangement of pores in cross-sectional view or rays in tangential view.

EDGE-GRAINED (QUARTERSAWN, RADIALLY CUT, VERTICALLY GRAINED)
Pieces of wood in which the growth rings form an angle of 45° or more (approaching 90°) with the wood surface or lumber face; approaching or coinciding with a radial surface. *See also* Rift grain.

ELECTRON MICROSCOPE
An instrument that achieves magnification by using beams of electrons rather than ordinary light and magnetic fields rather than glass lens systems. A transmission electron microscope (TEM) is used to examine ultrathin sections. A scanning electron microscope (SEM) magnifies opaque surfaces.

ELEMENT
Same as cell.

ENCASED KNOT (BLACK KNOT, LOOSE KNOT)
The dead portion of a branch embedded in the stem by subsequent tree growth. *See also* Intergrown knot.

END GRAIN
A cross-sectional surface.

END WALL
The portion of a cell wall that extends at a right angle or obliquely to the long axis of a cell, such as the tangential wall of a ray parenchyma cell.

EPITHELIUM (EPITHELIAL LAYER)
The layer of parenchyma (epithelial cells) that forms a resin canal or gum canal.

EVEN-GRAINED
Descriptive of wood with uniform or nearly uniform structure throughout the growth ring and little or no earlywood/latewood difference, as in basswood.

EVERGREENS
Trees whose foliage remains year-round.

EXCURRENT GROWTH
Growth form in which the tree has a dominant main stem, as is typical of most conifers. *See also* Deliquescent growth.

EXTENDED PIT APERTURE
A pit aperture that extends beyond the outline of the pit border. *See also* Included pit aperture.

EXTRACTIVES
Substances deposited in wood in association with the transition from sapwood to heartwood, often imparting significant color and decay resistance to the wood.

EYEPIECE (OCULAR)
The lens system at the viewer's end of the body tube in a compound microscope.

FACE GRAIN
The figure or pattern on the face side of a plywood panel or board.

FALSE RING(S)
More than one sequence of earlywood and latewood within a growth ring.

FAMILY
The classification level below orders and above genera. A family name usually ends in -aceae.

FEATHER CROTCH (FEATHER GRAIN)
The figure characterized by its featherlike appearance that is produced by cutting a longitudinal section through a tree crotch.

FENESTRIFORM PITTING
Same as windowlike pitting.

FIBER
1. A term applied collectively to separated wood cells, such as those used for making paper. 2. An elongated hardwood cell with pointed ends and thick walls that contributes greatly to the strength of the wood.

FIBER TRACHEID
A hardwood fiber with bordered pits and usually a fairly thick cell wall.

FIBRIL ANGLE
The angle between the longitudinal axis of a cell and the direction of the fibrils in the cell wall.

FIBRILS
Minute striations in the cell that are the smallest structural details visible with a light microscope.

FIDDLEBACK GRAIN
Same as curly figure.

FIELD (FIELD OF VIEW)
The area visible through a microscope or hand lens.

FIGURE
Any distinctive pattern or appearance on a longitudinal wood surface that results from routine or irregular anatomical structure, irregular coloration or defects.

FINE GRAIN
Same as close grain. Also sometimes used synonomously with "fine texture" in reference to woods with relatively small-diameter cells.

FINE-TEXTURED (FINE-GRAINED, FINE-GROWN)
Descriptive of hardwoods with small, closely spaced pores, or softwoods with small-diameter tracheids.

FLAME GRAIN
A figure produced on flatsawn boards or rotary-cut veneer.

FLATSAWN (FLAT-GRAINED, PLAINSAWN, SIDE-GRAINED, SLASH-GRAINED, TANGENTIALLY CUT)
Descriptive of wood machined approximately along a tangential plane so that growth rings intersect the surface at an angle of less than 45°.

FLECKS
See Pith flecks, Ray flecks.

FLOCCOSOIDS
Whitish spots that occur in the wood of western hemlock.

FLUORESCENCE
The absorption of invisible ultraviolet light ("black light") by a material that transforms the energy and its re-emission as visible light of a particular color.

FORAMINATE
Same as reticulate perforation plate.

FUNGI (SINGULAR: FUNGUS)
Simple forms of microscopic plants whose parasitic development in wood may cause mold, stain or decay.

FUSIFORM PARENCHYMA
See Longitudinal parenchyma.

FUSIFORM RAY
A ray that contains a horizontal resin canal, giving it a spindle shape when viewed in tangential section.

GASH PITS (GASHLIKE PITTING)
The irregular diagonal lines that sometimes appear on radial walls of latewood vessels in certain species of *Juglans* due to reticulate thickenings on the vessel walls.

GELATINOUS FIBERS
Fibers with modified cell walls that appear gelatinous in cross section. They are usually associated with reaction wood (tension wood) in hardwoods.

GELATINOUS LAYER
The abnormal innermost layer of the cell wall of gelatinous fibers.

GENUS (PLURAL: GENERA)
The level of classification below family but above and including one or more species. The genus name is the capitalized first word of the binomial term used to designate a species; e.g., in *Acer rubrum*, red maple, *Acer* is the genus and *rubrum* is the species.

GRAIN(ED)
A broad-ranging and therefore confusing term whose specific meaning is made apparent by the context or associated adjectives. It has several common meanings: 1.The direction or orientation of cells (e.g., along the grain, spiral-grained). 2.The surface appearance or figure (ribbon grain). 3. Growth-ring placement (vertically grained). 4. Plane of cut (end grain). 5. Growth rate (narrow-grained). 6. Earlywood/latewood contrast (uneven-grained). 7. Relative cell size (open-grained, meaning coarse-textured). 8. Machining defects (chipped grain). 9. Artificial decorative effects (graining, grain-painted).

GRAIN DIRECTION
The direction of the long axes of the dominant longitudinal cells in a piece of wood. *See also* Cross grain, Spiral grain, Straight grain.

GREEN
1. Descriptive of freshly cut and unseasoned wood. 2. Descriptive of wood with a moisture content above the fiber saturation point.

GROSS FEATURES
Physical features that can be perceived with the unaided eye. This term is sometimes meant to include macroscopic features, which can be seen with a hand lens.

GROUND MASS
The darker background mass of indistinguishable fibers in a microscopic cross-sectional surface of hardwoods against which rays, pores or arrangements of parenchyma are visibly distinct.

GROWTH RING (GROWTH INCREMENT, GROWTH LAYER)
The layer of wood or bark added to the stem in a given growth period. In the temperate zones, where one layer is added per yearly growth period, it is often called an annual ring.

GROWTH-RING BOUNDARY
The demarcation between the outer extreme of one growth increment and the inner extreme of the next.

GUM DUCTS (GUM CANALS)
Tubular passageways in certain hardwoods that contain gum.

GUM SPOT
A small patch of accumulated gum that is often the result of bird pecks, insect damage or other injury to the growing wood.

GYMNOSPERM
A member of the naked-seed class of plants — plants whose seeds are not enclosed in ovaries. Within this group are all the trees that yield softwood lumber.

HALF-BORDERED PIT PAIR
A pairing of a bordered pit and a simple pit between two adjacent cells.

HAND LENS
A hand-held magnifying lens. Ten power (10x) is the most common magnification used in wood identification.

HARDWOODS (POROUS WOODS)
Woods from broad-leaved trees in the botanical group angiosperms. Since these woods have vessels, they are also called porous woods. (Note: the term hardwood does not describe the degree of hardness or softness of the wood.)

HEARTWOOD (DURAMEN)
The central core of wood in mature stems that was at one time sapwood but no longer conducts sap or has living cells. In most species, infiltration of extractives imparts a perceptibly darker color to this wood.

HELICAL THICKENINGS
Same as spiral thickenings.

HERBARIUM
A reference collection of dried and pressed plant material important in authenticating wood specimens. Also, the facility in which the collection is housed.

HETEROCELLULAR RAY (HETEROGENEOUS RAY)
In hardwoods, a ray composed of both upright and procumbent cells; in softwoods, a ray composed of both ray parenchyma and ray tracheids.

HOMOCELLULAR RAY (HOMOGENEOUS RAY)
A ray having only one type of cell.

HORIZONTAL RESIN CANAL
See Resin canal.

HYPHAE
The microscopic filaments of a fungus that digest and absorb material from its host. *See also* Incipient decay.

INCIPIENT DECAY
The early stage of decay in which hyphae have invaded the cell structure, sometimes discoloring the wood, but have not perceptibly softened the wood.

INCLUDED PIT APERTURE
A pit aperture within the outline of the pit border. *See also* Extended pit aperture.

INCLUDED SAPWOOD (INTERNAL SAPWOOD)
Areas of light-colored wood, apparently sapwood, found within the portion of the stem that has become heartwood.

INCREMENT
The growth added in a given period. *See* Growth ring.

INDENTED RINGS
V-shaped dips in the growth rings, as viewed on cross-sectional surfaces, that result from narrow depressions of the growth increment extending up to several inches along the grain.

INDENTURE
A depression in the horizontal walls of ray parenchyma at the junction of the tangential end wall as viewed in radial section.

INITIAL PARENCHYMA
See Marginal parenchyma.

INNER BARK
See Bark.

INTERCELLULAR CANAL
Same as resin canal.

INTERCELLULAR LAYER (MIDDLE LAMELLA)
The layer, primarily lignin, between the primary walls of adjacent cells. *See also* Compound middle lamella.

INTERCELLULAR SPACE
A space between cells typically seen at the corners of cells when viewed in cross section. Intercellular spaces are normal in only a few woods, *Juniperus*, for example. However, they commonly occur as an abnormal (for the species) feature in compression wood. Resin canals and gum ducts are technically intercellular spaces within the sheath of epithelium.

INTERGROWN KNOT (RED KNOT, TIGHT KNOT)
A portion of a branch that was alive and formed growth layers continuous with those of the surrounding stem. *See also* Encased knot.

INTERLOCKED GRAIN (INTERWOVEN GRAIN)
Repeated alternation of left- and right-hand spiral grain, with each reversal usually distributed over several growth rings.

INTERMEDIATE WOOD
Wood that is transitional between sapwood and heartwood.

INTERNAL SAPWOOD
Same as included sapwood.

INTERVESSEL PITTING (INTERVASCULAR PITTING)
Pitting, usually bordered pitting, between laterally adjacent vessel elements.

INTERWOVEN GRAIN
Same as interlocked grain.

ISOTROPIC
Having equal properties, for example, strength, permeability and thermal conductivity, in all structural directions. *See also* Anisotropic.

JUVENILE WOOD (CORE WOOD, CROWN-FORMED WOOD, PITH WOOD)
Wood formed near the pith of the tree that is often characterized by wide growth rings of lower density and abnormal properties. *See also* Adult wood.

KNOT
The portion of a branch that becomes overgrown by the expanding girth of the bole or larger branch. *See also* Intergrown knot, Encased knot, Pin knot.

LATEWOOD (SUMMERWOOD)
The later-formed portion of the growth ring that is often characterized by smaller cells and/or higher density. *See also* Earlywood.

LATTICED PERFORATION PLATE
Same as scalariform perforation plate.

LEAF GRAIN
A term for the figure on flat-grained surfaces.

LIBRIFORM FIBER
A hardwood fiber with simple pits. Libriform fibers are difficult to distinguish from fiber tracheids in cross section.

LIGHT MICROSCOPE
See Microscope.

LIGULE
A pointed extension at the end of a vessel element.

LONGITUDINAL (AXIAL)
An adjective describing the direction parallel to the stem axis of the tree or branches or describing the axial direction of the dominant cell structure; along the grain.

LONGITUDINAL GRAIN
Same as longitudinal plane.

LONGITUDINAL PARENCHYMA (AXIAL PARENCHYMA)
Parenchyma cells that are elongated vertically in the stem. They most commonly occur as a vertical series of short cells and are sometimes referred to as strand parenchyma. They occur less commonly as elongated cells without cross walls, in which case they are called fusiform parenchyma.

LONGITUDINAL PLANE
Any plane — radial, tangential or intermediate between the two — cut parallel to the grain direction of wood.

LONGITUDINAL RESIN CANAL
See Resin canal.

LONGITUDINAL STRAND PARENCHYMA
See Longitudinal parenchyma.

LONGITUDINAL TRACHEIDS
Vertically elongated, nonliving wood cells that are the principal conductive and supportive cells of softwoods, comprising 90% or more of the wood volume.

LOOSE KNOT
Same as encased knot.

LUMEN (PLURAL: LUMINA) (CELL LUMEN)
Same as cell cavity.

LUSTER
The light reflectiveness of a surface.

MACROSCOPIC FEATURES
Anatomical details visible with no greater magnification than a 10x hand lens.

MARGINAL CELLS
Ray cells along the upper or lower margin of the ray.

MARGINAL PARENCHYMA
Cells of longitudinal parenchyma along the growth-ring boundary. These are called initial parenchyma if they occur at the inner edge of the growth ring or terminal parenchyma if they occur at the outer edge.

MATURE WOOD
Same as adult wood.

MEDULLA
Same as pith.

MEDULLARY RAYS (PITH RAYS)
Rays connected to the pith. This term is sometimes used loosely to refer to any rays.

METATRACHEAL PARENCHYMA
See Diffuse parenchyma.

**MICROMETER
(ABBREVIATION: μm)**
1/1000 of a millimeter, or approximately 1/25,000 of an inch. (Replaces the equivalent term micron.)

MICROSCOPE
An optical instrument used to magnify details too small to be seen with the unaided eye. The unqualified term microscope usually implies a compound light microscope. A light microscope utilizes ordinary light passing through glass lenses; a compound microscope has a two-stage magnification system consisting of an objective and an eyepiece. *See also* Electron microscope.

MICROSCOPE SLIDE
A glass plate, nominally 1 in. by 3 in., upon which tissue sections are mounted for microscopic examination. *See also* Prepared slide.

**MICROSCOPIC FEATURES
(MINUTE FEATURES)**
Anatomical details that cannot be seen without the aid of a compound microscope.

MICROTOME
An instrument for cutting thin sections of tissue for examination with a microscope.

MIDDLE LAMELLA
Same as intercellular layer. *See also* Compound middle lamella.

MIL
1/1000 of an inch.

MINERAL STAIN (MINERAL STREAK)
Dark grey, black or greenish-black discolorations in wood sometimes associated with injury but not directly due to fungi, as commonly seen in soft maple and other hardwoods.

MINUTE FEATURES
Same as microscopic features.

**MONOCOTS
(MONOCOTYLEDONS)**
A class of plants within the angiosperms characterized by a single cotyledon or seed leaf. *See also* Dicots.

MOONSHINE CROTCH FIGURE
A swirling figure produced on crotch-cut surfaces of certain species, as in mahogany.

MOTTLED FIGURE
A type of broken-stripe figure with irregular interruptions of curly figure.

MULTIPLE
See Pore multiple.

MULTIPLE PERFORATION PLATE
A vessel-element end wall with more than one opening. *See also* Scalariform perforation plate.

MULTISERIATE RAY
A ray that is two or more cells wide tangentially.

**NANOMETER
(ABBREVIATION: nm)**
1/1,000,000 of a millimeter, 1/1,000 of a micrometer or 10 angstroms.

NARROW-GRAINED
Same as close-grained.

NEEDLEPOINT GRAIN
Same as rift grain.

NODED RAYS
Rays that appear swollen at the growth-ring boundary, as seen on transverse surfaces or sections.

NODULAR END WALL
The common end wall between two longitudinal or ray parenchyma cells that appears to have beadlike thickenings (nodules) in sectional view.

NONPOROUS WOODS
See Softwoods.

NOSEPIECE
A revolving holder with two or more objectives at the lower end of the body tube of a compound microscope.

OBJECTIVE
The lens system at the lower end of the body tube of a compound microscope.

OCULAR
Same as eyepiece.

OIL CELL
A specialized parenchyma cell containing oil, usually evident as a prominent rounded globule.

OPEN-GRAINED
1. Descriptive of wood with relatively large pores, and used especially in reference to its finishing characteristics. Coarse-textured is a preferred term.
2. Descriptive of wood having widely spaced growth rings, in contrast to close-grained or dense-grained.

OPPOSITE PITTING
Two or more pits aligned horizontally on vessel element or tracheid walls. The pits are sometimes rectangular because of crowding. *See also* Alternate pitting.

ORDER
A classification level ranking below class and above family.

OUTER BARK
See Bark.

OUTER WOOD
Same as adult wood.

PARATRACHEAL PARENCHYMA
Longitudinal parenchyma that is in contact or close association with pores as viewed in cross section. *See also* Aliform parenchyma, Confluent parenchyma, Scanty parenchyma, Vasicentric parenchyma.

PARENCHYMA (SOFT TISSUE)
Thin-walled wood cells (living when in the sapwood) that are involved mainly in food storage and distribution. Groupings of parenchyma may appear as light-colored areas on cross sections as seen with a hand lens. *See also* Prosenchyma.

PEANUT-SHELL FIGURE
A tangential figure resulting from closely spaced, oblong blister-grain formation. It is commonly produced in Japanese ash (tamo).

PECKY DRY ROT (PECK, PECKINESS)
Advanced decay in living trees occurring as elongated pockets of rot. It is most familiar in baldcypress and incense-cedar.

PERFORATION (VESSEL PERFORATION)
A single or multiple opening in the common end wall joining two consecutive elements in a vessel.

PERFORATION BARS
See Scalariform perforation plate.

PERFORATION PLATE
The common end wall between two consecutive vessel elements in which perforations occur.

PERFORATION RIM
The periphery of the perforation plate at the edge of a simple perforation.

PHLOEM
The tissue of the inner bark that conducts food upward through the tree.

PHOTOMICROGRAPH
A photograph taken of a microscopically enlarged image.

PHYLUM (PLURAL: PHYLA)
See Division.

PICEOID PITTING (PICIFORM PITTING)
Cross-field pitting in the earlywood of some conifers, e.g., *Picea*, *Larix* and *Pseudotsuga*, in which the pits are round or oval with a narrow or slitlike aperture that may be either included or extended. *See also* Extended pit aperture, Included pit aperture.

PIGMENT FIGURE
Figure resulting from the irregular deposition of extractive coloration.

PIN KNOT
A knot ½ in. to ¼ in. or less in diameter.

PINOID PITTING
Cross-field pitting in the earlywood of many hard pines that typically consists of multiple, variably sized oval to football-shaped pits that are elongated diagonally on the cross field. They can be either simple pits or bordered pits.

PIT
A void in the secondary cell wall through which fluids pass from cell to cell. A pit is usually paired with a pit in an adjacent cell and is separated from it by the remaining compound middle lamella (pit membrane). Pits are classified as simple or bordered, depending on the shape of the cavity. *See also* Bordered pit, Bordered pit pair, Simple pit, Simple pit pair.

PIT APERTURE
The opening from the cell lumen into the cavity of a pit.

PIT BORDER
The portion of the cell wall that overarches the pit cavity in a bordered pit.

PIT CAVITY
The void in the cell wall forming a pit.

PITCH
The material formed in the resin canals of softwoods; also, the accumulation of resin as seen in pockets, streaks or seams.

PITCH POCKET
A flattened round or oval tangential separation in the wood of conifers that contains or contained liquid or solidified resin.

PITH (MEDULLA)
The small core of soft spongy tissue that forms the central longitudinal axis of a tree stem, branch or twig.

PITH FLECKS
Longitudinal streaks of wound tissue caused by the larvae of small flies tunneling along the cambium during wood formation.

PITH RAY
Same as medullary ray.

PITH WOOD
Same as juvenile wood.

PIT MEMBRANE
See Pit, Torus.

PIT PAIR
Pits of adjacent cells located opposite one another and sharing a pit membrane.

PLAINSAWN (PLAIN-GRAINED)
See Flatsawn.

POCKET ROT
A localized, sharply delineated pocket of advanced decay surrounded by apparently sound wood.

PORE
The cross section of a hardwood vessel.

PORE ARRANGEMENT
The position of pores relative to one another in cross section.

PORE CHAIN
A distinct radial series or alignment of several to many solitary pores or pore multiples, as in *Ilex*.

PORE CLUSTER
Few to many pores grouped or nested together.

PORE MULTIPLE
Two or more pores that are in close contact, and, typically, in radial alignment. They may assume the appearance of a single large pore with subdivisions.

POROUS WOODS
Same as hardwoods.

PREPARED SLIDE
A standard glass microscope slide upon which a specimen, usually stained, has been permanently mounted under a cover glass.

PRIMARY WALL
The thin outer layer of the cell wall that is initially present during cell division. *See also* Secondary wall.

PROCUMBENT RAY CELLS
Ray cells that are elongated horizontally (radially). *See also* Upright ray cells.

PROSENCHYMA
A term — becoming obsolete — for nonliving wood cells, such as tracheids, vessel elements and fibers, that perform conduction and support functions. *See also* Parenchyma.

PROTOPLAST
The mass of protoplasm within a living cell that includes the nucleus and cytoplasm.

QUARTERSAWN (QUARTERED, QUARTER-GRAINED)
Same as edge-grained.

QUILTED FIGURE
A figure sometimes seen in bigleaf maple that is characterized by crowded bulges in the grain.

R
The abbreviation or symbol for a radial section or surface.

RADIAL
A direction in wood perpendicular to the (longitudinal) grain direction and following the orientation of the rays.

RADIALLY CUT (RADIALLY GRAINED)
Same as edge-grained.

RADIAL MULTIPLE
A pore multiple in radial alignment.

RADIAL PLANE (RADIAL-LONGITUDINAL PLANE)
A longitudinal plane in wood that coincides with the radial direction.

RADIAL SECTION
A thin section cut along the grain in a radial plane. Also a surface exposed by such a cut.

RAISED GRAIN
The elevation of latewood above the earlywood on the pith side of longitudinal surfaces of flatsawn boards that have been dressed by machine. This elevation may also result from weathering or wearing away of the softer earlywood.

RATE OF GROWTH
The relative rate of increase in girth of a tree, usually expressed as rings per inch.

RAY CROSSING
See Cross field.

RAY CELLS
See Rays.

RAY FLECK
The conspicuous appearance of rays on an edge-grain surface.

RAY MARGIN
See Marginal cells.

RAY PARENCHYMA
Any cells of the ray that are parenchyma; this term usually does not include epithelial cells.

RAYS
Flattened bands of tissue composed of ray cells extending horizontally in a radial plane through the tree stem. See also Medullary rays.

RAY TRACHEID
A nonliving cell type with small bordered pits that occurs in the rays of certain softwood species. See also Dentate ray tracheids.

RAY-VESSEL PITTING
Pitting between a ray cell and a vessel element.

REACTION WOOD
Abnormal wood formed in leaning stems and branches in trees. In softwood trees, it forms on the lower side of the stem and is called compression wood; this wood is denser but more brittle, has greater than normal longitudinal shrinkage, and has rounded tracheids with pronounced spiral checking and numerous intercellular spaces. In hardwood trees, it forms typically on the upper side of the stem and is called tension wood. It is characterized by woolly surfaces when machined, greater than normal longitudinal shrinkage and gelatinous fibers.

RED KNOT
Same as intergrown knot.

REDWOOD
Same as compression wood. See Reaction wood.

RESIN
Material secreted into resin canals by epithelial cells in softwood trees. See also Pitch.

RESIN CANALS (INTERCELLULAR CANALS, RESIN DUCTS)
Tubular passageways in the wood of certain softwoods (Pinus, Larix, Picea, Pseudotsuga) formed as intercellular spaces encircled by epithelial cells. The epithelial cells secrete resin into the canal. Longitudinal resin canals are larger and run along the grain, transverse or horizontal resin canals are smaller and occur within fusiform rays. See also Traumatic resin canals.

RESOLUTION (RESOLVING POWER)
The measure of the closeness of points that can be distinguished from one another; it describes the sharpness of focus of magnified images.

RETICULATE PARENCHYMA
Banded parenchyma of approximately the same width and spacing as the rays, which together form a netlike pattern on a cross-sectional surface, as in Carya. If the rays are wider, forming a ladderlike pattern with the connecting banded or diffuse-in-aggregates parenchyma, the pattern may be called scalariform, as in obeche.

RETICULATE (FORAMINATE) PERFORATION PLATE
A perforation plate with multiple openings resulting in a netlike pattern.

RETICULATE THICKENINGS (RETICULUM)
Meshlike thickenings on the inner surface of the secondary wall in some vessel elements. See also Gashlike pitting.

RIBBON FIGURE (RIBBON GRAIN, STRIPE FIGURE)
Figure on an edge-grain surface of wood with interlocked grain that is characterized by vertical bands of varying luster and vessel markings.

RIFT GRAIN (COMB GRAIN, NEEDLEPOINT GRAIN)
The surface or figure produced by a longitudinal plane of cut at approximately 45° to both rays and growth rings. The term is used especially for white oak with its large rays. The term comb grain is used where the vessel lines are parallel to the board edge and the rays produce a uniform pencil stripe.

RING-POROUS WOOD
Hardwood in which relatively large pores are concentrated in the earlywood and distinctly smaller pores are found in the latewood. See also Diffuse-porous wood.

RING SHAKE (RING FAILURE, SHELL SHAKE)
A separation of the wood structure parallel to the growth rings, often through the first layer(s) of earlywood cells.

RIPPLE MARKS
Fine striations perpendicular to the grain that are most apparent on tangential surfaces of wood with storied rays.

ROEY GRAIN (ROE FIGURE)
Stripes less than 1 ft. long formed on a radial surface by irregular interlocked grain.

ROT
Same as decay.

ROUND KNOT
The round or oval exposed section of a knot cut more or less crosswise to the limb axis.

SAPSTAIN
Same as blue stain.

SAPWOOD (ALBURNUM)
The active sap-conductive wood containing some living cells in mature stems. It comprises one to many of the outermost growth rings of the wood, and is usually lighter in color than the central heartwood.

SCALARIFORM PARENCHYMA
See Reticulate parenchyma.

SCALARIFORM PERFORATION PLATE (LATTICED PERFORATION PLATE)
A perforation plate with multiple perforations that are elongated transversely and are parallel. The remnants of the plate between openings are called perforation bars.

SCALARIFORM PITTING
Intervessel pitting in which transversely elongated pits are arranged in ladderlike series. Scalariform pitting may intergrade with opposite pitting.

SCANTY (PARATRACHEAL) PARENCHYMA
Parenchyma cells in contact with pores that occur singly or as an intermittent sheath around the pores.

SECONDARY CELL WALL
The cell-wall layer added to the cell-cavity side of the primary wall. The bulk of the cell wall visible with a light microscope is secondary wall. Pits are voids in this layer.

SECTION
A thin slice of wood tissue; also the cellular surface exposed by cutting.

SEMI-DIFFUSE-POROUS WOOD (SEMI-RING-POROUS WOOD)
Hardwood with fairly evenly distributed pores whose size gradually decreases from the earlywood to latewood portion of the growth ring.

SERIATION
The width of a ray as measured in cells.

SHAKE
See Ring shake.

SHELL SHAKE
Same as ring shake.

SIDE-GRAINED
1. Flatsawn (side-cut). 2. Referring to any longitudinal surface, as opposed to an end-grained surface.

SILVER GRAIN
Figure produced by showy or lustrous ray fleck on an edge-grained surface.

SIMPLE PERFORATION
A single, large, rounded opening in the perforation plate.

SIMPLE PIT
A pit in which the pit cavity remains approximately constant in size from the pit membrane to the cell cavity.

SIMPLE PIT PAIR
An intercellular pairing of two simple pits that share the same pit membrane.

SLASH-GRAINED (SLASH-SAWN)
Same as flatsawn.

SLIDE
Same as microscope slide.

SOFT TISSUE
Same as parenchyma.

SOFTWOODS (CONIFEROUS WOODS)
Woods produced by coniferous trees in the botanical group gymnosperms. Since these woods lack vessels, they are sometimes referred to as nonporous woods. (Note: The term softwood does not describe the degree of hardness of the wood.)

SOLITARY PORE
A pore that does not come in contact with other pores but is surrounded by other cell types as seen in cross section.

SOUND WOOD
Wood that is decay-free.

SPALTED WOOD
Partially decayed wood characterized by irregular discolorations appearing as dark to black zone lines on the surface.

SPECIES (PLURAL: SPECIES)
A kind of plant (or animal); the lowest basic unit of classification, just below genus. All members of a species can interbreed and produce fertile offspring, even though varieties are sometimes recognized within a species. Also, the binomial (genus name plus specific epithet); e.g., *Acer rubrum*, red maple.

SPECIFIC EPITHET (SPECIFIC ADJECTIVE, SPECIFIC NAME)
The second word of the binomial term used to name a species.

SPECIFIC GRAVITY
The ratio of the weight of a body to the weight of an equal volume of water; relative density.

SPIRAL CHECKING (SPIRAL CAVITIES)
Separations (checks) of the cell wall following the helical fibril orientation of the secondary wall. This checking may develop as extensions of pit apertures. A common characteristic of compression-wood tracheids.

SPIRAL GRAIN
Cross grain indicated by grain deviation from the edge of a tangential surface. Spiral grain may result naturally from helical grain direction in a tree or may be produced artificially in a piece by misaligned sawing.

SPIRAL THICKENINGS (HELICAL THICKENINGS)
Helical ridges along the inner surface of the cell walls in tracheids or vessel elements in certain species. On longitudinal sections, such cells have the appearance of coil springs. *See also* Reticulate thickenings.

SPRINGWOOD
Same as earlywood.

STAGE
The tablelike platform of a compound microscope directly beneath the objective lens upon which the slide-mounted specimen is positioned for examination.

STAGE MICROMETER
A microscope slide with a precise scale etched or imprinted on it.

STAIN
1. A discoloration in wood caused by stain fungi, metals or chemicals. 2. A material used intentionally to change the color of wood.

STEM-FORMED WOOD
Same as adult wood.

STORIED STRUCTURE
An arrangement of longitudinal cells or rays in horizontal tiers as viewed tangentially. *See also* Ripple marks.

STRAIGHT GRAIN
Grain direction parallel to the axis or edges of a piece of wood.

STRAND PARENCHYMA
See Longitudinal parenchyma.

STRIPE FIGURE (STRIPE GRAIN)
Same as ribbon figure.

STUMP WOOD
Wood from the flared basal region of the tree stem near ground level.

SUMMERWOOD
Same as latewood.

SWIRL CROTCH
A figure that is transitional between center crotch figure and normal stem wood.

T
The abbreviation or symbol for a tangential section or surface.

TANGENTIAL
The direction in wood that is perpendicular to the grain and to the rays, and therefore is tangent to the growth ring.

TANGENTIALLY CUT (TANGENTIALLY GRAINED)
Same as flatsawn.

TANGENTIAL PLANE (TANGENTIAL-LONGITUDINAL PLANE)
A longitudinal plane perpendicular to the rays and more or less parallel to the growth layer.

TANGENTIAL SECTION
A thin section cut along the grain in the tangential plane. Also a surface exposed by such a cut.

TAXODIOID PITTING
A type of cross-field pitting in the earlywood of some conifers (e.g., *Sequoia, Taxodium, Abies*). The pits are oval to circular in outline, with a wide, oval, included pit aperture and narrow border.

TAXONOMY
The science of classification. The descending levels of classification of the plant kingdom are division, class, order, family, genus, species and variety.

TENSION WOOD
See Reaction wood.

TERMINAL PARENCHYMA
See Marginal parenchyma.

TEXTURE
Relative cell size described by adjectives ranging from fine to coarse. In softwoods, texture is determined by relative tracheid diameter; in hardwoods, it is determined by relative pore diameter. The term is also used to indicate evenness of grain — e.g., uniform texture — and uniformity in size and distribution of pores — e.g., even texture.

TIGER GRAIN
Same as curly figure.

TIGHT KNOT
Same as intergrown knot.

TISSUE
A group or mass of cells that have a similar function or a common origin.

TORUS
The thickened central portion of the pit membrane in the bordered pit pairs of softwoods.

TRACHEIDS
Nonliving cell types with bordered pits that function in food conduction and tree support. These are the principal cell type in softwoods but are a minor cell type in hardwoods. *See also* Longitudinal tracheid, Ray tracheid, Vasicentric tracheid.

TRANSVERSE
The direction perpendicular to the grain.

TRANSVERSE PLANE
A cross-sectional plane perpendicular to the grain.

TRANSVERSE RESIN CANALS
See Resin canals.

TRANSVERSE SECTION
Same as cross section.

TRAUMATIC RESIN CANALS
Resin canals formed in response to injury to the living tree. Although they are rarely found, their occurrence is usually in the wood of species that do not normally have resin canals.

TRUNK
The main stem of a tree. *See also* Bole.

TYLOSES
Bubblelike structures that form in the vessel elements of certain hardwood species, usually in conjunction with heartwood formation.

TYLOSOIDS
The walls of epithelial cells that have bulged out into the passageway of a resin canal. They may occur in pines that have thin-walled epithelial cells.

ULMIFORM PORE ARRANGEMENT
Same as wavy bands of pores.

ULTRASTRUCTURE
Anatomical detail too small to be resolved with a light microscope.

ULTRAVIOLET LIGHT
Invisible light with wavelengths beyond the lowest visible light wavelength of about 380 nanometers (violet). Ultraviolet light, in the range of 320-380 nanometers, is called longwave ultraviolet or "black light," and can be used safely to determine the fluorescent properties of wood. *See also* Fluorescence.

UNEVEN-GRAINED
Describing wood with growth rings that show a pronounced difference in appearance between their earlywood and latewood, as in southern yellow pine or white ash.

UNISERIATE RAY
A ray that is only one cell wide tangentially.

UPRIGHT RAY CELLS
Ray cells whose longest axes are parallel to the longitudinal direction of the wood. These cells commonly occur as the marginal cells of heterocellular rays in hardwoods. *See also* Procumbent ray cells.

VASICENTRIC PARENCHYMA
Paratracheal parenchyma that forms a complete sheath one to many cells thick around a vessel.

VASICENTRIC TRACHEID
An irregularly shaped tracheid with numerous bordered pits that is found in association with the large earlywood vessels of certain hardwoods such as oaks and chestnut.

VERTICAL GRAIN
Same as edge grain.

VESSEL
A longitudinal conductive passageway formed by an aligned series of vessel elements.

VESSEL ELEMENT (VESSEL CELL)
A hardwood cell type with a relatively large diameter, thin cell wall and perforate (open) ends; one of the cellular components of a vessel. The cross section of a vessel element is called a pore.

VESSEL LINES
The visible lines produced on longitudinal surfaces of hardwoods with fairly large-diameter vessels wherever the plane of cut opens the vessels lengthwise.

WAVY BANDS OF PORES (ULMIFORM PORE ARRANGEMENT)
Pores arranged in undulating bands approximately parallel to the growth rings, as in the latewood of elm and hackberry.

WAVY GRAIN (CURLY GRAIN)
Undulations of the grain direction creating horizontal corrugations on radially, and sometimes tangentially, split surfaces.

WHITE ROT
Decay caused by a type of fungus that attacks both cellulose and lignin, leaving a whitish spongy or stringy residue. *See also* Brown rot.

WINDOWLIKE PITTING (FENESTRIFORM PITTING)
A type of cross-field pitting in the earlywood of soft pines and certain other pines in which the pits are large and rectangular or rounded and without apparent borders. They occupy a large percentage of the cross-field.

WOOD (XYLEM)
The cellular tissue of the tree, exclusive of the pith, enclosed by the cambium.

WOOD SUBSTANCE
The solid material of which wood is composed, principally cellulose and lignin, exclusive of extractives and sap.

WOUND HEARTWOOD
Regions of discolored sapwood resembling normal heartwood formed in association with wounds and other cambial damage.

X
The abbreviation or symbol for cross section or a cross-sectional surface.

XYLARIUM (PLURAL: XYLARIA)
A formal reference collection of wood samples.

XYLEM
Same as wood.

XYLEM PARENCHYMA (WOOD PARENCHYMA)
Parenchyma occurring in the xylem (in contrast to that occurring in the phloem) usually in either of two systems, axial or radial (ray parenchyma).

XYLOTOMY
The microscopic study of wood structure.

ZONATE PARENCHYMA
See Confluent parenchyma.

ZONE LINES
See Spalted wood.

BIBLIOGRAPHY

REFERENCES BY CATEGORY

I. CURRENTLY AVAILABLE BOOKS ON WOOD IDENTIFICATION
Core, Côté and Day
Edlin
Panshin and deZeeuw
Wheeler, Pearson, LaPasha, Zack and Hatley
White

II. FIBER ANATOMY AND IDENTIFICATION
Côté
Parham and Gray
Strelis and Kennedy

III. TECHNICAL REFERENCES, PRIMARILY WOOD ANATOMY AND/OR IDENTIFICATION
Barefoot and Hankins
Brazier and Franklin (Forest Products Research Laboratory Bulletin No. 46)
Dechamps
Forest Products Research Laboratory Bulletin Nos. 25 and 26
Friedman
Greguss
Harrar
Kribs
Miles
Phillips (Forest Products Research Laboratory Bulletin No. 22)
Schweingruber
Wheeler, Baas and Gasson (IAWA Bulletin No. 10)

IV. TECHNICAL REFERENCES CONTAINING SECTIONS ON ANATOMY/IDENTIFICATION
Desch
Jane
Johnston
Spencer and Luy
Summit and Sliker
Tsoumis

V. NONTECHNICAL BOOKS CONTAINING INTERESTING MATERIAL ON IDENTIFICATION
Bramwell
Constantine
Lincoln
Rendle
Titmuss
United States Forest Products Lab. (U.S.D.A. Agr. Handb. No. 101)

VI. NAMES OF TREES AND WOODS
British Standards Institution (BS No. 881/589)
Kartesz and Kartesz
Little (U.S.D.A. Agr. Handb. No. 541)

VII. TREE IDENTIFICATION
Elias

VIII. MISCELLANEOUS
Fowells (range maps)
Gregory (bibliography)
International Association of Wood Anatomists (glossary)
Stern (wood collections)
Timber Research and Development Association (examination/techniques)

Barefoot, A.C., and F.W. Hankins. 1982. *Identification of Modern and Tertiary Woods.* Oxford: Oxford University Press. 189 pp.

An identification guide for modern and fossil woods for use with multiple-entry key-punch cards. Presents a well-illustrated catalog of wood-identification features, with excellent supplementary sections on variability in wood anatomy and on specimen preparation.

Bramwell, Martyn, ed. 1976. *The International Book of Wood.* New York: Simon and Schuster. 276 pp.

An enjoyable and informative full-color book about all aspects of wood, from its anatomical structure to its use in building, transportation, crafts and art. A chapter on world timbers describes 144 species, each accompanied by a small color photo of the wood and a low-power transverse view of cell structure.

Brazier, J.D., and G.L. Franklin. 1961. *Identification of Hardwoods: A Microscope Key* (Forest Products Research Bulletin No. 46). London: Her Majesty's Stationery Office. 96 pp.

A major reference describing microscopic identification features of about 380 commercial hardwoods representing some 800 species. Data are presented in coded form appropriate for use with the Princes Risborough Laboratory perforated card system. The bulletin is currently out of print, but under revision.

British Standards Institution. 1974. *Nomenclature of Commercial Timbers, Including Sources of Supply* (British Standards No. BS 881 & 589). 87 pp.

This publication combines British Standards No. BS 881, *Nomenclature of Hardwoods*, and BS 589, *Nomenclature of Softwoods*, to provide a checklist of standard trade names and scientific names for commercial timbers as used in the United Kingdom.

Constantine, Albert, Jr. 1975. *Know Your Woods.* Rev. ed. New York: Charles Scribner's Sons. 360 pp.

Fascinating facts and lore about wood. The second half of the book describes over 300 woods "from abura to zebrano." Although the book was not designed as an identification book, photographs show the characteristic figure of many of the species discussed.

Core, H.A., W.A. Côté and A.C. Day. 1979. *Wood Structure and Identification.* 2d ed. Syracuse, New York: Syracuse University Press. 182 pp.

A multipurpose manual useful to wood-identification students as well as woodworkers. Wood structure is presented at gross, microscopic and ultrastructural levels, and the book features outstanding photomicrographs and electron micrographs. The anatomy of commercially important American species is analyzed individually. Keys for identification are based on features visible with a hand lens or with a microscope.

Côté, Wilfred A. 1980. *Papermaking Fibers: A Photomicrographic Atlas.* Syracuse, New York: Syracuse University Press. 186 pp.

Vivid scanning-electron micrographs show how cells are separated chemically from wood and reunited as a layer of paper. Photomicrographs illustrate details of cells of 64 species of wood and of many other natural and synthetic materials. This book is invaluable for fiber identification.

Dechamps, Roger. 1973. *How To Understand the Structure of Hardwood.* Rev. and trans. P. Baas. Tervuren, Belgium: Musée Royal de l'Afrique Centrale. 71 pp.

A concise and inexpensive booklet on hardwood anatomy whose aim is to explain current terminology and anatomical descriptions used in the literature. Visual and microscopic anatomical features are explained and illustrated in photographs, mostly of tropical species.

Desch, H.W. 1973. *Timber — Its Structure and Properties.* 5th ed. New York and London: Macmillan Press Ltd. 424 pp.

A general text on the cellular anatomy, gross features, properties and technology of timber. Chapters on wood identification include only a brief anatomical chart and key. For each of 36 commercial hardwoods, mostly European and tropical, a macroscopic description accompanies an accurate drawing of transverse cell structure.

Edlin, Herbert L. 1969. *What Wood Is That? — A Manual of Wood Identification.* New York: The Viking Press. 160 pp.

Tree structure and gross anatomical features form the basis for a unique key system for recognizing 40 woods; detailed macroscopic or microscopic anatomy is not used. A fold-out chart has 40 mounted veneer specimens, each specimen measuring 7/8 in. by 3 in.

Elias, Thomas S. 1980. *The Complete Trees of North America: Field Guide and Natural History.* New York: Van Nostrand Reinhold Company, Outdoor Life/Nature Books. 948 pp.

A comprehensive field identification guide based on both summer and winter keys that is written in nontechnical language. Covers 652 native and 100 introduced species, and includes range maps and commentary on wood use that are helpful in wood identification.

Forest Products Research Laboratory. 1960. *Identification of Hardwoods: A Lens Key* (Forest Products Research Bulletin No. 25). London: Her Majesty's Stationery Office. 126 pp.

A concise survey of hand-lens features developed for use with the Princes Risborough Laboratory perforated-card identification system. A guide to the identification of some 450 worldwide hardwoods, including principal species of North America.

Forest Products Research Laboratory. 1961. *An Atlas of End-Grain Photomicrographs for the Identification of Hardwoods* (Forest Products Research Bulletin No. 26). London: Her Majesty's Stationery Office. 88 pp.

This atlas comprises 396 photomicrographs at 10x magnification to illustrate the woods described in F.P.R.B. No. 25, *Identification of Hardwoods: A Lens Key.*

Fowells, H.A. 1965. *Silvics of Forest Trees of the United States* (U.S.D.A. Forest Service Agr. Handb. No. 721). Washington, D.C.: U.S. Government Printing Office. 762 pp.

Although intended primarily as a comprehensive summary of silvical information on American forest trees, this book contains detailed maps (many by county) of natural ranges of trees, excellent for evaluating probable sources of American woods.

Friedman, Janet. 1978. *Wood Identification by Microscopic Examination* (Heritage Record Series No. 5). Victoria: British Columbia Provincial Museum. 84 pp.

Written as a guide for archaeologists on the northwest coast of North America, this book contains descriptions of microscopic features of 42 woods, including lesser-known species of trees and shrubs.

Gregory, M. 1980. *Wood Identification: An Annotated Bibliography.* (International Association of Wood Anatomists Bulletin [new series] 1[1/2]:3-41). Leiden, The Netherlands: International Association of Wood Anatomists.

An extensive compilation of world literature on the subject since 1900. References are grouped by geographic region and indexed by author.

Greguss, Pál. 1955. *Identification of Living Gymnosperms on the Basis of Xylotomy.* Budapest: Akadémiai Kaidó. Trans. 263 pp.

A comprehensive volume with 350 plates embracing virtually every important softwood species of the world. Detailed anatomical descriptions are accompanied by photomicrographs (30x, 100x, 300x) and interpretive drawings of microscopic detail. Includes keys to families, genera and species, as well as charts summarizing anatomical features.

Harrar, Ellwood S. 1957. *Hough's Encyclopaedia of American Woods.* New York: Robert Speller & Sons.

This multiple-volume set is a treasured classic. Each two-part volume covers 25 species. The first part is a ring binder. On each page are mounted transverse, radial and tangential slices (1.6 in. x 3.7 in. x 0.01 in.) of actual wood tissue. The second part is a bound edition of species descriptions.

International Association of Wood Anatomists. 1964. *Multilingual Glossary of Terms used in Wood Anatomy.* Zurich: IAWA Committee on Nomenclature. 186 pp.

A principal reference for terminology of anatomical features used in describing wood for identification and anatomical study. Some 206 terms are defined in English, French, German, Italian, Portuguese, Spanish and Serbo-Croatian.

Jane, F.W. 1970. *The Structure of Wood.* 2nd ed., rev. London: Adam and Charles Black, Ltd. 478 pp.

Authoritatively discusses wood anatomy from ultrastructure to figure. A third of the book contains concise keys to genera and a detailed discussion of species differences. Includes important commercial timbers of the north temperate zone, especially the U.S. and Europe. There is a separate chapter on commercial mahogany, walnut and teak.

Johnston, David. 1983. *The Wood Handbook for Craftsmen.* New York: Arco Publishing, Inc. 168 pp.

Describes 106 hardwoods with hand-lens identification features and a cross-sectional micrograph (25x, 50x or 60x) for each; includes excellent color photographs of longitudinal surfaces of 30 species. Major conifers are described, but identification information is not given.

Kartesz, John T., and Rosemarie Kartesz. 1980. *A Synonymized Checklist of the Vascular Flora of the United States, Canada and Greenland, Volume 2, the Biota of North America.* Chapel Hill: University of North Carolina Press. 498 pp.

An exhaustive and authoritative listing by scientific name of all vascular plants. Arranged alphabetically by family and by genus, and indexed by family and genus. Common and trade names are not given.

Kribs, David A. 1968. *Commercial Foreign Woods on the American Market.* New York: Dover Publications, Inc. 241 pp.

A most useful guide to foreign woods. The general properties, anatomy and uses are described for each of more than 350 species. Each listing is accompanied by photographs of transverse sections (10x and 80x) and tangential sections (100x). Includes an illustrated glossary, bibliographies and keys for identification.

Lincoln, William A. 1986. *World Woods in Color.* New York: Macmillan Publishing Company. 320 pp.

Presents full-scale color photographs of 231 hardwoods and 31 softwoods representing the most important commercially available world timbers. A general description of gross features is given for each, but specific anatomical identification features are lacking. Scientific, commercial and local names as well as technical and working properties are included.

Little, Elbert L., Jr. 1980. *Checklist of United States Trees (Native and Naturalized)* (U.S.D.A. Agr. Handb. No. 541). Washington, D.C.: U.S. Government Printing Office. 375 pp.

This revised checklist supersedes the 1953 edition and is generally accepted as a principal authority on scientific and common names. It compiles scientific names and current synonyms, approved common names and other names in use, and the geographic ranges of 679 native and 69 naturalized tree species of the United States (except Hawaii).

Miles, Anne. 1978. *Photomicrographs of World Woods.* Building Research Establishment. London: Her Majesty's Stationery Office. 233 pp.

This atlas of 1,353 high-quality micrographs ranging from 25x (transverse) to 250x (radial) provides a valuable illustrative supplement to verbal microscopic descriptions presented in Forest Products Research Bulletin No. 22 for softwoods (Phillips, 1979) and No. 46 for hardwoods (Brazier and Franklin, 1961).

Panshin, A.J., and Carl deZeeuw. 1980. *Textbook of Wood Technology.* 4th ed. New York: McGraw-Hill Book Company. 722 pp.

Probably the most widely used text for wood anatomy and identification. Essentially two books in one, Part 1 discusses the anatomy and defects of commercially important woods, and Part 2 includes identification features and keys. For each of 75 woods, there are photomicrographs with descriptions of gross and minute features.

Parham, Russell A., and Richard L. Gray. 1982. *The Practical Identification of Wood Pulp Fibers.* Atlanta: Technical Association of the Pulp and Paper Industry, Inc. 212 pp.

Cell-wall features are utilized in a comparison method of fiber identification. Typical features of 33 softwoods and 43 hardwoods are illustrated with both phase-contrast light microscopy and scanning-electron microscopy.

Phillips, E.W.J. 1979. *Identification of Softwoods by Their Microscopic Structure* (Forest Products Research Bulletin No. 22). London: Her Majesty's Stationery Office. 56 pp.

A concise survey of microscopic features. Developed for use with the Princes Risborough Laboratory perforated-card identification system, it provides a guide to the identification of softwoods, including those of principal commercial importance in North America.

Rendle, B.J. 1969, 1970. *World Timbers, Volume 1: Europe & Africa, Volume 2: North & South America, Volume 3: Asia & Australia & New Zealand.* London: Ernest Benn Ltd., and

Downsview, Ontario: University of Toronto Press. 516 pp.

This comprehensive work includes more than 200 timbers of the world. Each wood is illustrated by a high-quality color photograph of a longitudinal surface that is useful for visual comparison in identification work. Properties and characteristics are described with a minimum of technical terms.

Schweingruber, Fritz H. 1978. *Microscopic Wood Anatomy: Structural Variability of Stems and Twigs in Recent and Subfossil Woods from Central Europe.* Zurcher, Zug, Switzerland: Swiss Federal Institute of Forestry. 226 pp.

Text in German, French and English. A high-quality atlas of micrographs of European woods. Anatomical descriptions emphasize distinguishing features important to identification. Includes special sections on abnormal wood such as reaction wood, decayed wood and charcoal.

Spencer, Albert G., and Jack A. Luy. 1975. *Wood and Wood Products.* Columbus, Ohio: Charles E. Merrill Publishing Co. 246 pp.

An informative book covering a wide scope of wood technology and product information. A chapter on identification emphasizes macroscopic features as a basis for a key to 30 hardwoods and 17 softwoods, which are accompanied by low-power photographs of transverse surfaces.

Stern, William L. 1988. *Index Xylariorum. Institutional Wood Collections of the World, 3* (International Association of Wood Anatomists Bulletin 9[2]:203-252). Leiden, The Netherlands: IAWA.

For each of 137 institutional wood collections, this guide lists location, address, staff, size of collection, areas of specialization and availability of material.

Strelis, I., and R.W. Kennedy. 1967. *Identification of North American Commercial Pulpwoods and Pulp Fibres.* Downsview, Ontario: University of Toronto Press. 117 pp.

Concise descriptions and excellent photographs of microscopic features for the identification of both solid wood tissue and wood fibers. Includes identification summaries for 22 softwoods and 14 hardwoods and a clearly illustrated guide to microscopic identification of nonwoody and man-made fibers.

Summit, Robert, and Alan Sliker. 1980. *CRC Handbook of Materials Science, Volume IV: Wood.* Boca Raton, Florida: CRC Press, Inc.

Anatomical, physical and chemical properties of wood. For each of the 172 species described, micrographs ranging from 10x (transverse) to 150x (radial) are accompanied by a systematic listing of gross, hand-lens and microscopic identification features. Summary identification charts are included.

Timber Research and Development Association. 1971. *Examination of Timbers* (Teaching Aid No. 7). High Wycombe, Buckinghamshire, England: Timber Research and Development Association. 20 pp.

An easy-to-read and informative summary of approaches and techniques for preparing and examining wood for identification, including both freehand methods and microtome sectioning.

Titmuss, F.H. 1971. *Commercial Timbers of the World.* 4th ed. London: The Technical Press Ltd. 351 pp.

General discussions of each of 252 woods that focus on commercial importance, properties, workability and uses. For about two-thirds of the woods, a paragraph describing macroscopic identification is included, but, regrettably, only 38 woods are accompanied by low-power photographs of their transverse surfaces.

Tsoumis, George. 1968. *Wood as Raw Material.* Long Island City, New York: Pergamon Press. 276 pp.

Anatomical, physical, chemical and mechanical properties of wood are discussed, including wood formation, abnormalities and degradation. A chapter on identification includes 13 separate keys to North American and European hardwoods and softwoods. The appendix describes techniques for the microscopic investigation of wood.

United States Forest Products Laboratory. 1956. *Wood — Colors and Kinds* (U.S.D.A. Agr. Handb. No. 101). Washington, D.C.: U.S. Government Printing Office. 36 pp.

A pictorial guide to the identification of the species most commonly found in retail lumber markets. For each of 18 hardwoods and 14 softwoods, full-scale color photographs of wood surfaces are provided. The range, properties, uses and gross features of each species are described.

Wheeler, E.A., P. Baas and P.E. Gasson, eds. 1989. *IAWA List of Microscopic Features for Hardwood Identification.* (International Association of Wood Anatomists Bulletin [new series] 10[3]:219-332). Leiden, The Netherlands: IAWA.

This concise list of 163 anatomical features useful in identification includes definitions, explanatory commentary and appropriate illustrations. The appendix defines 58 miscellaneous features, including color, odor and fluorescence.

Wheeler, E.A., R.G. Pearson, C.A. LaPasha, T. Zack and W. Hatley. 1986. *Computer-Aided Wood Identification* (North Carolina Agricultural Research Service Bulletin No. 474). Raleigh: North Carolina State University. 160 pp.

Designed for use with computerized identification programs based on multiple-entry punch cards originally compiled at Oxford and Princes Risborough, and expanded at NCSU. The bulk of the manual contains excellent micrographs of anatomical characters, which are ideal for learning identification features. Data bases and programs are available for IBM-PC, Macintosh and Apple II computers.

White, Marshall S. 1980. *Wood Identification Handbook: Commercial Woods of the Eastern United States.* Falls Church, Virginia: Colonial Hardwoods, Inc. 88 pp.

Describes gross and macroscopic features of 7 softwoods and 25 hardwoods. Contains simplified identification keys accompanied by transverse macrophotographs.

INDEX

Editor: Steve Marlens
Designer: Deborah Fillion
Layout artist: Greta Sibley
Illustrator: Marianne Markey
Copy editor: Peter Chapman
Production editor: Victoria W. Monks
Indexer: Harriet Hodges

Typeface: R Stempel Garamond
Paper: 70-lb. Warren Patina, neutral pH
Printer and binder: Ringier America, New Berlin, Wisconsin